RELIABILITY AND OPTIMIZATION OF STRUCTURAL SYSTEMS

PROCEEDINGS OF RELIABILITY AND OPTIMIZATION OF STRUCTURAL SYSTEMS,
TUM, MÜNCHEN, GERMANY, 7–10 APRIL 2010

Reliability and Optimization of Structural Systems

Editor

D. Straub
Engineering Risk Analysis Group, Technische Universität München (TUM), Germany

CRC Press
Taylor & Francis Group
Boca Raton London New York Leiden

CRC Press is an imprint of the
Taylor & Francis Group, an **informa** business
A BALKEMA BOOK

CRC Press/Balkema is an imprint of the Taylor & Francis Group, an informa business

Typeset by MPS Ltd. (A Macmillan Company), Chennai, India

Published by: CRC Press/Balkema
P.O. Box 447, 2300 AK Leiden, The Netherlands
e-mail: Pub.NL@taylorandfrancis.com
www.crcpress.com – www.taylorandfrancis.co.uk – www.balkema.nl

ISBN: 978-0-415-88179-1 (Hbk)
ISBN: 978-0-203-84141-9 (eBook)

Reliability and Optimization of Structural Systems – Straub (ed)
© 2010 Taylor & Francis Group, London, ISBN 978-0-415-88179-1

Table of Contents

Participants of the 15th IFIP WG7.5 Working Conference at Max-Joseph-Platz, Munich

Reliability and Optimization of Structural Systems – Straub (ed)
© 2010 Taylor & Francis Group, London, ISBN 978-0-415-88179-1

Preface

These are the proceedings of the fifteenth working conference of the International Federation of Information Processing (IFIP) Working Group 7.5 on Reliability and Optimization of Structural Systems, which took place at the Oskar von Miller Forum in Munich, Germany, on April 7–10, 2010. This volume contains 28 papers presented at the conference and a brief summary of one of two plenary discussions.

The 15th IFIP WG7.5 working conference in Munich was dedicated to Prof. Rüdiger Rackwitz of Technische Universität München. Professor Rackwitz's influence on the development of the field of structural reliability and optimization is second to none. Many of the achievements of our community, including the use of semi-probabilistic safety formats and probabilistic methods in Eurocode, are a direct consequence of his pioneering works. Besides being the originator of many ideas and methods that are now commonly used in our community, Prof. Rackwitz has made no less impact during conferences and workshops, such as the IFIP WG7.5 working conferences, through his informative and uncompromising way of discussing and asking questions. Professor Rackwitz is deeply esteemed by the members of this community for his contributions and personal integrity, which, in the opinion of this editor, make him a role-model for any scientist. During the general assembly of the working group, it was unanimously decided to honor Prof. Rackwitz by offering him Honorary Membership of IFIP WG7.5.

The conference was supported by IFIP TC-7, IFIP's technical committee on Modeling and Optimization, the Technische Universität München (TUM) and the Oskar von Miller Forum in Munich, which provided the excellent conference location; this support is greatly appreciated. Financial support from TUM is also acknowledged.

The purpose of the Working Group 7.5 is to promote modern structural system reliability and optimization theory and its applications, to stimulate research, development and application of structural system reliability and optimization theory, to assist and advance research and development in these fields, to further the dissemination and exchange of information on reliability and optimization of structural systems, and to encourage education in structural system reliability and optimization theory.

The main themes of the conference were structural reliability methods and applications, engineering risk analysis and decision making, new optimization techniques and various applications in civil engineering; a significant number of papers were related to the assessment and optimization of existing structural systems. The conference was marked by lively discussions, which culminated in two discussion sessions, and the noteworthy participation of various young researchers who presented valuable contributions to the field and brought new viewpoints to the discussions.

IFIP Working Group 7.5 has the following membership (dated April 25 2010). Many of the researchers and practitioners listed below played an active role in the scientific programme of the conference.

A.H.-S. Ang, USA
G. Augusti, Italy
F. Biondini, Italy
F. Bontempi, Italy
A. Borri A., Italy
C. Bucher, Austria
J. Casas, Spain
H.N. Cho, South Korea
M. Ciampoli, Italy
R. Corotis, USA
D. De Leon, Mexico
A. Der Kiureghian, USA
O. Ditlevsen, Denmark
M. Dogaki, Japan
H. Ellis, USA
L. Esteva, Mexico (Former chair)
M.H. Faber, Switzerland (Chair)
D.M. Frangopol, USA
G. Fu, USA
H. Furuta, Japan
P. Geyskens, USA
M. Gioffre, Italy

M. Grigoriu, USA
C. Guedes-Soares, Portugal
E. Heredia-Zavoni, Mexico
H. Hong, Canada
J. Li, China
M. Kawatani, Japan
C.W. Kim, Japan
N. Kogiso, Japan
M. Lemaire, France
M.A. Maes, Canada
K. Marti, Germany
R. Melchers, Australia
Y. Murotsu, Japan
K. Nishijima, Switzerland
A. Nowak, USA
M.D. Pandey, Canada
R. Rackwitz, Germany
J.O. Royset, USA
S. Ruiz, Mexico
M. Sakano, Japan
T. Sato, Japan
N. Shetty, UK
W. Shiraki, Japan
P. Sniady, Poland
J. Song, USA
J. Sørensen, Denmark
D. Straub, Germany (Vice-chair)
B. Sudret, France
M.M. Szerszen, USA
W. Tang, Hong Kong
N.J. Tarp-Johansen, Denmark
P. Thoft-Christensen, Denmark
P. Waarts, The Netherlands
W.-K. Wen, USA
Z. Zembaty, Poland
Y. Zhao, Japan

Members of the technical committee were:

Alfredo H.-S. Ang, USA
Christian Bucher, Austria
Armen Der Kiureghian, USA
L. Esteva, Mexico (co-chair)
M.H. Faber, Switzerland (co-chair)
Hitoshi Furuta, Japan
Ernesto Heredia Zavoni, Mexico
Marc A. Maes, Canada
Kurt Marti, Germany
Robert Melchers, Australia
Rüdiger Rackwitz, Germany
Junho Song, USA
John D. Sørensen, Denmark
Bruno Sudret, France
Wilson Tang, Hong Kong

Members of the local organization were:

Panagiota Papakosta
Anneliese Spitzauer
Daniel Straub

Results from previous conferences have been published in the following proceedings:

1. Reliability and optimization of structural systems. Proceedings of the first IFIP WG 7.5 working conference, Aalborg, Denmark, May 6–8, 1987 (ed. P. Thoft-Christensen). Lecture notes in engineering, Vol. 33, Springer-Verlag Berlin Heidelberg New York.
2. Reliability and optimization of structural systems. Proceedings of the second IFIP WG 7.5 working conference, London, UK, Sept. 26–28, 1988 (ed. P. Thoft-Christensen). Lecture notes in engineering, Vol. 48, Springer-Verlag Berlin Heidelberg New York.
3. Reliability and optimization of structural systems. Proceedings of the third IFIP WG 7.5 working conference, Berkeley, California, USA, Mar. 26–27, 1990 (eds. A. Der Kiureghian and P. Thoft-Christensen). Lecture notes in engineering, Vol. 61 , Springer-Verlag Berlin Heidelberg New York.
4. Reliability and optimization of structural systems. Proceedings of the fourth IFIP WG 7.5 working conference, Munich, Germany, Sept. 11–13, 1991 (eds. R. Rackwitz and P. Thoft-Christensen). Lecture notes in engineering, Vol. 76, Springer-Verlag Berlin Heidelberg New York.
5. Reliability and optimization of structural systems. Proceedings of the fifth IFIP WG 7.5 working conference, Takamatsu, Kagawa, Japan, March 22–26, 1993 (eds. P. Thoft-Christensen and H. Ishikawa). IFIP transactions B-12, Applications in technology, Elsevier Science Publishers B. V, Amsterdam.
6. Reliability and optimization of structural systems. Proceedings of the sixth IFIP WG 7.5 working conference, Assisi, Italy, Sept. 7–9, 1994 (eds. R. Rackwitz, G. Augusti , A. Bori), 1995 Chapman & Hall, London.

7. Reliability and optimization of structural systems. Proceedings of the seventh IFIP WG 7.5 working Conference, Boulder, CO, USA, April 2–4, 1996 (eds. D.M. Frangopol, R.B. Corotis, and R. Rackwitz).
8. Reliability and optimization of structural systems. Proceedings of the eigth IFIP WG 7.5 working conference, Krakow, Poland, May 11–13, 1998 (eds. A.S. Nowak and M.M. Szerszen), University of Michigan, Ann Arbor, MI, USA.
9. Reliability and optimization of structural systems. Proceedings of the ninth IFIP WG 7.5 working conference, University of Ann Arbor, Michigan, Sept. 25–27, 2000 (eds. A.S. Nowak and M.M. Szerszen), University of Michigan, Ann Arbor, MI, USA.
10. Reliability and optimization of structural systems. Proceedings of the tenth IFIP WG 7.5 working conference, Kansai University, Osaka, Japan, Mar. 25–27, 2002 (eds. H. Furuta, M. Dogaki and M. Sakano), A.A. Balkema Publishers.
11. Reliability and optimization of structural systems. Proceedings of the eleventh IFIP WG 7.5 working conference, Banff, Canada, Nov. 2–5, 2003 (eds. M. Maes and L. Huyse), A. A. Balkema Publishers.
12. Advances in Reliability and Optimization of Structural Systems. Proceedings of the twelfth IFIP WG 7.5 working conference, Aalborg, Denmark, May 22–25, 2005 (eds. J.D. Sørensen and D.M. Frangopol), Taylor and Francis/Balkema.
13. Reliability and Optimization of Structural Systems. Proceedings of the thirteenth IFIP WG 7.5 working conference, Kobe, Japan, Oct 11–14, 2006 (eds. D.M. Frangopol, M. Kawatani and C.-W. Kim), Taylor and Francis/Balkema.
14. Reliability and Optimization of Structural Systems. Proceedings of the fourteenth IFIP WG 7.5 working conference, Toluca, Mexico, Aug 6–9, 2008 (CD version; ed. L. Esteva, Institute of Engineering, National University of Mexico, lestevam@iingen.unam.mx).

The editor wishes to thank his team, under the guidance of Anneliese Spitzauer, which helped him in the successful organization of the workshop. Special thanks are also due to Patty Papakosta for her superb work in compiling these proceedings. Finally, the authors of the various papers are acknowledged for the excellent contribution they made to the conference. Without each of them, the Munich conference would not have been successful.

The Editor,
Daniel Straub
TU München, Germany

Reliability and Optimization of Structural Systems – Straub (ed)
© 2010 Taylor & Francis Group, London, ISBN 978-0-415-88179-1

Bayesian network modeling of system performance

M.T. Bensi & A. Der Kiureghian
University of California, Berkeley, USA

D. Straub
Engineering Risk Analysis Group, TU München, Germany

ABSTRACT: Bayesian Networks (BNs) provide an excellent framework for modeling system performance, particularly in near-real time applications when it is necessary to update models in light of observations. However, BNs can be very demanding of computer memory and inference can become intractable if care is not taken to optimize their topology. In this paper, efficient BN formulations for modeling system performance are presented. First, formulations are developed for series and parallel systems. Then, results are extended to general systems for which the minimal link and/or cut sets are known. Finally, an optimization algorithm is developed to automate the generation of efficient BN formulations for modeling system performance.

1 INTRODUCTION

Civil infrastructure systems are backbones of modern societies, yet in many regions of the world they are vulnerable to seismic hazards. The authors are currently working toward the development of a probabilistic decision-support system for seismic infrastructure risk assessment and management using a Bayesian Network (BN) formulation. Components of the BN framework for seismic risk assessment and management include: (1) a seismic demand model of ground motion intensity as a spatially distributed Gaussian random field accounting for finite fault rupture and directivity effects; (2) models of component performance; (3) models of system performance; and (4) the extension of the BN to include decision and utility nodes to aid pre- and post-earthquake decision-making. This paper focuses on the third component of the proposed framework, modeling of system performance, with particular emphasis on issues related to computational efficiency.

A BN is a probabilistic graphical model that represents a set of random variables and their probabilistic dependencies. The ease with which BNs facilitate information updating makes them ideally suited for the proposed application. However, like all computational methods, BNs have limitations. In particular, calculations in BNs can be highly demanding of computer memory. Space constraints do not permit the inclusion of an introduction on BNs here. Comprehensive coverage is available in textbooks (Jensen & Nielson 2007; Kjaerulff & Madsen 2008) and overviews are presented in many papers (e.g. Friis-Hansen 2004; Bayraktarli et al. 2006; D. Straub 2005; Bensi et al. 2010).

2 MODELING SYSTEM PERFORMANCE VIA BAYESIAN NETWORK

Consider a system of n components. Each component $i \in [1, n]$ has d_i discrete component states. Therefore, the number of distinct configurations of the system is $\prod_{i=1}^{n} d_i$. Bensi et al. (2009) refer to the BN formulation that corresponds to the joint consideration of all combinations of component and system states as the *naïve BN formulation*. In this formulation, a single node S_{sys}, representing the state of the system, is defined as a child of all nodes representing the states of the components (a converging structure) as shown in Figure 1a. For a system with n binary components and binary system state (Pr(survive) = 1 − Pr(failure)), node S_{sys} has a conditional probability table (CPT) of size 2^n. (The CPT provides the conditional probabilities of the states of a node for given states of its parent nodes.) For systems with a large number of components, the size of the CPT in the naïve formulation quickly causes the BN to become computationally intractable.

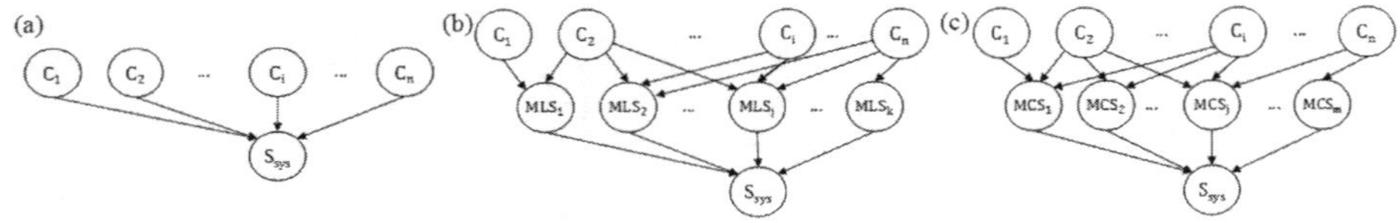

Figure 1. (a) Naïve formulation; (b) MLS formulation; and (c) MCS formulation.

However, the naïve formulation is useful in applications where the number of components is small, e.g. in the reliability assessment of simple structural systems. Mahadevan et al. (2001) utilize a formulation as in Figure 1a with components representing limit-state functions corresponding to failure modes of structural systems. The number of limit-state functions in their example is sufficiently small so that the size of the CPT associated with the system node is not prohibitively large. However, for realistic infrastructural systems such an approach to modeling is impractical.

As an alternative to the naïve approach to modeling system performance, Bensi et al. (2009) offer four additional BN formulations. We first review two of the formulations presented in that paper, which are based on use of minimal link and cut sets. Then, we adapt these two formulations with the goal of minimizing computational demands. We focus first on series and parallel systems followed by consideration of general systems. Due to page limitations, we limit the scope of this paper to systems with binary states. However, it is noted that the formulations presented here can be extended to systems with multi-state components.

2.1 *Modeling performance of series and parallel systems with binary component states*

A common approach to making systems analysis more methodical is through the use of minimal link sets and minimal cut sets. A minimal link set (MLS) is a minimum set of components whose joint survival constitutes survival of the system. The *minimal link set BN formulation* introduces intermediate nodes between the component and system nodes which correspond to the MLSs. Torres-Toledano and Sucar (1998) use a MLS-based BN formulation for modeling system performance, though with less formality and generality than is described here. Figure 1b shows an example of the MLS BN formulation. The binary states of the MLS nodes are defined such that each MLS node is in the survival state only if all its constituent components have survived; otherwise it is in the failure state. The system node is in the survival state if any MLS node is in the survival state. With binary component and system states, the size of the CPT for each MLS is 2 to the power of the number of its constituent components, and the size of the system node CPT is 2 to the power of the number of MLSs. As a result, when the number of MLSs is large, the size of the CPT associated with the system node S_{sys} becomes large. A similar problem occurs for an MLS nodes when the number of its constituent components is large.

The dual of the MLS formulation is the *minimal cut set BN formulation*. A minimal cut set (MCS) is a minimum set of components whose joint failure constitutes failure of the system. In this formulation, the system node is a child of nodes representing MCSs, and each MCS node is a child of nodes representing the states of its constituent components. Figure 1c shows a conceptual BN employing the MCS formulation. The system node is a series system of all the MCS nodes (i.e., the system node is in the failure state if at least one MCS node is in the failure state), whereas each MCS is a parallel system of its parent nodes (i.e. all constituent components must be in the failure state for the MCS node to be in the failure state). As with the MLS formulation, the CPTs in this formulation become large as the number of MCSs increases and/or the number of components in an individual MCS becomes large.

In general, BN structures with nodes arranged in chains are significantly more efficient than the converging configurations characterizing the naïve and MLS/MCS formulations described above. Consider the two equivalent BN structures shown in Figure 2. The figure on the left shows a converging structure and the one on the right illustrates a chain structure. Both BNs model systems whose component states are dependent on a common demand D. A formal description of the construction of the BN in Figure 2b is presented later in this paper. Based on an investigation of growth in computational complexity for both structures, it is found that the converging structure is associated with exponentially increasing memory demand, while the complexity of the chain structure grows linearly. However, the converging structure is more efficient than the chain structure for systems with less than four components.

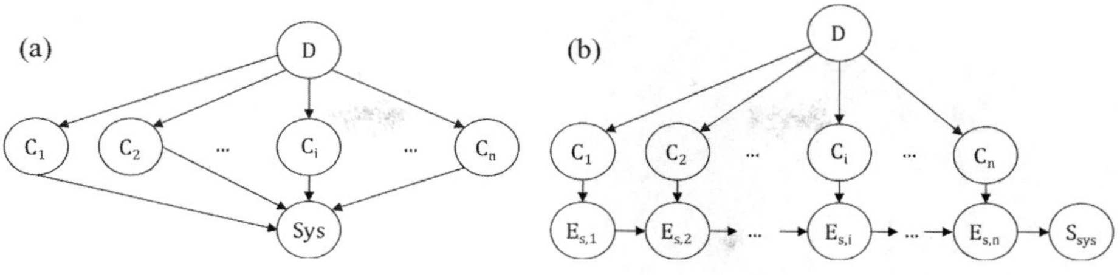

Figure 2. Illustration of BNs with (a) converging structure; and (b) chain structure.

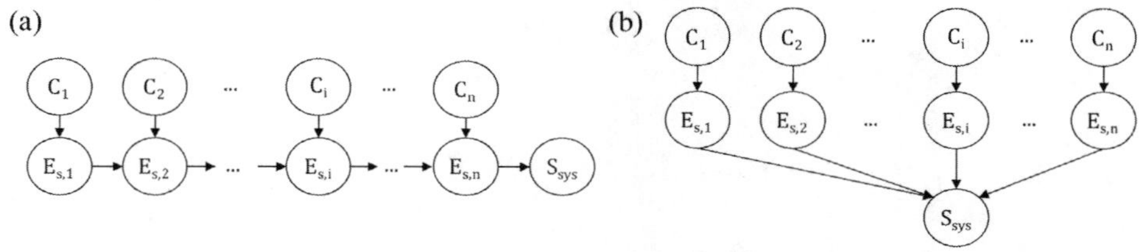

Figure 3. BN using SPEs to define performance of (a) a series system and (b) a parallel system.

We next describe how system performance can be modeled with BNs having the chain topology. Define a *survival path sequence* (SPS) as a chain of events, corresponding to a MLS, in which the terminal event in the sequence indicates whether or not all the components in the MLS are in the survival state. Note that the term "sequence" does not have any time implications. A series system has one MLS and a parallel system has *n* MLSs. It follows that a series system has one SPS and a parallel system has *n* SPSs. A SPS is comprised of a chain of *survival path events* (SPEs), each of which describes the state of the sequence up to that event. SPEs are represented in the BN by nodes labeled $E_{s,i}$, the subscript *i* indicating that the particular SPE is associated with component *i*. The state of $E_{s,i}$ is defined as

$$\begin{aligned} E_{s,i} &= 1 \text{ if } E_{s,\mathrm{Pa}(i)} = 1 \cap C_i = 1 \\ &= 0 \text{ otherwise} \end{aligned} \tag{1}$$

where $E_{s,\mathrm{Pa}(i)}$ defines the state of the SPE node that is parent to $E_{s,i}$; $E_{s,i} = 1$ indicates that the node (or any other node considered in this paper) is in the survival state and $E_{s,i} = 0$ indicates its failure. C_i denotes the state of component *i* with $C_i = 1$ ($C_i = 0$) indicating the survival (failure) state. Thus, for a series system, the BN formulation takes the form shown in Figure 3a. The state of node $E_{s,1}$ is equal to the state of node C_1. $E_{s,2}$ is in survival state only if $E_{s,1}$ is in survival state and C_2 is in survival state. This pattern continues such that $E_{s,n}$ is in the survival state only if both $E_{s,n-1}$ and C_n are in the survival state. Consequently, the state of $E_{s,n}$ describes the state of the entire SPS (i.e. it indicates whether all components in the MLS have survived) and, therefore, that of the system.

A parallel system has a SPS corresponding to each component. The resulting BN formulation is shown in Figure 3b. The system node indicates system survival if any node $E_{s,i}$ is in the survival state. Like the naïve formulation, the exponential growth in the size of the CPT associated with node S_{sys} renders the BN intractable when the number of components in the system is large.

Define a *failure path sequence* (FPS) as a chain of events, corresponding to a MCS, in which the terminal event in the sequence indicates whether or not all components in the MCS are in the failure state. For a parallel system, there is only one MCS and thus one FPS. For a series system with *n* components, there are *n* FPSs, one corresponding to each component. A FPS is comprised of a chain of *failure path events* (FPEs), each of which gives the state of the sequence up to that event. Let $E_{f,i}$ be the FPE associated with component *i*. The state of $E_{f,i}$ is expressed as

$$\begin{aligned} E_{f,i} &= 0 \text{ if } E_{f,\mathrm{Pa}(i)} = 0 \cap C_i = 0 \\ &= 1 \text{ otherwise} \end{aligned} \tag{2}$$

where $E_{f,\mathrm{Pa}(i)}$ defines the state of the FPE node that is parent to $E_{f,i}$. Thus, for a parallel system, the BN formulation takes the chain form shown in Figure 4a. The Boolean logic used to construct

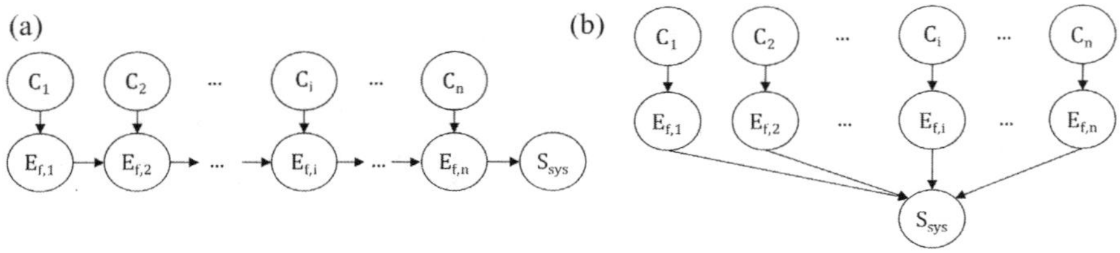

Figure 4. BN using FPEs to define (a) a parallel system and (b) a series system.

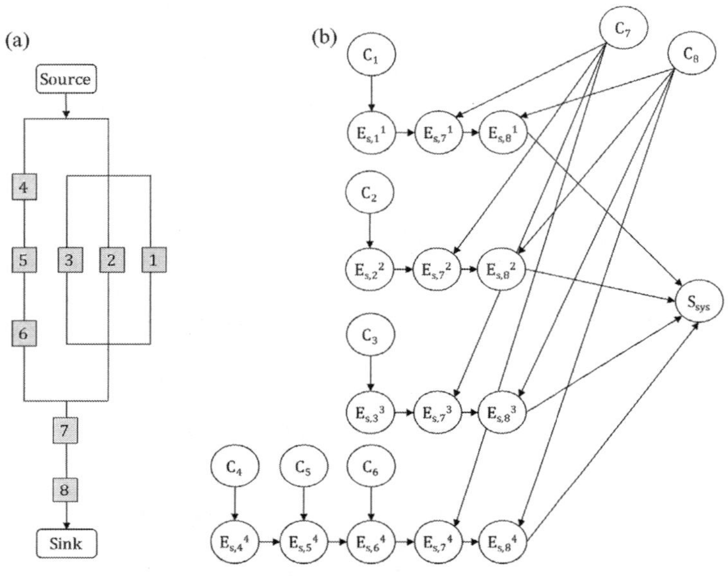

Figure 5. (a) Example system; (b) Efficient MLS formulations for system in (a) with distinct SPSs.

the CPTs in this BN is dual of that used for SPSs, i.e. $E_{f,i}$ is in the failure state only if the parent FPE is in the failure state and C_i is also in the failure state. For a series system, the BN formulation using FPSs is shown in Figure 4b. The size of the CPT associated with S_{sys} is 2^n and there is no computational advantage to this approach over the naïve formulation for series systems. These findings suggest that a combination of SPS and FPS formulations can be used to efficiently model general systems. This approach is described in the next section.

2.2 *Modeling performance of general systems with binary component states*

A MLS is a series system of its constituent components. Therefore, based on the above discussion, one can construct a SPS to describe each MLS. Consider the example system in Figure 5a, which has four MLSs: $MLS_1 = \{1, 7, 8\}$, $MLS_2 = \{2, 7, 8\}$, $MLS_3 = \{3, 7, 8\}$ and $MLS_4 = \{4, 5, 6, 7, 8\}$. In Figure 5b each MLS is modeled as an individual SPS. The SPEs, $E^j_{s,i}$, in each SPS are indexed by a subscript corresponding to the associated component *i* and a superscript corresponding to SPS/MLS number *j*. The dependence between the SPEs corresponding to the same component is modeled through a common parent node. The system node is in the survival state if the terminal node of any SPS is in the survival state. For reference, the BN formulation in which the MLSs are arranged in chain structures is named *efficient MLS BN formulation*. Similar logic leads to the creation of an *efficient MCS BN formulation*, whereby strings of FPSs are constructed corresponding to each MCS. The dependence between SPEs or FPEs sharing a component increases the computational demand when performing inference in the BN. By coalescing common SPEs/FPEs that appear in multiple SPSs/FPSs, the number of nodes and links in the BN, and hence the computational demand, are reduced. In the example system, components 7 and 8 appear in all SPSs. We take advantage of this observation and introduce only one "instance" of the SPEs associated with these

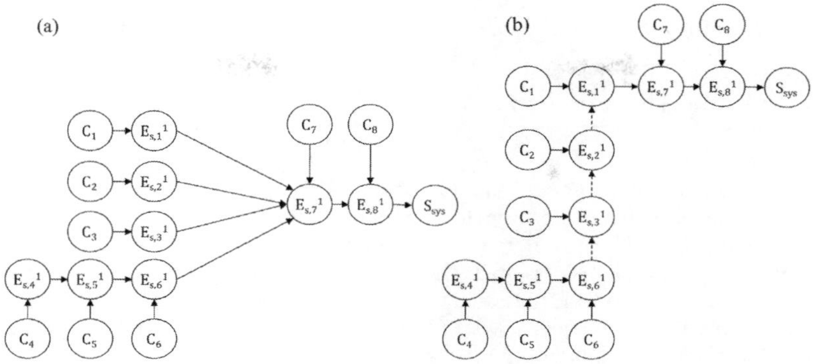

Figure 6. Efficient MLS formulations for the example system with coalesced SPEs associated with components 7 and 8 using (a) a converging structure (b) a chain structure.

components. The resulting BN is shown in Figure 6a. The states of SPE nodes having multiple SPEs as parents (e.g. node $E_{s,7}$ in Figure 6a) are specified using the Boolean relation

$$\begin{aligned} E_{s,i} &= 1 \text{ if } \left[\cup\{E_{s,\mathrm{Pa}(i)} = 1\}\right] \cap C_i = 1 \\ &= 0 \text{ otherwise} \end{aligned} \tag{3}$$

A notational change has been introduced in Figure 6a: the superscript associated with each SPE node, which previously indicated the MLS number, now represents the instance of the SPE, i.e. if multiple SPEs are associated with the same component, then they are recognized as different instances of the SPE and are distinguished through the superscript. Because, for this system, each component is associated with only one SPE, all superscripts in Figure 6a are 1.

It is noted that node $E_{s,7}$ in Figure 6a has more than 3 parents. Earlier, it was indicated that chain structures are more efficient than converging structures when the number of parents is greater than 3. Thus, the BN in Figure 6a is further modified by replacing the parallel SPE nodes associated with components 1, 2, 3, and 6 with nodes arranged in a chain, resulting in the BN in Figure 6b, with CPTs defined using the relation

$$\begin{aligned} E_{s,i} &= 1 \text{ if } \left[(C_i = 1) \cap \left(\cup\{E_{s,Pa(i)'} = 1\}\right)\right] \cup \{E_{s,Pa(i)''} = 1\} \\ &= 0 \quad \text{otherwise} \end{aligned} \tag{4}$$

where $E_{s,Pa(i)'}$ are the SPE nodes that are parent to $E_{s,i}$ before the addition of the chain modification and which remain parents after it; $E_{s,Pa(i)''}$ are the SPE nodes that become parents to $E_{s,i}$ after the chain structure is added (identified by dashed links in Figure 6b).

Thus far, the SPEs in a SPS (FPEs in FPSs) corresponding to a particular MLS (MCS) have been arranged in an arbitrary order. However, for complex systems, the arrangement of the SPEs in the SPSs may strongly influence our ability to coalesce SPEs in multiple SPSs (and analogously for FPEs in FPSs). The order in which SPEs appear can be optimized such that SPEs in as many SPSs as possible are coalesced. As mentioned earlier, this reduces the number of nodes and links in the BN. This optimization problem is described next. For brevity, only the formulation employing SPSs is presented; a dual formulation applies to FPSs and, in fact, the example at the end of this paper will use MCSs/FPSs.

Let $L(i^m, j^n) = 1$ indicate the existence of a directed link from $E^m_{s,i}$ to $E^n_{s,j}$ in the efficient MLS BN formulation and $L(i^m, j^n) = 0$ indicate its absence, where i and j are component indices and m and n are indices denoting the instances of these SPE nodes in the BN. Similarly, let $C^m_i = 1$ indicate a directed link between the node representing component i and node $E^m_{s,i}$ and $S^m_i = 1$ indicate a directed link between $E^m_{s,i}$ and the system node (with $C^m_i = 0$ and $S^m_i = 0$ respectively denoting their absences). $L(i^m, j^n)$, C^m_i and S^m_i are the decision variables in the optimization problem. Formulation of the optimization problem assumes the use of only SPE nodes and a converging structure at the system node. To further increase computational efficiency of the resulting BN, the converging structure at any node with more than 3 SPE nodes as parents is replaced by a chain structure in the manner described in Figure 6b.

The objective of the optimization problem is to minimize the number of links in the BN, i.e.

$$\min\left[\sum_{i=1}^{N_c}\sum_{j=1}^{N_c}\sum_{m=1}^{N_I}\sum_{n=1}^{N_I}L(i^m,j^n)+\sum_{i=1}^{N_c}\sum_{m=1}^{N_I}C_i^m+\sum_{i=1}^{N_c}\sum_{m=1}^{N_I}S_i^m\right] \tag{5}$$

where N_C is the number of components in the system and N_I is the maximum number of instances of any SPE. It is desirable that N_I be as small as possible, but its value is not known prior to solving the optimization problem. Thus, an iterative procedure is used to find the smallest N_I value under which the optimization problem is feasible. The existence of links between the component and SPE nodes as well as between the SPE nodes and the system node are controlled by the arrangement of SPE nodes in the BN. Specifically, $C_i^m = 1$ if node $E_{s,i}^m$ exists in the BN, which occurs if the decision variables indicate a link going into or out of node $E_{s,i}^m$. (A node without links going into or out of it can be removed from the BN.) Mathematically, this constraint is written as

$$\left[\sum_{j=1}^{N_C}\sum_{n=1}^{N_I}L(i^m,j^n)+L(j^n,i^m)\geq 1\right]\Rightarrow C_i^m=1 \tag{6}$$

Techniques are available for modeling "if-then" and "k-out-of-n" (which are needed later) constraints in numerical optimization (Sarker & Newton 2008). The decision variable $S_i^m = 1$ if node $E_{s,i}^m$ is a terminal node in a SPS, i.e. $E_{s,i}^m$ exists and has no other SPE node as a child. Mathematically, this is written as

$$\left[\sum_{j=1}^{N_C}\sum_{n=1}^{N_I}L(j^n,i^m)\geq 1\right]\cap\left[\sum_{j=1}^{N_C}\sum_{n=1}^{N_I}L(i^m,j^n)=0\right]\Rightarrow S_i^m=1 \tag{7}$$

There are two constraints governing the arrangement of the SPE nodes in the BN: (1) each MLS must be represented by a SPS; and (2) no SPS may exist that is not strictly a MLS. If the first constraint is violated, then one or more MLSs are excluded resulting in overestimation of the system failure probability. If the second constraint is violated, then the BN will include one or more fictitious MLSs and thus underestimate the system failure probability.

The first constraint requires that each MLS be represented as a SPS, i.e. at least one *permutation* of the SPEs associated with the components in each MLS must be connected as a chain. Define MLS_i to be the set of components contained in the ith MLS and let $N_{MLS,i}$ be the number of components in MLS_i. For the system in Figure 5a: $N_{MCS,1} = N_{MCS,2} = N_{MCS,3} = 3$ and $N_{MCS,4} = 5$. Let P_i be the set of permutations, without replacement, of the components in MLS_i and define $p_i^\alpha = \{p_{i,1}^\alpha, p_{i,2}^\alpha, \ldots, p_{i,N_{MCS,i}}^\alpha\}$ as the αth permutation contained in the set P_i. As an example, for the system in Figure 5a, $P_1 = [p_1^1 = \{8,7,1\}, p_1^2 = \{8,1,7\}, p_1^3 = \{7,8,1\}, p_1^4 = \{7,1,8\}, p_1^5 = \{1,7,8\}, p_1^6 = \{1,8,7\}]$.

Next, let Q_i be the set of permutations with replacement of $N_{MCS,i}$ draws from the index set $\{1,\ldots,N_I\}$. Define $q_i^\beta = \{q_{i,1}^\beta, q_{i,2}^\beta, \ldots, q_{i,N_{MLS,i}}^\beta\}$ as the set of instance indices ordered according to the βth member of Q_i. Using the same example and assuming $N_I = 2$, we have $Q_1 = \{q_1^1 = (1,1,1), q_1^2 = (1,1,2), q_1^3 = (1,2,1), q_1^4 = (1,2,2), q_1^5 = (2,1,1), q_1^6 = (2,1,2), q_1^7 = (2,2,1), q_1^8 = (2,2,2)\}$. Note that P_i has $N_{MLS,i}!$ members, while Q_i has $N_I^{N_{MLS,i}}$ members.

Define a set $r_i^{\alpha,\beta} = [r_{i,1}^{(\alpha,\beta)}, r_{i,2}^{(\alpha,\beta)}, \ldots, r_{i,N_{MLS,i}}^{(\alpha,\beta)}]$ which combines the elements of p_i^α and q_i^β. Specifically, $r_i^{\alpha,\beta}$ includes the set p_i^α with superscripts given by the set q_i^β. For the example system, $r_1^{1,1} = \{8^1, 7^1, 1^1\}$, $r_1^{1,2} = \{8^1, 7^1, 1^2\}$, $r_1^{2,4} = \{8^1, 1^2, 7^2\}$, etc. Overall, for this specific MLS, there are $3! * 2^3 = 48$ possible ways to arrange the component numbers given by p_i^α and the subscripts given by q_i^β.

For convenience, define the sum $X_i^{(\alpha,\beta)} = \sum_{l=1}^{N_{MLS,i}-1} L[r_{i,l}^{(\alpha,\beta)}, r_{i,l+1}^{(\alpha,\beta)}]$, where $r_{i,l}^{(\alpha,\beta)}$ is the lth element of $r_i^{(\alpha,\beta)}$. $X_i^{(\alpha,\beta)} = N_{MLS,i} - 1$ only if the SPEs corresponding to the components in MLS_i form a SPS in the order specified by p_i^α and with instance indices assigned by q_i^β. For a required SPS to exist in the BN, $X_i^{(\alpha,\beta)} = N_{MLS,i} - 1$ for *at least one* component/instance index ordering from the set $r_i^{(\alpha,\beta)}$. The constraint is written as

$$\max_{\alpha,\beta} X_i^{(\alpha,\beta)} \geq N_{MLS,i}-1 \qquad \forall i, \qquad \alpha = 1,\ldots,N_{MLS,i}!, \qquad \beta = 1,\ldots,N_I^{N_{MCS,i}} \tag{8}$$

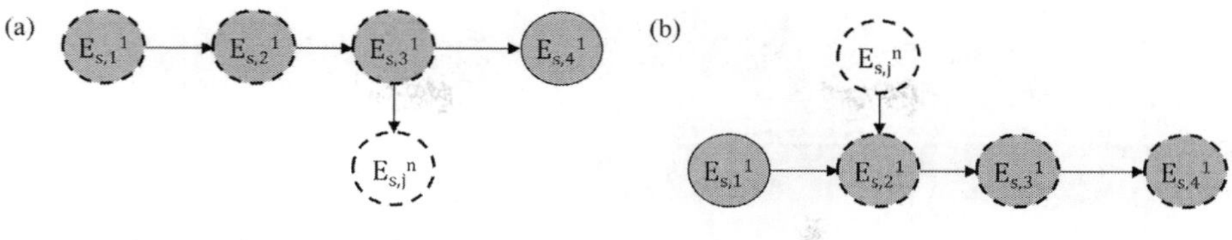

Figure 7. BN used to illustrate constraint (2).

The second constraint requires that no SPS exist in the BN which does not correspond to a MLS. Consider the BN shown in Figure 7a. Let the shaded nodes ($E_{s,1}^1 \rightarrow E_{s,2}^1 \rightarrow E_{s,3}^1 \rightarrow E_{s,4}^1$) represent a particular permutation of components/instance indices $r_i^{(\alpha,\beta)} = \{1^1, 2^1, 3^1, 4^1\}$ resulting in a valid SPS. Constraint (2) must prohibit a SPE $E_{s,j}^n$, for any n, from "branching-off" the SPS at any point (i.e. being a child of any node in the chain) unless the component j exists in a MLS with all preceding components in the sequence. For example, in Figure 7a, $E_{s,j}^n$ cannot exist as a child of $E_{s,3}^1$ unless components 1,2,3, and j all exist together in a MLS. If components 1,2,3, and j do not exist in a MLS, then the false survival path shown by nodes with dashed edges is introduced into the BN. The associated constraint is written as

$$\begin{gathered}\left[X_i^{(\alpha,\beta)} = N_{MLS,i} - 1, \forall i, \alpha, \beta\right] \Rightarrow \\ L\left[r_{i,l}^{(\alpha,\beta)}, j^n\right] = 0, \qquad \forall j: \left\{p_{i,1}^{(\alpha)}, \ldots, p_{i,l}^{(\alpha)}, j\right\} \notin MLS_m, \forall n, \forall m, \forall l\end{gathered} \tag{9}$$

Furthermore, the constraint must prohibit SPE $E_{s,j}^n$, for any n, from being a parent to any node in a valid SPS, unless component i exists in a MLS with all subsequent components in the sequence. For example, in Figure 7b, $E_{s,j}^n$ cannot be a parent of $E_{2,1}$ unless components 2,3,4, and j all exist together in a MLS. The second constraint takes the form

$$\begin{gathered}\left[X_i^{(\alpha,\beta)} = N_{MLS,i} - 1, \forall i, \alpha, \beta\right] \Rightarrow \\ L\left[j^n, r_{i,l}^{(\alpha,\beta)}\right] = 0 \quad \forall j: \left\{j, p_{i,l}^{(\alpha)}, \ldots, p_{i,N_{MLS,i}}^{(\alpha)}\right\} \notin MLS_m, \forall n, \forall m, \forall l\end{gathered} \tag{10}$$

The combination of these two requirements along with the objective function (the minimization of which ensures links that are not necessary for constructing the required SPSs are not in the BN) prohibits invalid SPSs in the BN. Combining (9) and (10) results in the constraint

$$\begin{gathered}\left[X_i^{(\alpha,\beta)} = N_{MLS,i} - 1, \forall i, \alpha, \beta\right] \Rightarrow \\ \sum_{m=1}^{N_{MLS}} \sum_{n=1}^{N_I} \sum_{l=1}^{N_{MLS,i}} \sum_{\forall j: \left\{p_{i,1}^{(\alpha)}, \ldots, p_{i,l}^{(\alpha)}, j\right\} \notin MLS_m} L\left[r_{i,l}^{(\alpha,\beta)}, j^n\right] + \\ \sum_{m=1}^{N_{MLS}} \sum_{n=1}^{N_I} \sum_{l=1}^{N_{MLS,i}} \sum_{\forall j: \left\{j, p_{i,l}^{(\alpha)}, \ldots, p_{i,N_{MLS,i}}^{(\alpha)}\right\} \notin MLS_m} L\left[j^n, r_{i,l}^{(\alpha,\beta)}\right] = 0\end{gathered} \tag{11}$$

The integer optimization problem described above requires consideration of permutations of components. Consequently, it becomes difficult to solve this problem in practice for large systems. To overcome this problem, several heuristics have been developed to reduce the size of the optimization problem that must be considered. Specifically, groups of components are considered as single "super-components," thus reducing the number of components, and measures are taken to eliminate unnecessary permutations of component indices. These heuristics will be described in an extended version of this paper.

To briefly illustrate the computational advantages of using the proposed efficient BN formulation, consider the structural system in Figure 8a consisting of 10 labeled components which can fail. The system has 11 MCSs: {1,2}, {3,4}, {1,3,10}, {1,4,10}, {2,3,10}, {2,4,10}, {5}, {6}, {7}, {8}, and {9}. The BN obtained using the optimization algorithm (including the use of heuristics) is shown in Figure 8b. An example of a heuristic adaption is observed when the single component MCSs are arranged as a series system using SPE nodes on the left of Figure 8b. Because this BN uses a mixture of FPE and SPE nodes, the relations between nodes require modifications of the equations given above. The total clique table size associated with this BN obtained using the optimization scheme is 164. The total clique table size of the MLS BN formulation with the converging structure is 5,140. Thus, the optimized BN is over an order of magnitude computationally more efficient.

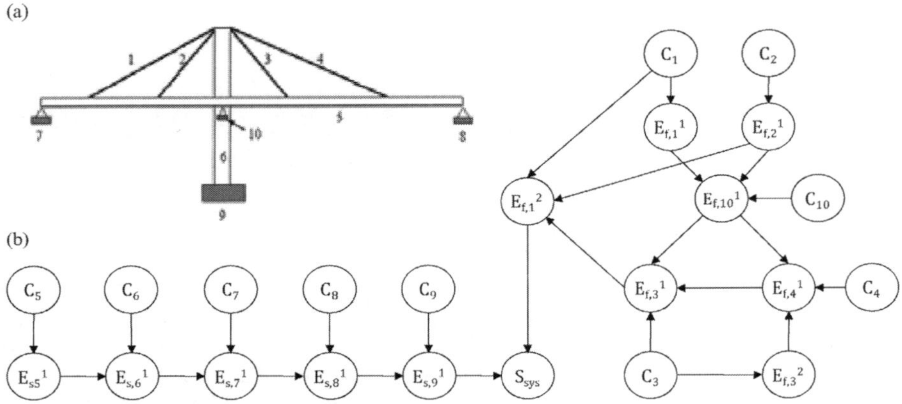

Figure 8. (a) Example structural system; (b) system performance BN.

3 CONCLUSIONS

This paper develops efficient BN formulations for modeling the performance of general systems for which the MLSs or MCSs are known. We show that BNs with nodes in chain structures are more efficient than when nodes are arranged in converging topologies. We demonstrate how BNs with nodes arranged in chain structures are constructed for series and parallel systems. This idea is then extended to develop similar topologies for general systems. Finally, a binary integer optimization program is developed with the goal of automatically constructing the BN such that the number of links is minimized. An example highlights the advantage gained from the efficient BN formulation.

REFERENCES

Bayraktarli, Y.Y., Yazgan, U., Dazio, A. & Faber, M., 2006. Capabilities of the Bayesian probabilistic networks approach for earthquake risk management. In *Proceedings of First European Conference on Earthquake Engineering and Seismology*. Geneva, Switzerland.

Bensi, M., Der Kiureghian, A. & Straub, D., 2010. Bayesian network for seismic infrastructure risk assessment: challenges and opportunities. In *Joint Conference Proceedings: 7th International Conference on Urban Earthquake Engineering and 5th International Conference on Earthquake Engineering*. Tokyo Institute of Technology, Tokyo, Japan.

Bensi, M., Straub, D., Friis-Hansen, P. & Der Kiureghian, A., 2009. Modeling infrastructure system performance using BN. In *Proceedings of the 10th International Conference on Structural Safety and Reliability*. Osaka, Japan.

Friis-Hansen, P., 2004. Structuring of complex systems using Bayesian Networks (with additional material from the unpublished extension to the paper, 2008). In O. Ditlevsen & P. Friis-Hansen, eds. *Proceedings of Two Part Workshop at DTU*. Technical University of Denmark.

Jensen, F.V. & Nielson, T.D., 2007. *Bayesian Networks and Decision Graphs* second., New York: Springer Science Business Media|Springer.

Kjaerulff, U.B. & Madsen, A.L., 2008. *Bayesian Networks and Influence Diagrams: A Guide to Construction and Analysis*, New York: Springer Science+Business Media, LLC.

Mahadevan, S., Zhang, R. & Smith, N., 2001. Bayesian networks for system reliability reassessment. *Structural Safety*, 23, 231–251.

Sarker, R.A. & Newton, C.S., 2008. *Optimization Modelling: a Practical Approach*, Boca Raton, Florida: CRC Press, Taylor and Francis Group, LLC.

Straub, D., 2005. Natural hazards risk assessment using Bayesian networks. In *Proceedings of 9th International Conference on Structural Safety and Reliability (ICOSSAR 05), Rome, Italy*.

Torres-Toledano, J. & Sucar, L., 1998. Bayesian Networks for Reliability Analysis of Complex Systems. In *Proceedings of the 6th Ibero-American conference on AI (IBERAMIA 98). Lecture notes in artificial intelligence, vol. 1484*. Springer-Verlag Berlin Heidelberg, pp. 195–206.

Reliability and Optimization of Structural Systems – Straub (ed)
© 2010 Taylor & Francis Group, London, ISBN 978-0-415-88179-1

Reliability analysis of a pressurized water reactor vessel using sparse polynomial chaos expansions

Géraud Blatman
EDF R&D, Site des Renardières, Moret-sur-Loing cedex, France

Bruno Sudret
Clermont Université, IFMA, Laboratoire de Mécanique et Ingénieries, BP 10448, Clermont-Ferrand, France
Phimeca Engineering S.A., Centre d'affaires du Zénith, Cournon, France

ABSTRACT: Polynomial chaos expansion have received much attention in the past ten years in the domain of computational stochastic mechanics. One keypoint in this context is the problem of the *curse of dimensionality*. In order to bypass this difficulty, the present paper proposes the use of *sparse* polynomial chaos basis built from a prior choice of monomials by the so-called hyperbolic q-norms and a selection algorithm known as *Least Angle regression*. The approach is illustrated by solving a structural reliability problem of a nuclear reactor vessel submitted to a pressurized thermal shock.

1 INTRODUCTION

Polynomial Chaos (PC) expansions allow one to represent explicitly the random response of a mechanical system whose input parameters are modelled by random variables. The PC coefficients may be efficiently computed using non intrusive techniques such as projection (Ghiocel and Ghanem, 2002) or regression (Berveiller et al., 2006). However, the required number of model evaluations (*i.e.* the computational cost) increases with the PC size, which itself dramatically increases with the number of input variables when the common truncation scheme of the PC expansion is applied (*i.e.* retain all the multivariate polynomials of total degree not greater than a prescribed p). To circumvent this problem, a truncation strategy based on the use of q-quasi-norms with $0 < q < 1$ is proposed. It is motivated by the so-called *sparsity-of-effects principle* (Montgomery, 2004), which states that most models are principally governed by main effects and low-order interactions. The related truncated PC expansions contain a low number of likely important terms compared to the full representation.

Using these truncation strategies, an adaptive method based on *Least Angle Regression* (LAR) (Efron et al., 2004) is proposed in order to retain progressively a small number of significant PC coefficients, leading to a *sparse* PC representation. Beside the adaptivity in terms of PC basis, the experimental design is systematically complemented such that the various regression problems are well-posed (Blatman and Sudret, 2008, 2010a). This may be achieved using *sequential sampling* techniques, *e.g. quasi-random numbers* (Niederreiter, 1992).

2 POLYNOMIAL CHAOS REPRESENTATION OF THE MODEL RESPONSE

2.1 *Mathematical framework*

Consider a mechanical system described by a numerical model $\mathcal{M}$ which can be analytical or more generally algorithmic (*e.g.* a finite element model). Suppose that this model has M uncertain input parameters which are represented by *independent* random variables $\{X_1, \dots, X_M\}$ gathered into a random vector $\boldsymbol{X}$ of prescribed joint probability density function $f_{\boldsymbol{X}}(\boldsymbol{x})$. Hence the model response denoted by $Y = \mathcal{M}(\boldsymbol{X})$ is also random. For the sake of simplicity, Y is assumed to be scalar throughout the paper (in case of a vector response $\boldsymbol{Y}$, the following derivations hold component-wise). Provided that the random variable Y has a finite variance, it may be recast as follows (Soize and Ghanem, 2004):

$$Y = \mathcal{M}(\boldsymbol{X}) = \sum_{\boldsymbol{\alpha} \in \mathbb{N}^M} a_{\boldsymbol{\alpha}} \psi_{\boldsymbol{\alpha}}(\boldsymbol{X}) \tag{1}$$

This expansion is referred to as the *polynomial chaos* (PC) *representation* of Y. The $a_{\boldsymbol{\alpha}}$'s are unknown deterministic coefficients and the $\psi_{\boldsymbol{\alpha}}$'s are multivariate polynomials which are orthonormal with respect to the joint PDF $f_{\boldsymbol{X}}$ of the input random vector $\boldsymbol{X}$, *i.e.* $\mathbb{E}[\psi_{\boldsymbol{\alpha}}(\boldsymbol{X})\psi_{\boldsymbol{\beta}}(\boldsymbol{X})] = 1$ if $\boldsymbol{\alpha} = \boldsymbol{\beta}$ and 0 otherwise. For instance, if $\boldsymbol{X}$ is a standard normal random vector, the $\psi_{\boldsymbol{\alpha}}$ are normalized multivariate Hermite polynomials.

2.2 *Computation of the coefficients by regression*

Let us consider a set of realizations of $\boldsymbol{X}$ denoted by $\mathcal{X} = \{\boldsymbol{x}^{(1)}, \ldots, \boldsymbol{x}^{(N)}\}$ and referred to as the *experimental design* (ED). Let us denote by $\mathcal{Y}$ the associated set of model response quantities, say $\mathcal{Y} = \{\mathcal{M}(\boldsymbol{x}^{(1)}), \ldots, \mathcal{M}(\boldsymbol{x}^{(N)})\}$.

The PC coefficients may be estimated by ordinary least square regression (Berveiller et al., 2006). This method requires the choice of a truncation of the PC *ab initio*, *i.e.* a non empty finite set $\mathcal{A} = \{\boldsymbol{\alpha}_0, \ldots, \boldsymbol{\alpha}_{P-1}\} \subset \mathbb{N}^M$ which contains the multi-indices of the retained basis polynomials $\{\psi_{\boldsymbol{\alpha}_0}, \ldots, \psi_{\boldsymbol{\alpha}_{P-1}}\}$. $\mathcal{A}$ is referred to as the *truncation set* in the sequel. The corresponding PC approximation is denoted by $Y_{\mathcal{A}} \equiv \mathcal{M}_{\mathcal{A}}(\boldsymbol{X})$.

The PC expansion is commonly truncated by retaining those polynomials whose total degree is not greater than p, which leads to the truncation set $\mathcal{A}^{M,p} = \{\boldsymbol{\alpha} \in \mathbb{N}^M \ : \ \|\boldsymbol{\alpha}\|_1 \equiv \sum_{i=1}^M \alpha_i \le p\}$. Accordingly, the number of PC terms is given by $P = \binom{M+p}{p}$. Hence it dramatically increases with both p and M. Consequently, the minimal size of the ED (*i.e.* the computational cost) that is required for an accurate solution of the regression problem blows up. This problem is tackled in Section 3.

2.3 *Post-processing*

The statistical moments of the response PC expansion can be analytically derived from its coefficients. In particular, the mean and the variance respectively read:

$$\widehat{\mu}_Y \equiv a_0 \tag{2}$$

$$\widehat{\sigma}_Y^2 \equiv \sum_{\boldsymbol{\alpha} \in \mathcal{A} \setminus \{\boldsymbol{0}\}} a_{\boldsymbol{\alpha}}^2 \tag{3}$$

Besides, it is also possible to compute the so-called *Sobol' sensitivity indices* (Sobol', 1993) at a negligible cost, see details in Sudret (2008). The Sobol' indices aim at quantifying the relative importance of each input parameter in the response variance:

$$S_i = \frac{\mathbb{V}\left[\mathbb{E}\left[\mathcal{M}(\boldsymbol{X})|X_i\right]\right]}{\mathbb{V}\left[\mathcal{M}(\boldsymbol{X})\right]} \tag{4}$$

where $\mathbb{E}[\cdot]$ and $\mathbb{V}[\cdot]$ are the mathematical expectation and variance operators, respectively. S_i is referred to as the *first order sensitivity index* of the model response to the input random variable X_i. This measure may be extended to each subset $\{X_{i_1}, \ldots, X_{i_s}\}$ of input random variables to quantify the interaction effects. Let us define by $\mathcal{I}_{i_1,\ldots,i_s}$ the set of $\boldsymbol{\alpha}$-tuples in $\mathcal{A}$ such that only the indices $\{i_1, \ldots, i_s\}$ are nonzero:

$$\mathcal{I}^p_{i_1,\ldots,i_s} = \left\{ \boldsymbol{\alpha} \in \mathcal{A}^p \ : \ \alpha_k = 0 \iff k \notin (i_1, \ldots, i_s) \quad , \forall k = 1, \ldots, M \right\} \tag{5}$$

It is easy to derive sensitivity indices from the PC representation:

$$\widehat{S}_{i_1,\ldots,i_s} = \frac{1}{\widehat{\sigma}_Y^2} \sum_{\boldsymbol{\alpha} \in \mathcal{I}^p_{i_1,\ldots,i_s}} a_{\boldsymbol{\alpha}}^2 \tag{6}$$

It is also possible to carry out inexpensively reliability analysis by intensive Monte Carlo simulation of the PC metamodel, as shown in Blatman (2007).

3 ADAPTIVE POLYNOMIAL CHAOS EXPANSIONS

Solutions for breaking the *curse of dimensionality* due to the classical truncation scheme introduced in Section 2.2 are now proposed.

3.1 *Hyperbolic polynomial chaos expansions*

Any truncation strategy of the PC representation corresponds to a specific choice of a non empty finite set $\mathcal{A}$ of indices $\boldsymbol{\alpha}$. For instance the common scheme is equivalent to the choice:

$$\mathcal{A} = \mathcal{A}^{M,p} \equiv \{\boldsymbol{\alpha} \in \mathbb{N}^M : \|\boldsymbol{\alpha}\|_1 \leq p\} \tag{7}$$

where $\|\boldsymbol{\alpha}\|_1 = \sum_{i=1}^{M} \alpha_i$

As an alternative, the use of the following index sets based on q-quasi-norms[1], $0 < q < 1$ is proposed:

$$\mathcal{A}_q^{M,p} \equiv \left\{ \boldsymbol{\alpha} \in \mathbb{N}^M : \|\boldsymbol{\alpha}\|_q \equiv \left(\sum_{i=1}^{M} \alpha_i^q \right)^{1/q} \leq p \right\} \tag{8}$$

Such quasi-norms penalize the high-rank indices all the more since q is low, as shown in Figure 1 for a two-dimensional case. This leads to favor the main effects and low-order interactions, which are more likely to be significant than the high-order interactions in the governing equations of the model according to the *sparsity-of-effects principle* (Montgomery, 2004).

Note that setting q equal to 1 corresponds to the usual truncation scheme, *i.e.* $\mathcal{A}_1^{M,p} = \mathcal{A}^{M,p}$. When using $q < 1$, the retained basis polynomials are located under an hyperbola-like surface, hence the name *hyperbolic index sets*. It may be shown that decreasing q results in a significant reduction of the number of terms compared to the usual truncated PC expansion (*i.e.* $q = 1$). For instance if $M = p = 10$, and say $q = 0.4$, the cardinality of $\mathcal{A}_q^{M,p}$ is 3 orders of magnitude smaller than that of $\mathcal{A}^{M,p}$. Note that $\mathcal{A}_q^{M,p}$ contains by construction univariate polynomials up to degree p and that $\lim_{p \to +\infty} \mathcal{A}_q^{M,p} = \mathbb{N}^M$ whatever the value of q: the enumeration scheme associated with the q-quasi-norm does not introduce any spurious truncation.

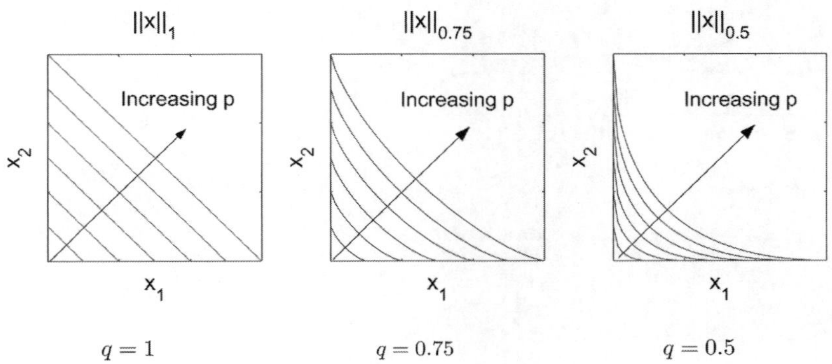

Figure 1. Principle of the truncation strategy based on q-quasi-norms ($0 < q \leq 1$).

[1] The quantity $\|\cdot\|_q$ defined in Eq.(8) is not rigorously a norm since it does not satisfy the triangular inequality.

3.2 *Adaptive algorithm based on Least Angle Regression*

Least Angle Regression (LAR) (Efron et al., 2004) is an efficient procedure for variable selection. In other words, it is aimed at selecting those predictors (*i.e.* the basis polynomials ψ_α in our case) that have the greatest impact on the model response $Y \equiv \mathcal{M}(\boldsymbol{X})$, among a possibly large set of candidates. LAR eventually provides a *sparse* PC approximation, *i.e.* which only contains a small number of terms compared to a classical full representation. A limitation of LAR lies in the requirement of an *a priori* truncation set $\mathcal{A}$. To circumvent this difficulty, one proposes a procedure for enriching iteratively the truncation set of the PC approximation, *i.e.* the set of active basis functions.

LAR is first applied to a first-order PC approximation corresponding to $\mathcal{A}_q^{M,1}$. The selection of the best LAR metamodel is performed using a modified cross-validation scheme (Blatman, 2007; Blatman and Sudret, 2010b). The associated optimal subset of indices (resp. error estimate) is stored in the variable $\mathcal{A}^{(1)}$ (resp. $\varepsilon^{(1)}$). Note that $\varepsilon^{(1)}$ has been already computed in order to select the optimal LAR-based PC approximation ($\varepsilon^{(1)}$ is the error estimate of the metamodel $\mathcal{M}_{\mathcal{A}^{(1)}}(\boldsymbol{X})$). One sets $\varepsilon^* \equiv \varepsilon^{(1)}$ and one denotes by $\mathcal{A}^*$ the corresponding set of indices. If ε^* is less than a prescribed target error ε_{tgt}, one stops the algorithm. Otherwise one iterates with the second-order PC approximation related to $\mathcal{A}_q^{M,2}$ and sets $\varepsilon^* \equiv \min(\varepsilon^*, \varepsilon^{(2)})$, and so on. Thus the sparse PC approximation is sought among the best LAR metamodels for each degree $p = 1, \ldots, p_{max}$.

It is possible that the approximation error $\varepsilon^{(p)}$ increases from a given value of the degree p. This might be due to an *overfitting* situation, in which the number of accepted predictors gets too important with respect to the size N of the experimental design $\mathcal{X}$. In order to avoid this phenomenon, the following heuristic criterion is introduced: if the approximation error $\varepsilon^{(p)}$ increases twice in a row (say $\varepsilon^{(p)} \geq \varepsilon^{(p-1)} \geq \varepsilon^{(p-2)}$), then the algorithm is stopped, returning a warning message of possible overfitting.

The optimal subset $\mathcal{A}^*$ is eventually retained. The coefficients of the related sparse PC approximation $\mathcal{M}_{\mathcal{A}^*}(\boldsymbol{X})$ are computed by ordinary least-square regression. As soon as overfitting is detected in the adaptive LAR iterations (*i.e.* when the error estimate increases twice in a row), the experimental design is enriched. Note that the number of additional points is fixed *a priori*. Then the procedure is restarded from the PC degree p that yielded the best metamodel prior to complementing the design. The computational flowchart of the algorithm is sketched in Figure 2.

The proposed procedure revealed accurate and efficient for conducting uncertainty, sensitivity and reliability analysis of various academic test models (Blatman, 2007). In particular, the algorithm allowed one to tackle problems with a high nominal dimension (*i.e.* up to 100 random variables)

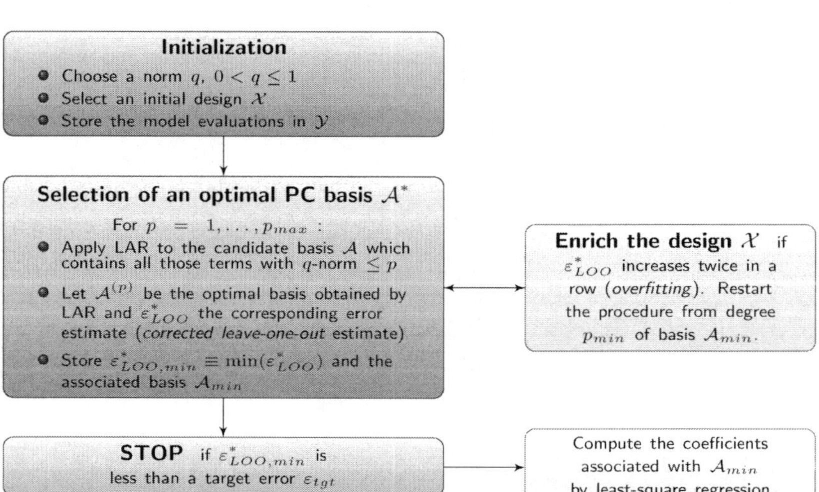

Figure 2. Computational flowchart of the basis-adaptive LAR procedure for building up an adaptive sparse polynomial chaos expansion.

using a relatively small number of model evaluations, whereas it would have been unaffordable using the classical full PC expansions.

4 ANALYSIS OF THE INTEGRITY OF THE REACTOR PRESSURE VESSEL OF A NUCLEAR POWERPLANT

4.1 *Problem statement*

The Reactor Pressure Vessel (RPV) is an essential component which could potentially limit the lifetime duration of Pressurized Water Reactor (PWR) nuclear power plants. The vessel core is subjected to time-varying thermal and mechanical loadings, called *transients*, resulting from the operation of the reactor in the normal situation, but also in an incidental or accidental situation. Certain types of transients, like those induced by a *loss of coolant accident* (LOCA), may induce rapid cooldown of the RPV with relatively high or increasing system pressure, which is called the *pressurized thermal shock* (PTS). The structural integrity of the RPV during PTS is assessed by assuming the presence of a crack as sketched in Figure 3. In this paper only longitudinal underclad cracks in the base metal are considered.

The structural integrity of the RPV is assessed by estimating the risk of crack propagation in the structure. To this end one computes the following safety margin:

$$z_{min} \equiv \min_{t \in [0,T]} (K_{IC}(t) - K_{\beta}(t)) \tag{9}$$

where $K_{IC}(t)$ (resp. $K_{\beta}(t)$) is the material fracture toughness (resp. the *plastic stress intensity factor*) at the crack edge in the base metal (Figure 3), at instant t of the transient loading, and T is the total duration of the transient.

Computing the safety margin in Eq.[9] is not an easy task since it is affected by uncertainty. Indeed, it depends on several uncertain variables. In our probabilistic framework, the latter are represented by random variables, which are listed in Table 1 together with their distribution type. Numerical values obtained from measurements have been omitted for the sake of confidentiality.

4.2 *Global sensitivity analysis*

From now on, the mean value of the fluence parameter, which corresponds to a given age of the RPV, is denoted by μ_{Φ}. The evolution in time of the global sensitivity of the random variable $Y \equiv \mathcal{M}_{\mu_{\Phi}}(\boldsymbol{X})$ to each input parameter $\{X_i, i = 1, \ldots, M\}$ and interactions thereof is considered. To this end, a parametric study is carried out varying the mean value μ_{Φ} of the fluence parameter Φ. This corresponds to values of the age of the powerplant varying from t_1 to t_5. The related safety margins are then denoted by $\mathcal{M}_{\mu_{\Phi}}(\boldsymbol{X})$. In this case, $\boldsymbol{X}$ is of size 10.

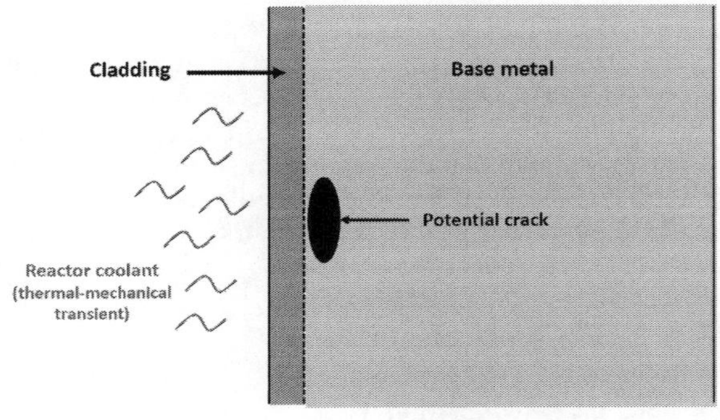

Figure 3. RPV structural integrity - Scheme of a RPV subjected to a thermal-mechanical transient and featuring an underclad crack.

Table 1. RPV structural integrity - Definition of the random variables.

Random variable	Symbol	Type
Crack heigth	h	Weibull
Crack aspect ratio	c	Lognormal
Cladding thickness	e	Uniform
Copper mass concentration	Cu	Normal
Phosphorus mass concentration	P	Normal
Nickel mass concentration	Ni	Normal
Fluence parameter	Φ	Normal
Transition temperature parameter	ξ_1	Normal
Initial transition temperature	RT_{NDT}^{ini}	Normal
Toughness parameter	ξ_2	Normal

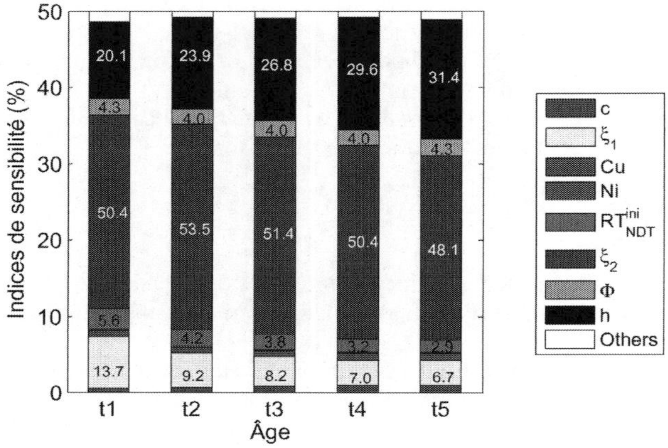

Figure 4. RPV structural integrity - Evolution in time of the global sensitivity indices.

Each model response $\mathcal{M}_{\mu_\Phi}(\boldsymbol{X})$ is approximated by a generalized PC representation. The PC basis is thus made both of normalized Legendre (for the uniform random variables) and Hermite (for the other random variables) polynomials. An hyperbolic truncation scheme is adopted using a $q = 0.4$-quasi-norm. Sparse PC expansions are determined using the adaptive LAR procedure. The target accuracy ε_{tgt} is set equal to 0.01. Lastly, the PC coefficients are computed using a sequential experimental design made of Sobol' quasi-random numbers. The initial size N_{ini} and the number of additional points when detecting overfitting N_{add} are both set equal to 100.

The total number of model evaluations for building the various PC approximations is equal to $1,800$. Estimates of the global sensitivity indices are obtained by elementary algebraic operations on the PC coefficients. The results are presented in Figure 4.

It appears that the most important variables are, in descending order, the intrinsic variability of the steel toughness, the crack height and the intrinsic variability of the transition temperature offset. The effects of the cladding thickness, the mass concentration of phosphorus as well as all the interaction effects are negligible. No significant evolution in time of the values of the Sobol' indices is observed. It may be noticed though that the importance of the crack height keeps increasing in time up to a value greater than 30% at t_5. The impact of all other variables decreases accordingly, except the one of the fluence parameter which remains rather constant.

4.3 *Reliability analysis*

The integrity of the pressure vessel is now assessed. The limit state function of the associated reliability problem is exactly the safety margin itself:

$$g(\boldsymbol{x}) = \mathcal{M}_{\mu_\Phi}(\boldsymbol{x}) \leq 0 \tag{10}$$

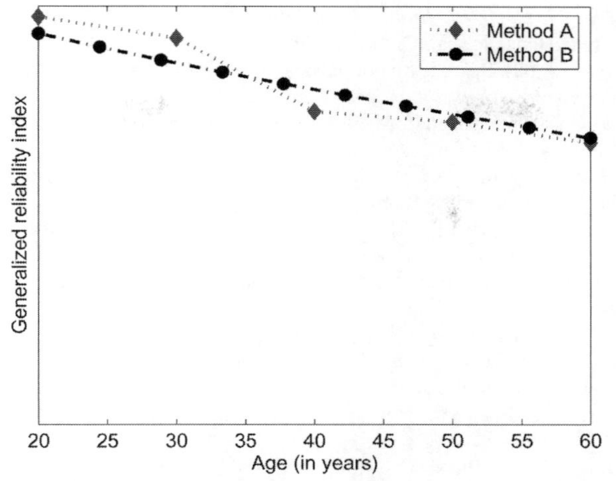

Figure 5. RPV structural integrity - Evolution in time of the generalized reliability index (labels have been discarded from the y-axis for the sake of confidentiality).

Two methods for computing the probabilities of failure at ages $t_1, \ldots, t_5$ are considered:

- *Method A*: as in the previous section, the various probabilities of failure are computed by post-processing separated sparse PC approximations for each age t_i;
- *Method B*: parameter μ_Φ is regarded as an additional input variable of the model $\mathcal{M}$ for computing the safety margin (μ_Φ is modelled by a *random* variable uniformly distributed over $[t_1, t_5]$); hence the various probabilities of failure are obtained by post-processing a single PC approximation.

Whatever the approach, a target error of $\varepsilon_{tgt} = 0.005$ is prescribed for building up the PC metamodels.

The results are depicted in Figure 5 in terms of generalized reliability indices $\widehat{\beta} \equiv -\Phi^{-1}(\widehat{P_f})$, together with the estimates obtained in the parametric study. It appears that both methods yield similar results, with a maximal relative discrepancy less than 10%. The required number or runs of the model is $N = 1,400$ when using Method B, instead of 2,900 when using Method A. Hence Method B allows the analyst to half the required number of model evaluations. Note that when using Method B, any number of vessel ages could be considered for the same global computational cost (*e.g.* 1,400 runs of the model), in contrast to Method A for which the computational cost is roughly proportional to the number of ages considered.

5 CONCLUSION

An adaptive LAR procedure is proposed to build up a sparse PC representation of the random response of a model with random input parameters. In order to reduce the number of unknown PC coefficients to identify and hence the required number of computer experiments (*i.e.* the computational cost), an adaptive algorithm is proposed for automatically detecting the significant PC terms (sparse PC representation). The procedure has been successfully applied to the sensitivity and reliability analysis of a vessel of a nuclear powerplant. An efficient method for carrying a parametric study at a reduced computational cost was also experimented in this study. Therefore, this method will be used in future investigations when parametric studies will be of interest.

REFERENCES

Berveiller, M., B. Sudret, and M. Lemaire (2006). Stochastic finite elements: a non intrusive approach by regression. *Eur. J. Comput. Mech. 15*(1–3), 81–92.

Blatman, G. (2007). *Adaptive stochastic finite element methods and metamodels for uncertainty propagation*. Ph. D. thesis, Université Blaise Pascal, Clermont-Ferrand. in preparation.

Blatman, G. and B. Sudret (2008). Sparse polynomial chaos expansions and adaptive stochastic finite elements using a regression approach. *Comptes Rendus Mécanique 336*(6), 518–523.

Blatman, G. and B. Sudret (2010a). An adaptive algorithm to build up sparse polynomial chaos expansions for stochastic finite element analysis. *Prob. Eng. Mech. 25*(2), 183–197.

Blatman, G. and B. Sudret (2010b). Adaptive sparse polynomial chaos expansion based on Least Angle Regression. *J. Comput. Phys.* submitted for publication.

Efron, B., T. Hastie, I. Johnstone, and R. Tibshirani (2004). Least angle regression. *Annals of Statistics 32*, 407–499.

Ghiocel, D. and R. Ghanem (2002). Stochastic finite element analysis of seismic soil-structure interaction. *J. Eng. Mech. 128*, 66–77.

Montgomery, D. (2004). *Design and analysis of experiments*. John Wiley and Sons, New York.

Niederreiter, H. (1992). *Random number generation and quasi-Monte Carlo methods*. Society for Industrial and Applied Mathematics, Philadelphia, PA, USA.

Sobol', I. (1993). Sensitivity estimates for nonlinear mathematical models. *Math. Modeling & Comp. Exp. 1*, 407–414.

Soize, C. and R. Ghanem (2004). Physical systems with random uncertainties: chaos representations with arbitrary probability measure. *SIAM J. Sci. Comput. 26*(2), 395–410.

Sudret, B. (2008). Global sensitivity analysis using polynomial chaos expansions. *Reliab. Eng. Sys. Safety 93*, 964–979. In press.

Reliability and Optimization of Structural Systems – Straub (ed)
© 2010 Taylor & Francis Group, London, ISBN 978-0-415-88179-1

The maximum of a stationary non-differentiable Gaussian process

Karl Breitung
Munich

ABSTRACT: The most mathematical models for time varying influences in reliability theory are stationary Gaussian processes. For these processes the asymptotic distribution of the maximum is known only for the case of differentiable and Markovian processes. Here the exact asymptotic distribution of the maximum is derived.

1 INTRODUCTION

In reliability problems time variant influences are modelled by stochastic processes. The most used are Gaussian processes, where all finite-dimensional distributions are Gaussian. Here the process is completely specified by its first and second order moments. If the process is stationary and standardized (i.e. $IE(x(t)) = 1$ and $\text{var}(t) = 1$), the processes are differing only in the autocorrelation function $r(t) = IE(x(0)x(t))$. This autocorelation function has near zero an expansion of the form:

$$r(t) = 1 - C \cdot |t|^{\alpha} + o(|t|^{\alpha}), t \to 0 \,, \tag{1}$$

with $0 \leq \alpha \leq 2$ and $C > 0$. If we are interested in the distribution of the maximum of the process, it is sufficient to consider only the case $C = 1$, since with a suitable time scaling a process where $C \neq 1$ can be transformed into one with $C = 1$.

In this case now, for a given stationary and standardized Gaussian process $x(t)$ where the autocorrelation function $r(t)$ has the expansion

$$r(t) = 1 - |t|^{\alpha} + o(|t|^{\alpha}), t \to 0 \,, \tag{2}$$

with $0 < \alpha \leq 2$, the maximum $M(1) = \max_{0 \leq t \leq 1} x(t)$ has the limiting distribution

$$\Pr(M(1) > u) \sim H_{\alpha} u^{2/\alpha} \Phi(-u), \; u \to \infty \tag{3}$$

with $\Phi(x)$ the standard normal integral and H_{α} a constant depending only on α. This result was first shown in Pickands III (1969b) and Pickands III (1969a). It is derived in Leadbetter et al. (1983), chap. 12 and with a different and much shorter method in Piterbarg (1996), p. 16. It is known only that $H_1 = 1$ and $H_2 = \pi^{-1/2}$. The first case includes the Ornstein-Uhlenbeck process which is Markovian and the second case is the class of differentiable processes.

We consider here only the case that $\alpha > 0$. For $\alpha = 0$ higher order expansions would be necessary.

2 POINT PROCESSES

In the following some results about point processes are needed, which are summarized here.

A point process $N(t)$ is defined by a sequence of points $\ldots, t_{-1}, t_0, t_1, t_2, \ldots$. Then $N(t)$ denotes the number of points in the interval $[0, t]$ and $N(t, s)$ the number of points in the interval $[t, s]$. A point process $N(t)$ is called stationary if its probability distributions are invariant under time shifts, i.e.

$$\Pr[N(s_1, t_1) = k_1, \ldots, N(s_m, t_m) = k_m] = \Pr[N(s_1 + h, t_1 + h) = k_1, \ldots, N(s_m + h, t_m + h) = k_m] \tag{4}$$

for all sets of m intervals $[s_1, t_1], \ldots, [s_m, t_m]$, all non-negative integers $k_1, \ldots, k_m$ and all real h. For a stationary process the following limit always exists

$$\lambda = \lim_{h \to 0} \frac{\Pr[N(0,h) \geq 1]}{h}. \tag{5}$$

This limit λ is called the intensity of the process. This means that for small $h > 0$ approximately $\Pr(N(0,h) \geq 1) \approx h\lambda$. A stationary point process is called regular (or orderly) if

$$\lim_{h \to 0} \frac{\Pr[N(0,h) > 1]}{h} = 0 \tag{6}$$

It means that for small intervals the probability of more than one point in them becomes negligible as the length goes to zero and that for small $h > 0$ approximately $\Pr(N(0,h) \geq 1) \approx h\lambda$.

For a regular point process one has (Korolyook theorem)

$$I\!E(N(0,1)) = \lim_{h \to 0} \frac{\Pr[N(0,h) = 1]}{h}. \tag{7}$$

The mean value is equal to the intensity (Cramer and Leadbetter 1967), p. 56.

3 UPCROSSINGS AND ϵ-UPCROSSINGS

For a sample path of a differentiable stationary Gaussian process $x(t)$ it is said to have an upcrossing of level u at time t_0 if (see figure 1.a)

$$x(t_0) = u \text{ and } x'(t_0) \geq 0. \tag{8}$$

The upcrossings of a level u form a stationary point process N_u:

$$t_{-2} \leq t_{-1} \leq t_0 \leq 0 \leq t_1 \leq t_2 \ldots \tag{9}$$

where t_n = time of the n-th upcrossing after 0.

This point process is also regular, i.e. between different points there is a positive distance with probability one. This is due to the differentiability of the paths, since at an upcrossing the derivative

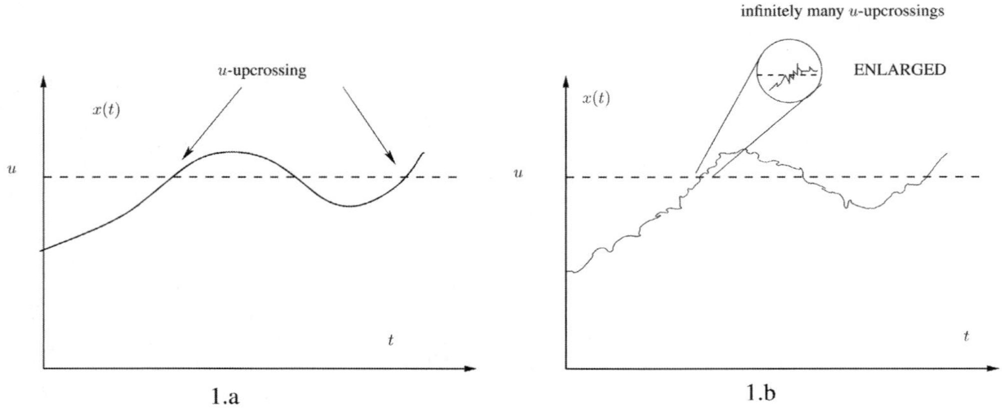

Figure 1. Upcrossings of differentiable processes (1.a) and of non-differentiable processes (1.b).

is positive with probability one, so the path will remain above u for some time. Therefore one can calculate its intensity

$$\mathbb{E}\left(N_u(0,1)\right) = \sqrt{\frac{-r''(0)}{2\pi}}\,\varphi(u) \tag{10}$$

with $\varphi(u)$ the standard normal density. For large u one obtains then

$$\Pr(M(0,1) > u) \approx \Pr(N(0,1) > 0) \approx \Pr(N(0,1) = 1) \approx \mathbb{E}(N(0,1)). \tag{11}$$

As $u \to \infty$ the point processes of u-upcrossings converge under scaling to a homogeneous Poisson process.

The behavior after an upcrossing is described by the Slepian model process (Leadbetter et al. (1983), p. 198–204). Under suitable scaling the path after an upcrossing has the form of a parabola. Let be defined

$$\zeta_u(t) = u(x(t/u) - u), \tag{12}$$

then if at $t = 0$ a u-upcrossing has occurred, then for $u \to \infty$

$$\zeta_u(t) \approx \frac{r''(0)t^2}{2} + \xi t. \tag{13}$$

where ξ has a Raleigh distribution with pdf $(-x/r''(0))\exp(x^2/(2r''(0))$.

This approach does not work if the process is not differentiable, i.e. if $r''(0)$ does not exist. In this case the limit goes to infinity (see figure 1.b). But this is possible to modify the concept such that it becomes applicable for these processes.

An ϵ-upcrossing above the level u of the process $x(t)$ (see figure 2) is a time point τ where

$$\max_{\tau-\epsilon\leq s\leq\tau} x(s) \leq u \quad \text{and for all } \eta > 0, x(t) > u \text{ for some } t \in (\tau, \tau+\eta). \tag{14}$$

The strictly stationary point processes $N_{\epsilon,u}$ of ϵ-upcrossings of level u converge as $u \to \infty$ under appropriate scaling towards a homogeneous Poisson process. From the definition of these processes follows that two points have always a distance not less than ϵ. Therefore they are regular point processes.

The intensities $\lambda_{\epsilon,u}$ of $N_{\epsilon,u}$ have the asymptotic form

$$\lambda_{\epsilon,u} \sim H_\alpha u^{2/\alpha}\Phi(-u),\ u \to \infty. \tag{15}$$

This follows from eq. (3).

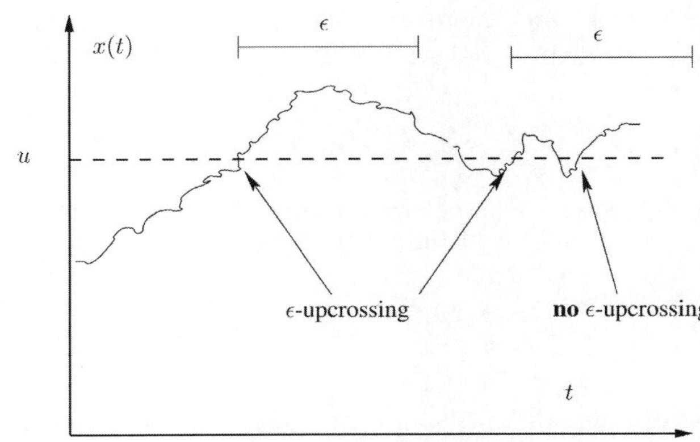

Figure 2. ϵ-upcrossings.

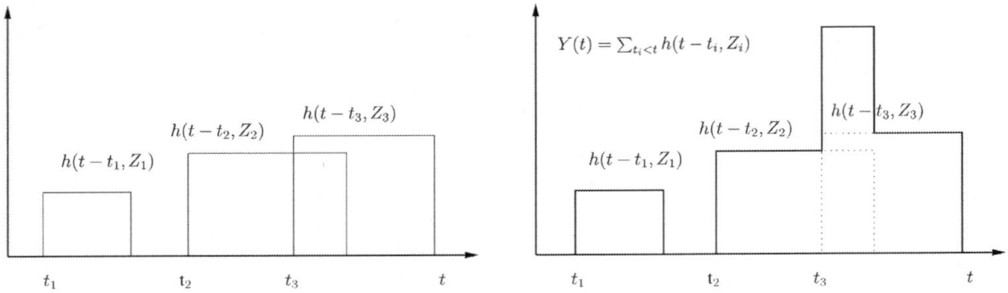

Figure 3. Filtered point process.

4 EXCEEDANCE MEASURES AND FILTERED POINT PROCESSES

To determine the values of H_α now we derive a relation between it and the time the process spends above the level u. For a stationary Gaussian process $x(t)$ we define the exceedance process $\eta_u(t)$ by

$$\eta_u(t) = \begin{cases} 1 & \text{if } x(t) > u, \\ 0 & \text{if } x(t) \leq u. \end{cases} \tag{16}$$

This process measures the time the process $x(t)$ spends above the level u. If $x(t)$ is a standardized stationary process, it follows (see Cramer and Leadbetter (1967), chap. 13.4, p. 291).

$$\int_0^T \eta_u(t)\, dt = T\mathbb{E}(\eta_u(t)) = T \cdot \Phi(-u). \tag{17}$$

To relate now the exceedance measures with the ϵ-upcrossing we have to introduce first the concept of filtered point processes. For a given stationary and regular point process $N(T)$ with points t_i a filtered point process is defined by

$$Y(t) = \sum_{t_i \leq t} h(t - t_i, Z_i). \tag{18}$$

Here $h(t, Z_i)$ is a random function depending on t and a rv Z_i. We assume that the function is zero for $t < 0$ so that the occurrence of the point t_i does not influence the past before t_i. The meaning of such a process is that each point t_i of the point process triggers a secondary process described by the influence function $h(t - t_i, Z_i)$. If the joint distributions are all stationary, also the resulting process $Y(t)$ is stationary.

For such processes there is a simple formula for calculating the mean of the process. Campbell's theorem says that the mean value of $Y(t)$ is given by

$$\mathbb{E}(Y(0)) = \lambda \times \int_0^\infty \lim_{x\to 0} \mathbb{E}(h(t, Z_1)|N(0,x) = 1)\, dt \tag{19}$$

with $\lambda = \mathbb{E}(N(0, 1))$ the intensity of the point process N. To understand this result, we approximate the process by a discretized version with time intervals of length Δ, it says then:

$$\begin{bmatrix} \text{Expected} \\ \text{output} \\ \text{at time } 0 \end{bmatrix} \approx \Delta^{-1} \sum_{i=0}^{\infty} \begin{bmatrix} \text{expected number of} \\ \text{points in } [-(i+1)\Delta, -i\Delta] \end{bmatrix} \times \begin{bmatrix} \text{expected influence of a} \\ \text{point in } [-(i+1)\Delta, -i\Delta] \\ \text{at time } 0 \end{bmatrix}. \tag{20}$$

Now the expected number of points in the interval Δ is $\lambda\Delta$. So in the limit as $\Delta \to 0$ we obtain Campbell's theorem. For the case that the underlying point process is a Poisson process this is shown in Papoulis (1965), p. 359. The general case is treated in Sigman (1995), p. 114.

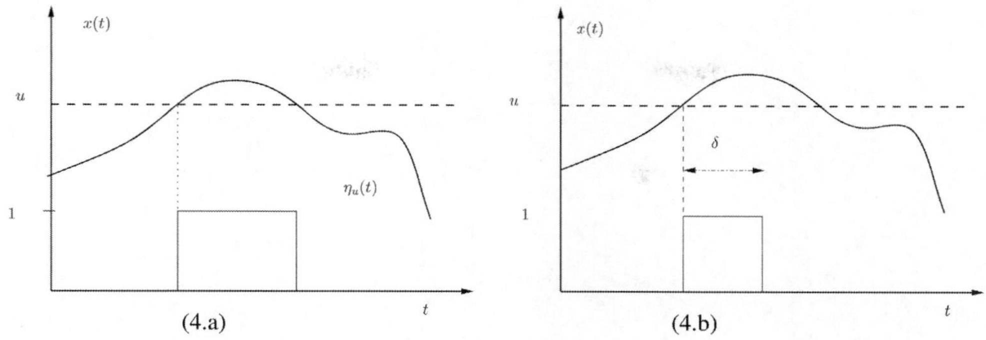

Figure 4. Exceedance process $\eta_u(t)$ (4.a) and $Y_{u,\epsilon,\delta}(t)$ (4.b).

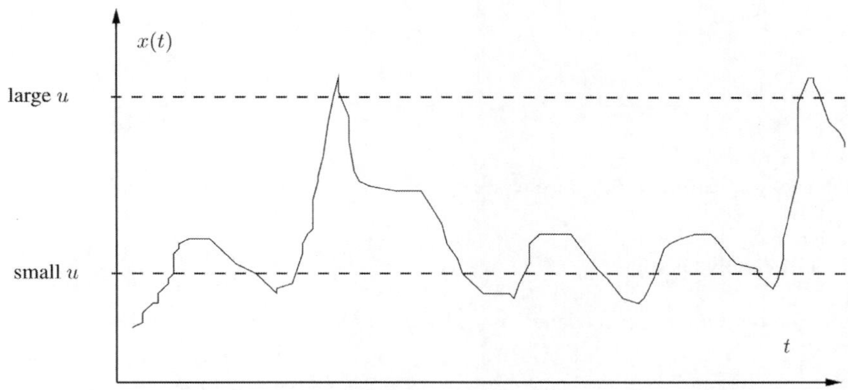

Figure 5. ϵ-upcrossings over small and large levels.

Given the point process $N_{\epsilon,u}$ of ϵ-upcrossings of u we define the influence function

$$h_{u,\delta}(t) = \begin{cases} 1, \text{ if } t \in [0,\delta] \text{ and } x(t) > u, \\ 0, \text{ elsewhere.} \end{cases} \tag{21}$$

With the point process of ϵ-upcrossings of the level u we define the filtered point process $Y_{u,\epsilon,\delta}(t)$

$$Y_{u,\epsilon,\delta}(t) = \sum_{t_i<t} h_{u,\delta}(t - t_i). \tag{22}$$

The term on the right side equal to unity for the time points after the ϵ-upcrossing at t_i where the path is still above u and the time points is not further than the distance δ after this crossing. This is not equal to the exceedance process $\eta_u(t)$. The difference is shown in figure 4. But it was shown in Leadbetter (1994) that asymptotically as $u \to \infty$, the exceedance process is approximated by the process $Y_{u,\epsilon,\delta}$. This is due to the fact that as $u \to \infty$ the ϵ-upcrossing become rarer and the length of the time the path stays above the crossed level becomes shorter (see figure 5). The distance between the ϵ-upcrossings goes to infinity and the length of stays above u to zero. Therefore

$$\Phi(-u) = I\!E(\eta_u(t)) \sim I\!E(Y_{u,\epsilon,\delta}(t)), \ u \to \infty \tag{23}$$

Remarkable is that the choice of ϵ and δ does not influence the asymptotic behavior.

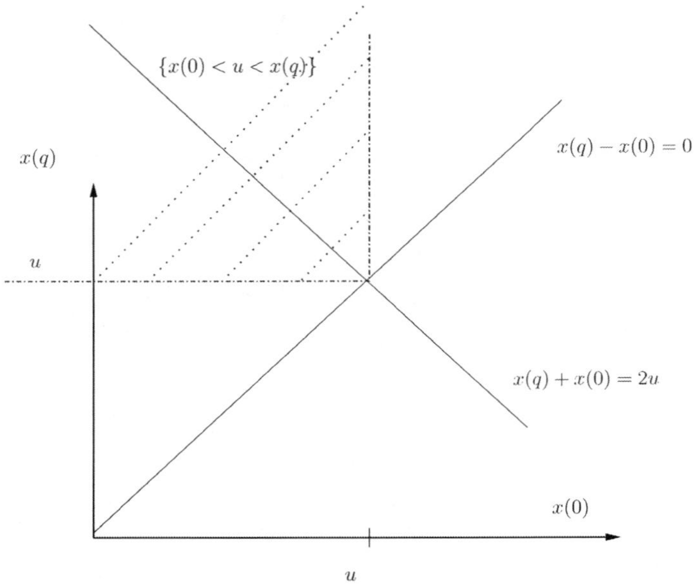

Figure 6. The event $\{x(0) < u < x(q)\}$ expressed by the rvs $x(0) + x(q)$ and $x(q) - x(0)$.

5 CALCULATION OF H_α

So the relation between the exceedance time above u and the filtered point processes is

$$\Phi(-u) \sim \lambda_{\epsilon,u} \int_0^\delta \Pr\left[x(t) > u | \ u\text{-upcrossing at } 0\right] \ dt \tag{24}$$

$$= u^{2/\alpha}\Phi(-u) \int_0^\delta \Pr\left[x(t) > u | \ u\text{-upcrossing at } 0\,\right] \ dt \tag{25}$$

It is sufficient to find the value of this integral for **one** process with $r(t) = 1 - |t|^\alpha + o(|t|^\alpha)$ to determine H_α. We use the process with $r(t) = \exp(-|t|^\alpha)$.

To evaluate the integral in eq. (24), we first derive the conditional distribution of $x(t)$ given an upcrossing of u at time 0. A discrete approximation for this event is

$$\{x(0) < u < x(q)\} \tag{26}$$

with q going to zero. This can be written in terms of the rvs $x(0) + x(q)$ and $x(q) - x(0)$ (see figure 6) which are independent as

$$\{x(0) < u < x(q)\} = \{x(q) - x(0) > 0, |x(0) + x(q) - 2u| < x(q) - x(0)\} \tag{27}$$

The correlation of $x(t)$ with these two rvs is

$$\text{corr}(x(t), x(0) + x(q)) = \frac{r(t) + r(t-q)}{\sqrt{2(1 + r(q))}} \to r(t) \text{as } q \to 0 \tag{28}$$

$$\text{corr}(x(t), x(q) - x(0)) = \frac{r(t-q) - r(t)}{\sqrt{2(1 - r(q))}} = K \cdot \lim_{q\to 0} q^{1-\alpha/2}. \tag{29}$$

The second limit is found using l'Hospital's rule; the constant $K > 0$ depends only on α and t. The second correlation goes to zero as $q \to 0$ if $0 < \alpha < 2$, for $\alpha = 2$ it goes to K. So for non-differentiable processes for $q \to 0$ the conditional distribution depends only on the value of $x(0)$

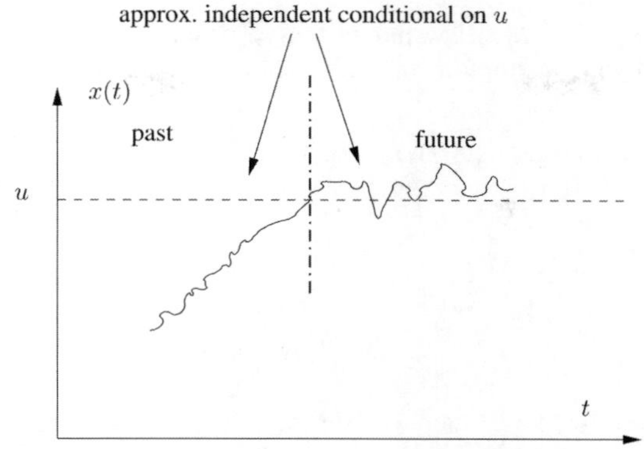

Figure 7. Approximate conditional independence of the near past and future of the process conditioned on an upcrossing.

which is u, since there is a u-upcrossing. From this we can derive by similar methods that the conditional distribution of $x(t)$ given an upcrossing at time 0 is a normal distribution with mean $r(t)u$ and variance $r^2(t)$ if $0 < \alpha < 2$. This is different from the case $\alpha = 2$, where the conditional behavior of the process after an upcrossing depends also on the derivative at the upcrossing time. This result was first shown for arbitrary autocorrelation functions by de de Maré (1977).

In a similar way we can show that conditional on an u-upcrossing at 0 the events $x(s) < u$ for all $s \in (-\epsilon, 0)$ and $x(t) > u$ for $t > 0$ become asymptotically independent as $u \to \infty$, i.e.

$$\Pr[x(s) < u \text{ for } -\epsilon < s < 0, x(t) > u | u\text{-upcrossing at } 0] \tag{30}$$

$$\approx \Pr[x(s) < u \text{ for } -\epsilon < s < 0 | u\text{-upcrossing at } 0] \times \Pr[x(t) > u | u\text{-upcrossing at } 0] \tag{31}$$

Using this, the integral in eq.24 simplifies to

$$\int_0^\delta \Pr[x(t) > u | u\text{-upcrossing at } 0]\, dt \sim \int_0^\delta \Pr[x(t) > u | x(0) = u]\, dt,\; u \to \infty \tag{32}$$

$$\Phi(-u) \;\sim\; u^{2/\alpha}\Phi(-u) H_\alpha \int_0^\delta \Phi\left(-\frac{-u(1-r(t))}{\sqrt{1-r^2(t)}}\right) dt \tag{33}$$

$$H_\alpha^{-1} \;\sim\; u^{2/\alpha} \int_0^\delta \Phi\left(-\frac{-u(1-r(t))}{\sqrt{1-r^2(t)}}\right) dt. \tag{34}$$

This can be evaluated by writing the integrand as the product of unity and the integrand and then integrating by parts:

$$\int_0^\delta 1 \cdot \Phi\left(-\frac{-u(1-r(t))}{\sqrt{1-r^2(t)}}\right) dt = \left[t \cdot \Phi\left(-\frac{-u(1-r(t))}{\sqrt{1-r^2(t)}}\right)\right]_0^\delta - \int_0^\delta t \cdot \varphi(-uh(t))\, uh'(t) dt \tag{35}$$

with $h(t) = -(1 - r(t))/(1 - r^2(t))$. The first term on the rhs is asymptotically negligible as $u \to \infty$. For $t \to 0$ one gets using the asymptotic form $r(t) \sim 1 - t^\alpha$ the following approximations for $h(t)$ and $h'(t)$

$$h(t) \;\sim\; \frac{t^{\alpha/2}}{\sqrt{2}} \tag{36}$$

Using the Laplace method for univariate integrals (see Erdélyi (1956), p. 37 or Breitung (1994), p. 48), one gets the following asymptotic approximation for the integral on the rhs of eq.(35)

$$-\int_0^\delta t \cdot \varphi(-uh(t))\, uh'(t)dt = u^{-2/\alpha} \frac{2^{1-2/\alpha}\sqrt{\pi}}{\Gamma(1/\alpha + 1/2)}. \tag{38}$$

This yields for H_α:

$$H_\alpha = \frac{2^{1-2/\alpha}\sqrt{\pi}}{\Gamma(1/\alpha + 1/2)}. \tag{39}$$

This is valid only for $0 < \alpha < 2$. At 2 there is a discontinuity, the value there is $\pi^{-1/2}$.

REFERENCES

Breitung, K. (1994). *Asymptotic Approximations for Probability Integrals*. Berlin: Springer. Lecture Notes in Mathematics, Nr. 1592.

Cramer, H. and M. Leadbetter (1967). *Stationary and Related Stochastic Processes*. New York: Wiley.

Erdélyi, A. (1956). *Asymptotic Expansions*. New York: Dover.

Leadbetter, M. (1994). Extremes and exceedance measures for continuously parameter stationary processes. In J. Galambos *et al.* (Ed.), *Extreme Value Theory and Applications*, Dordrecht, pp. 371–388. Kluwer. Proceedings of the Conference on Extreme Value Theory and Applications, Volume 1, Gaithersburg Maryland 1993.

Leadbetter, M., G. Lindgren, and H. Rootzén (1983). *Extremes and Related Properties of Random Sequences and Processes*. New York: Springer.

de Maré, J. (1977). The behaviour of a non-differentiable stationary Gaussian process after a level crossing. *Stochstic Process. Appl. 6*, 77–86.

Papoulis, A. (1965). *Probability, Random Variables and Stochastic Processes* (First ed.). Tokyo: McGraw-Hill Kogakusha, Ltd.

Pickands III, J. (1969a). Asymptotic properties of the maximum in a stationary Gaussian process. *Transactions of the American Mathematical Society 145*, 75–86.

Pickands III, J. (1969b). Upcrossing probabilities for stationary Gaussian processes. *Transactions of the American Mathematical Society 145*, 51–73.

Piterbarg, V. (1996). *Asymptotic methods in the Theory of Gaussian Processes and Fields*. Providence, RI: American Mathematical Society. Translations of mathematical monographs, vol. 148.

Sigman, K. (1995). *Stationary Marked Point Processes: An Intuitive Approach*. New York and London: Chapman and Hall.

Reliability and Optimization of Structural Systems – Straub (ed)
© 2010 Taylor & Francis Group, London, ISBN 978-0-415-88179-1

Introducing results from non-destructive testing into time variant safety analysis

Christian Bucher
Vienna University of Technology, Austria

ABSTRACT: Structures and infrastructure systems suffering from deterioration and damage due to corrosive actions and severe loads may eventually become unsafe. In order to prevent premature failure, monitoring can be implemented. Since structural monitoring essentially has to be non-destructive, direct information about the current strength or resistance of the structure is generally not available. Hence, resistance data must be inferred from observable or measurable data such as vibration characteristics or stiffness values, typically on a statistical basis. This introduces uncertainty into the process which can be modeled in terms of probability density functions. The paper presents a computational approach to combine Monte-Carlo simulation with analytical probabilistic models to assess the influence of observed stiffness values on the strength distribution model and on the prediction of expected failure rates for the future life time of the structure.

1 INTRODUCTION

Life-cycle considerations play an important role in the optimal design of structures and their maintenance planning (Frangopol, Kallen, and van Noortwijk 2004; Macke and Higuchi 2007). In this context it is essential to take into account the statistical uncertainty of loads and resistances in the expected life time of a structure. This is particularly important for overall life-cycle cost minimization including the important impact of maintenance planning and repair strategies. A very comprehensive survey on life-cycle oriented structural optimization utilizing a probabilistic basis is given by Frangopol and Maute 2003; Frangopol, Kallen, and van Noortwijk 2004). Further studies involving optimization under random actions and maintenance include (Higuchi and Bucher 2004; Bucher and Frangopol 2007; Bucher 2009b). In these approaches, it is necessary to have access to statistical information on the state of the structure, mainly regarding deterioration due to corrosive effects or damage due to possibly large loads. Therefore it is useful to introduce the results of structural monitoring on a probabilistic basis.

Due to the close relationship between structural properties such as stiffness and mass and vibration characteristics such as natural frequencies and mode shapes, it became quite popular to implement monitoring systems based on measurements of structural vibrations due to ambient excitations. One of the major purposes of such a monitoring system is to detect significant changes of the structural behavior early so that appropriate actions such as maintenance and repair can be taken (Carden and Fanning 2004).

Structural monitoring aims at being able to obtain better and more precise information on the actual state of the structure which is represented by the resistance R. Unfortunately, it is not possible to obtain the resistance directly without performing destructive tests. Hence, focus is on non-destructive testing and inferring resistance values R from the test results. As an important class, consider vibration monitoring concepts (see e.g. (Gul and Catbas 2008)). These strategies focus on the detection of changes in the dynamic behavior (i.e. the modal parameters (Peeters and de Roeck 1999; Chiuso and Picci 2001)) which are subsequently related to stiffness changes. Hence it is required to relate measured stiffness values K to actual strength values. Difficulties arise from the fact, that stiffness values cannot be measured directly but rather have to be inferred from modal response quantities such as natural frequencies and/or mode shapes.

The entire process involves several uncertain influences. The basic flow of information and actions is shown in Fig. 1. The initial vibration measurements are usually affected by measurement

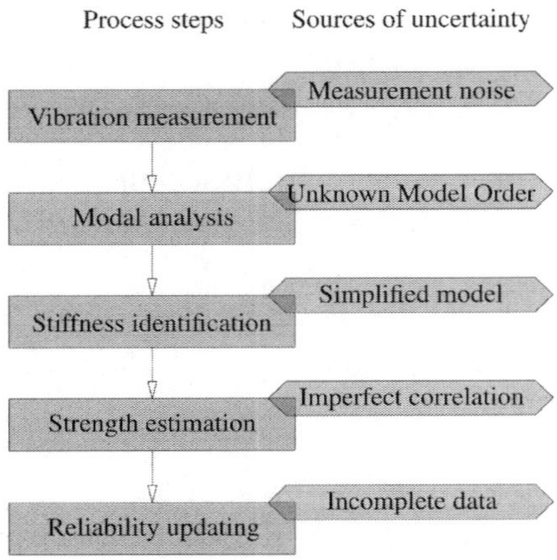

Figure 1. Reliability updating process based on vibration measurements.

noise due to random influences in the transducers and the recording equipment. In the next step, the essential modal characteristics of the structure are extracted from the measurement time series. This requires the establishment of a mathematical model whose properties are derived from certain properties of the measurements. As a result, natural frequencies and mode shapes can be obtained for a model of a certain order, in which the order of the model is arbitrary to some degree. In this context, it is referred to the literature on stochastic subspace identification (Peeters and de Roeck 2001). Then the modal changes are related to structural stiffness changes by means of a model updating procedure. Here, typically linear behavior of the structure is assumed. As at this stage the solution of an optimization procedure is required, there may be several problems associated with the specific nature of the optimization problem such as ill-posedness when many structural parameters have to be identified ((Bakir, Reynders, and de Roeck 2008)). Also, random fluctuations of structural properties as described by random fields may have an important impact on the identification results (Bucher, Huth, and Macke 2003). Hence the stiffness values may contain significant uncertainties. Furthermore, the essential quantity, i.e. the remaining strength of the structure or parts of the structure, is not perfectly correlated to the stiffness. Extensive tests comparing destructively measured strength values to non-destructively measured stiffness values indicate a statistical correlation of the order of 0.6 to 0.8 (see e.g. (Zhang and Wu 1997)). This, again, introduces further uncertainties. Finally, the prediction of future reliability involves assumptions on future loads, which may have statistical properties different from loads observed in the past.

The questions which follow from that are

- How does the stiffness value obtained from measurements affect the probabilistic description of the resistance?
- How does the updated probabilistic description of the resistance affect the reliability estimate?
- What are suitable computational procedures to perform the updating of the probabilistic information?

Conceptually, the answers to these questions rely on the notion of conditional probability. The present paper focusses on the aspects of relating uncertain information about stiffness values to probability-based predictions of future structural safety. The entire complex of system identification issues cannot be discussed here. A thorough survey can be found in e.g. in (Peeters and de Roeck 2001).

2 STATEMENT OF PROBLEM

In order to illustrate the problem from a basic viewpoint, consider a simple time-invariant structural reliability problem involving the random load $S > 0$ and the random resistance $R > 0$. Assume that

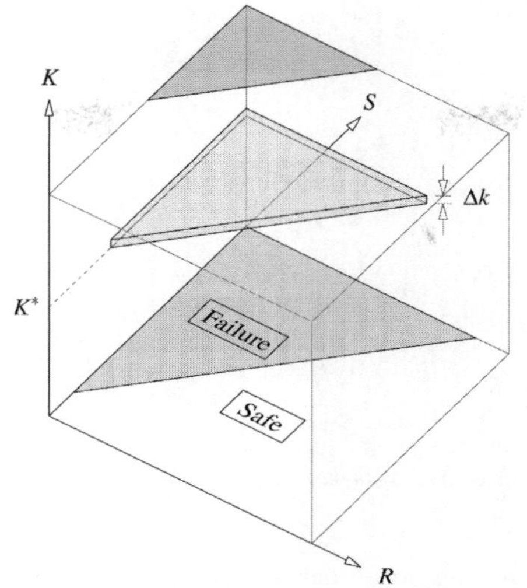

Figure 2. Failure domain depending on observed stiffness value K^*.

they are characterized by probability density function $f_S(s)$ and $f_R(r)$, respectively. The probability of failure P_F, i.e. the probability of $S \geq R$ can be computed from

$$P_F = \int_0^\infty \int_0^\infty \mathrm{H}(s-r) f_S(s) f_R(r) \mathrm{d}r \, \mathrm{d}s \tag{1}$$

in which H(.) denote the Heaviside unit step function which is zero for the argument being negative, and 1 for the argument being positive. If the influence of an observable additional variable (e.g. the stiffness K) should be included, then the event space of random variables must be extended to also include the variable K. We assume that the joint probability density function $f_{RK}(r,k)$ of R and K is known, typically in terms of the marginal densities $f_R(r), f_K(k)$ and the coefficient of correlation ρ_{RK}. Then the probability of failure can be written as

$$P_F = \int_0^\infty \int_0^\infty \int_0^\infty H f_S(s) f_{RK}(r,k) \mathrm{d}k \, \mathrm{d}r \, \mathrm{d}s \tag{2}$$

Once a particular value K^* of the stiffness has been observed, we can restrict the integration over k to a thin slice of thickness Δk around K^* (cf. Fig. 2) and renormalize by the probability of k falling into interval $[K^*, K^* + \Delta k]$.

$$P_F = \frac{1}{f_K(K^*)\Delta k} \cdot \int_0^\infty \int_0^\infty \int_{K^*}^{K^*+\Delta k} H(s-r) f_S(s) f_{RK}(r,k) \mathrm{d}k \, \mathrm{d}r \, \mathrm{d}s \tag{3}$$

which, upon taking the limit $\Delta k \to 0$, is equivalent to replacing the joint density of R and K in this equation by the conditional density:

$$f_{R|K} = \frac{f_{RK}(r,K^*)}{f_K(K^*)} \tag{4}$$

and from that

$$P_F = \int_0^\infty \int_0^\infty H(s-r) f_S(s) f_{R|K}(r,K^*) \mathrm{d}r \, \mathrm{d}s \tag{5}$$

Fig. 2 illustrates the shift of the failure domain due to different observed stiffness values.

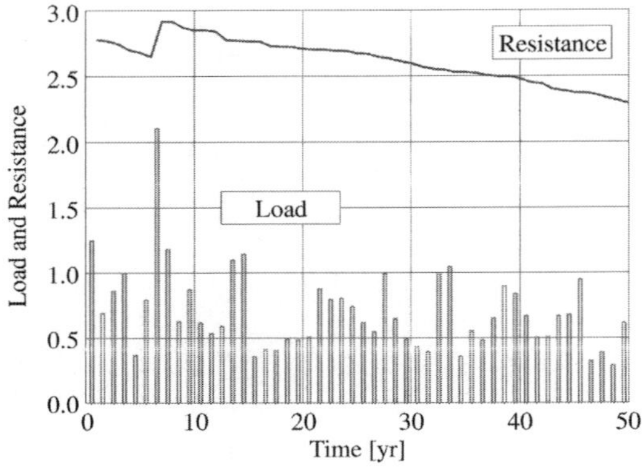

Figure 3. Sample history of load and resistance, repair due to large load at 7 years.

For the application to structures under the influence of deterioration and progressive damage, it is necessary to carry out a time-variant reliability analysis. In this case, it is more appropriate to use failure rates rather than failure probabilities. Nevertheless, the basic problem is exactly the same.

3 PROCESS MODEL

For the sake of simplicity, it will be assumed in the following that the loads acting on a structure or structural element are characterized by annual extreme events S_T. These loads are compared to the actual value of the resistance R_T. Resistance is assumed to deteriorate one one hand due to corrosive effects characterized by a deterioration rate ν_T. This rate depends mainly on environmental conditions. On the other hand, large loads below the actual capacity can introduce additional damage which should be compensated by proper maintenance or repair. This will result in a net change of resistance ΔR_T. The deterioration due to corrosive effects and small repeated loads is modeled by

$$\Delta R_T^c = R_T\left[1 - \exp\left(-\nu_T \Delta t\right)\right] \tag{6}$$

Here, and for the further analysis it is assumed that $\Delta t = 1$ year. Essential maintenance and repair are taken into account by applying improvement to the structural state after the occurrence of large loads S exceeding a repair threshold value S_{rep}. The repair results in an improvement of the resistance by the value ΔR_T^{rep}:

$$\Delta R_T^{rep} = R_T^{rep} \cdot \mathrm{H}(S_T - S_{rep}) \tag{7}$$

in which H(.) denotes the Heaviside unit step function. It is assumed that R_T^{rep} is a random variable. Hence, in each time interval T, the remaining resistance R_{T+1} is computed from

$$R_{T+1} = R_T\left[1 - \exp\left(-\nu_T \Delta t\right)\right] + R_T^{rep} \cdot \mathrm{H}(S_T - S_{rep}) \tag{8}$$

Failure at time $T+1$ is decided by comparing the resistance R_{T+1} to the load S_{T+1} to. Two sample time histories showing the development of structural resistance over time are given in Figs. 4 and 3. It can be seen from these figures that the simple maintenance strategy may or may not be effective.

4 NUMERICAL EXAMPLE

In this example, the probability density function of the initial resistance at time $i = 0$ is chosen to be lognormal with a mean value of $\bar{R}_0 = 3$ and a coefficient of variation of 0.15. The load S is assume to be Gumbel distributed with a mean value of $\bar{S} = 0.6$ and a coefficient of variation of 0.4. The

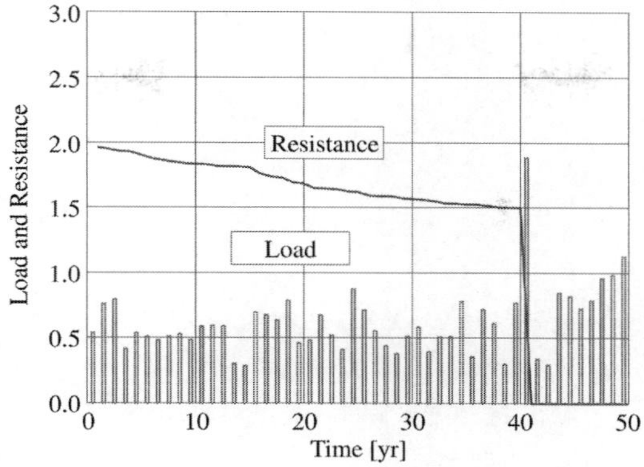

Figure 4. Sample history of load and resistance, failure due to large load at 41 years.

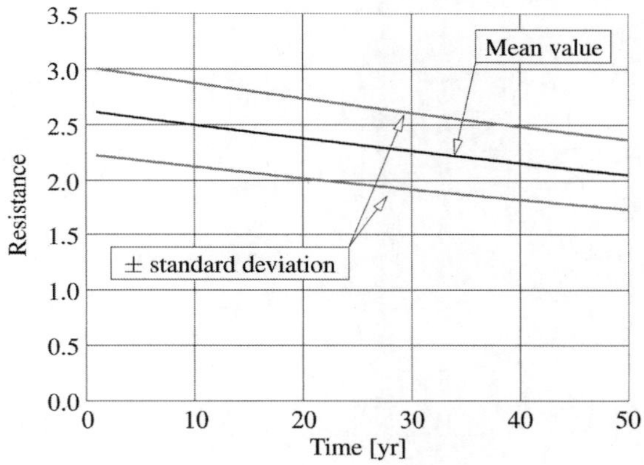

Figure 5. Statistics of resistance (mean value ± standard deviation).

annual deterioration rate is assumed to be exponentially distributed with a mean value $\bar{\nu} = 0.005$ and the maintenance effect is described by a repair threshold $S_{rep} = 2.0$ and a repair effect R^{rep} which is assumed to be a random variable uniformly distributed in the interval [0,0.5].

The temporal development of the structural resistance is analyzed using Monte-Carlo simulation of the load process. The statistics as obtained from 1,000.000 samples are shown in Fig. 5. There is a clear trend of the mean value toward smaller resistance values as the structural life time progresses. The overall uncertainty (as expressed by the standard deviation) is largely influenced by the initial uncertainty of the resistance, and not so much by the deterioration process. The corresponding failure rates as estimated directly from 1,000.000 samples are shown in Fig. 6. The increase of the failure rates over time is mainly due to the reduction of the mean resistance. This can also be seen from the temporary evolution of the probability density of the resistance as shown in Fig. 7. This figure shows the histograms of the resistance obtained from 1,000.000 samples at different values of T. These histograms are quite close to log-normal probability densities as can be seen from Figs. 8 and 9. The failure rates were re-computed based on the log-normal assumption, and the results are shown in Fig. 6 together with the direct simulation results. Although the log-normal densities match the simulation-based histograms very well, the agreement in the failure rates (cf. Fig. 6) was found to be somewhat less satisfactory.

Now stiffness measurements $K = K^*$ will be introduced, say at time $t = T^*$. In the following analysis, it will be assumed that the stiffness K_T at any time T is log-normally distributed with a coefficient of variation of 0.10. It is assumed to have a mean value proportional to the mean value of the strength R_T so that

$$\mathbf{E}[K_T] = A\,\mathbf{E}[R_T] \tag{9}$$

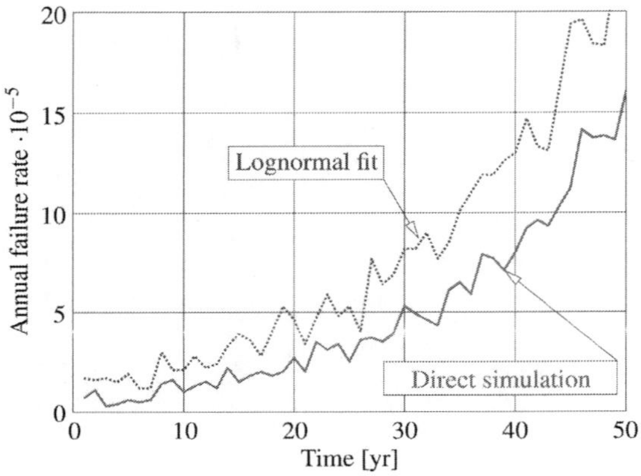

Figure 6. Failure rate vs. life time.

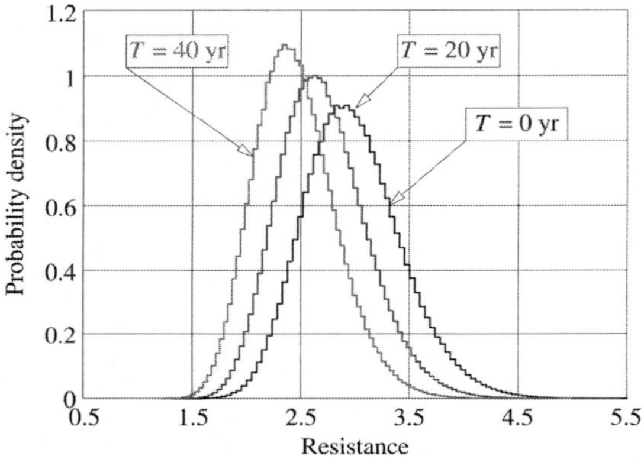

Figure 7. Histograms of resistance at different times T.

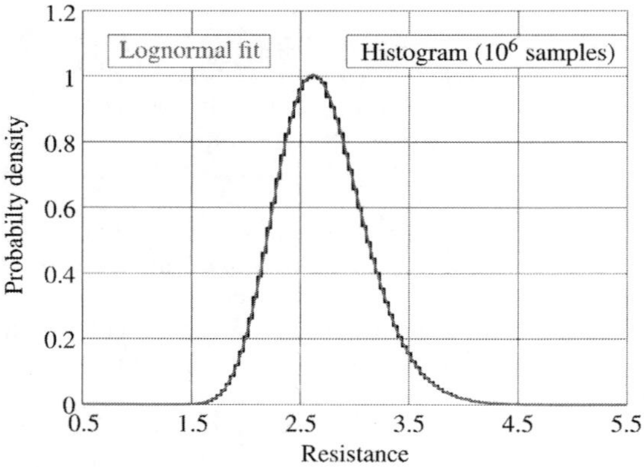

Figure 8. Histograms of resistance compared to log-normal density at $T = 20$ yr.

with a constant value of A. For the purpose of numerical analysis, it will be assumed that $A = 100$, although the actual value is immaterial in this context. Following the concept as outlined in section 2 it would be necessary to provide the joint probability density function of R_T and K_T. However, this is not an easy task since the temporal evolution of R is described only by a simulation procedure

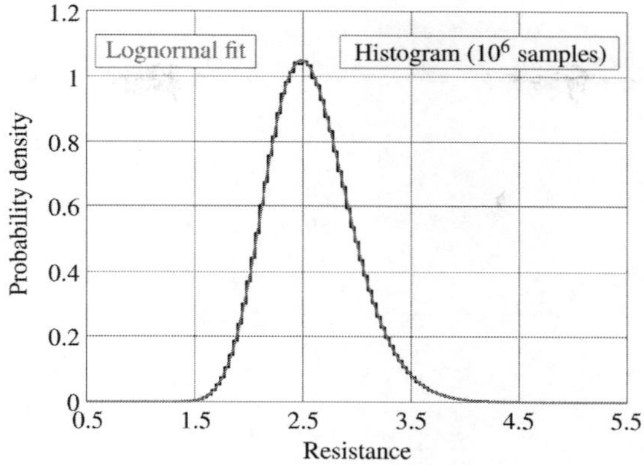

Figure 9. Histograms of resistance compared to log-normal density at $T = 30$ yr.

rather than an explicit formula. One possibility is to assume a semi-empirical copula (Genest and Rivest 1993) in which the marginal density of R is estimated from the simulation data and the marginal of K is assumed to be log-normal. This approach will not be pursued here. Instead, it will be assumed that the marginal of R is log-normal as well (with reasonable accuracy), and that the mean and standard deviation can be estimated from the simulation data on R_T. The accuracy of this assumption is discussed below.

The joint density function $f_{RK}(r,k)$ of resistance R and stiffness K is then assumed to follow the Nataf (Nataf 1962) model (i.e. a Gaussian copula is assumed). For two jointly log-normally distributed random variables, the joint pdf is given by (see e.g. (Bucher 2009a))

$$\begin{aligned} f_{RK}(r,k) &= \frac{1}{2\pi\, rk\; s_r s_k \sqrt{1-\rho'^2}} \cdot \\ &\cdot \exp\left[-\frac{v_r^2 - 2\rho' v_r v_k + v_k^2}{2(1-\rho'^2)}\right] \end{aligned} \tag{10}$$

$$v_r = \frac{\log r - \log \mu_r}{s_r}; \quad v_k = \frac{\log k - \log \mu_k}{s_k} \tag{11}$$

The parameters μ_r and s_r are related to the mean value $\bar{R} = \mathsf{E}[R]$ and the variance $\sigma_R^2 = \mathsf{E}[(R-\bar{R})^2]$ by

$$\begin{aligned} s_r &= \sqrt{\ln\left(\frac{\sigma_R^2}{\bar{R}^2} + 1\right)} \\ \mu_r &= \bar{R}\exp\left(-\frac{s_r^2}{2}\right) = \frac{\bar{R}^2}{\sqrt{\bar{R}^2 + \sigma_R^2}} \end{aligned} \tag{12}$$

Analogous relations hold for the stiffness K. The coefficient of correlation ρ' in $\mathbf{v}$-space can be computed as given by (Noh, Choi, and Du 2009)

$$\rho' = \frac{\log(1 + \rho s_r s_k)}{\sqrt{\log(1+s_r^2)\log(1+s_k^2)}} \tag{13}$$

Table 1. Measurement of stiffness at different times T^* resulting in different values K^*.

Case	T^*	K^*
A	20	280
B	30	280
C	20	230
D	30	230
0	–	–

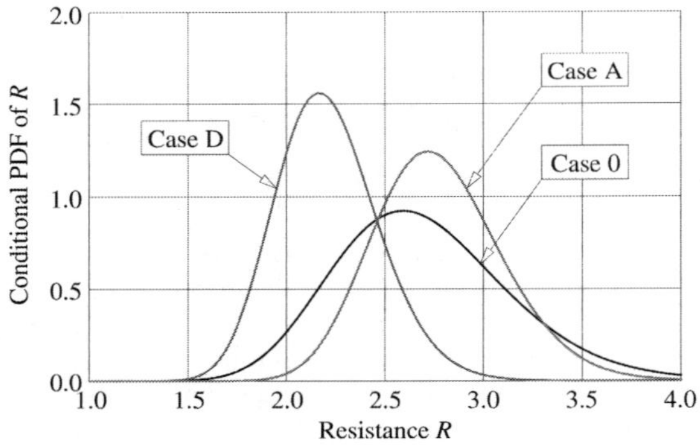

Figure 10. Conditional PDF of resistance given measured stiffness value K^*.

The marginal density $f_K(k)$ of the stiffness is given by

$$f_K(k) = \frac{1}{\sqrt{2\pi}k\ s_k} \cdot \exp\left(-\frac{v_k^2}{2}\right) \tag{14}$$

in which v_k is given by Eq. 11. Using the above relations, the conditional density $f_{R|K}(r, K^*)$ can be readily computed. It is a log-normal density

$$f_{R|K}(r, K*) = \frac{1}{\sqrt{2\pi}r\ s_r^*} \cdot \exp\left(-\frac{v_r^{*2}}{2}\right) \tag{15}$$

described by the parameters

$$s_r^* = \sqrt{1-\rho'^2} s_r; \quad v_r^* = \frac{\log r - \log \mu_r^*}{s_r^*}$$
$$\log \mu_r^* = \log \mu_r + \rho' \frac{s_r}{s_k} (\log k - \log \mu_k) \tag{16}$$

Based on these equations, the effect of a stiffness value K^* observed at time T^* on the (conditional) probability density function of R and on the prediction of expected failure rates can be readily quantified. For this purpose, four cases A–D as defined in Table 1 are considered. A fifth Case 0 without any observation is added for comparison. In all these cases it is assumed that the stiffness and resistance are correlated with $\rho = 0.7$.

Fig. 10 compares the conditional PDF of R at the same time time T^* for Cases A and D with observed stiffness values to Case 0 without observation. The mean values are shifted up or down

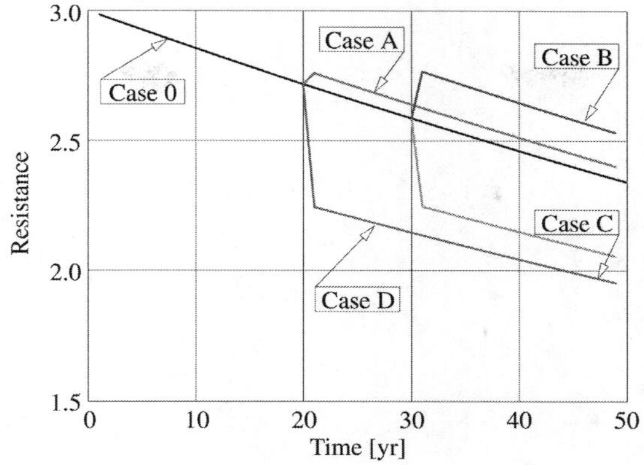

Figure 11. Conditional mean of resistance given measured stiffness value K^*.

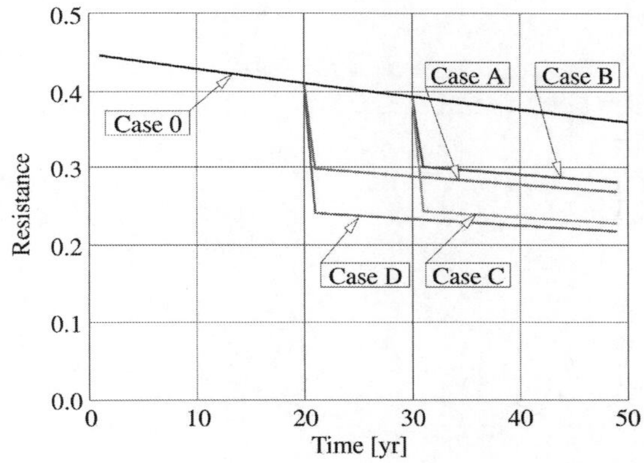

Figure 12. Conditional standard deviation of resistance given measured stiffness value K^*.

depending on the magnitude of the observed stiffness values, and the standard deviations are reduced in both Cases A and D.

After updating the PDF of R, the simulation is continued as described in section 3. The resulting second-order statistics of the resistance R are shown in Figs. 11 and 12. In principle, the mean values are shifted toward higher values if the observed stiffness is large, and vice versa. The standard deviations are reduced, as the uncertainty is reduced due to the observation. Finally, Fig. 13 shows the effect of different observed stiffness values K^* at different times T^* on the expected failure rates of the structure. The rates exhibit a similar effect of the observations as the mean values.

Obviously, the impact of the observed or measured values on the predicted resistance and failure rates significantly depends on the correlation between stiffness and resistance. For comparison, Fig. 14 shows results from Case A with correlation coefficients of 0.5 and 0.9 as well.

5 CONCLUDING REMARKS

An updating strategy for time-variant reliability analysis of deteriorating structures under non-destructive monitoring has been presented. The underlying analysis concept relies on a Monte-Carlo simulation approach to incorporate the effects of corrosion and load-induced damage together with essential maintenance into a numerical model. This simulation is combined with probabilistic modeling using appropriate joint density functions for observable stiffness and non-observable strength.

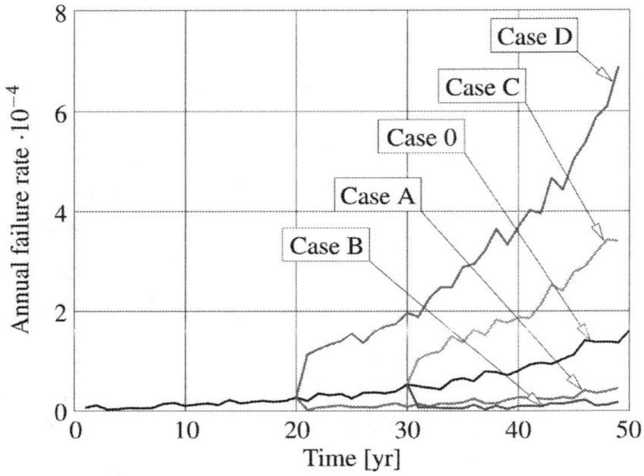

Figure 13. Conditional failure failure rate given measured stiffness value K^* at different times T^*.

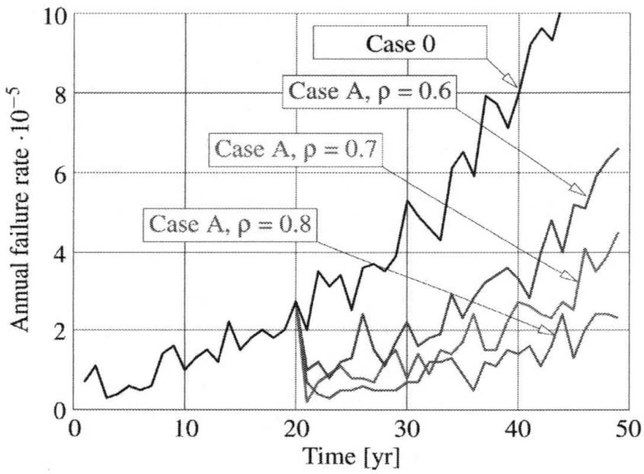

Figure 14. Conditional failure failure rate given measured stiffness value K^* for different correlations between stiffness and resistance.

The effect of introducing the observed values of the stiffness can be categorized as follows

- The mean value of the strength can be increased or decreased, depending on the magnitude of the observed stiffness. Values of the observed stiffness larger than its expected value will increase the updated mean of the strength, and vice versa.
- The standard deviation of the strength is reduced by the observation.
- The expected failure rates are significantly influenced by the observed stiffness values. The rates will increase if the observed stiffness is smaller than its expected value, and it will increase otherwise.

The deterioration and maintenance models are used in this paper are very simple, and should merely illustrate the basis principles of an updating procedure when the relevant quantities (strength) cannot be directly observed, but must rather be inferred from statistically correlated but physically different properties of the structure. Further studies will focus on the application to structural models involving more details regarding loading conditions, deterioration processes and failure criteria.

Another pint of future research is the application of a simulation.driven joint probability density model for stiffness and strength thus eliminating the need for analytical distribution models.

REFERENCES

Bakir, P. G., E. Reynders, and G. de Roeck (2008). An improved finite element model updating method by the global optimization technique 'coupled local minimizers'. *Computers and Structures 86*, 1339—1352.

Bucher, C. (2009a). *Computational analysis of randomness in structural mechanics*. Structures and Infrastructures Book Series, Vol. 3. London: Taylor & Francis.

Bucher, C. (2009b). Robustness analysis in structural optimization. *Structure and Infrastructure Engineering 5*(4), 287–294.

Bucher, C. and D. M. Frangopol (2007). Optimization of combined lifetime maintenance strategies. In G. Deodatis and P. D. Spanos (Eds.), *Computational Stochastic Mechanics 5*, pp. 103–107.

Bucher, C., O. Huth, and M. Macke (2003). Accuracy of system identification in the presence of random fields. In A. DerKiureghian, S. Madanat, and J. Pestana (Eds.), *Applications of Statistics and Probability in Civil Engineering*, pp. 427–433. Millpress.

Carden, E. P. and P. Fanning (2004). Vibration based condition monitoring: A review. *Structural Health Monitoring 3*(4), 355–377.

Chiuso, A. and G. Picci (2001). Some algorithmic aspects of subspace identification with inputs. *Int. J. Appl. Math. Comput. Sci. 11*(1), 55–75.

Frangopol, D. M., M.-J. Kallen, and J. M. van Noortwijk (2004). Probabilistic models for life-cycle performance of deteriorating structures: review and future directions. *Prog. Struct. Engng Mater. 6*, 197–212.

Frangopol, D. M. and K. Maute (2003). Life-cycle reliability-based optimization of civil and aerospace structures. *Computers and Structures 81*, 397–410.

Genest, C. and L.-P. Rivest (1993). Statistical inference procedure for bivariate archimedean copulas. *Journal of the American Statistical Association 88*(423), 1034–1043.

Gul, M. and F. N. Catbas (2008). Ambient vibration data analysis for structural identification and global condition assessment. *Journal of Engineering Mechanics 134*(8), 650–662.

Higuchi, S. and C. Bucher (2004). Performance-based optimal design of structures. In *Proceedings, 9th ASCE Specialty Conference on Probabilistic Mechanics and Structural Reliability*.

Macke, M. and S. Higuchi (2007). Optimizing maintenance interventions for deteriorating structures using cost-benefit criteria. *Journal of Structural Engineering 133*(7), 925–934.

Nataf, A. (1962). Détermination des distributions de probabilités dont les marges sont données. *Comptes rendus de l'Académie des sciences 225*, 42–43.

Noh, Y., K. Choi, and L. Du (2009). Reliability-based design optimization of problems with correlated input variables using a gaussian copula. *Struct. Multidisc. Optim. 38*, 1–16.

Peeters, B. and G. de Roeck (1999). Reference-based stochastic subspace identification for output-only modal analysis. *Mechanical Systems and Signal Processing 13*(6), 855–878.

Peeters, B. and G. de Roeck (2001). Stochastic system identification for operational modal analysis: A review. *Journal of Dynamic Systems, Measurement, and Control 123*(12), 659–667.

Zhang, B. and K. Wu (1997). Residual fatigue strength and stiffness of ordinary concrete under bending. *Cement and Concrete Research 27*(1), 115–126.

Reliability and Optimization of Structural Systems – Straub (ed)
© 2010 Taylor & Francis Group, London, ISBN 978-0-415-88179-1

Damage identification by Lévy ant colony optimization

G. Cottone
Engineering Risk Analysis Group, TU München, Germany
DISAG, Università di Palermo, Italy

A. Pirrotta & G. Fileccia Scimemi
Dipartimento di Ingegneria Strutturale, Aerospaziale e Geotecnica, Università di Palermo, Italy

E. Riva Sanseverino
Dipartimento di Ingegneria Elettrica, Elettronica e delle Telecomunicazioni, Università di Palermo, Italy

ABSTRACT: This paper deals with the identification of incipient damage in structural elements by non-destructive test based on experimentally measured structural dynamical response. By application of the Hilbert transform to the recorded signal the so-called phase of the analytical signal is recovered and a proper functional is constructed in such a way that its global minimum gives a measure of the damage level, meant as stiffness reduction. Minimization is achieved by applying a modified Ant Colony Optimization (ACO) for continuous variables, inspired by the ants' foraging behavior. The modification consists in the application of a new perturbation operator, based on alpha stable Lévy distribution. Numerical application to three degrees of freedom system is proposed to show the method and test the performance.

1 INTRODUCTION

Detecting and quantifying structural damage at early stage by means of non-destructive tests is of primary concern in many engineering applications. Most of non-destructive techniques are based on the correlation between measured data (in the majority of cases from dynamic tests) and the response of an analytical model of the damaged structure (model-based procedures). Different structural response characteristics can be used for this purpose including modal data (Adams et al. (1991), Davini et al. (1992) Gounaris et al. (1996), Cerri & Vestroni (2000)), curvature measures (Pandey et al. (1991), Luo & Hanagud (1997), Ratcliffe (1999)), frequency response functions (Wang et al. (1997), Thyagarajan et al. (1998), Sampaio et al. (1999), Lee & Shin (2002)) and strain energy (Hearn & Testa (1991), Shi et al. (1998), Cornwell et al. (1999)). However, frequency-based identification procedures often fail to give a satisfactory answer for very low damage levels. Indeed, especially for small damage extent, the response in terms of impulse response function and amplitude is almost insensible to the system perturbation. Conversely, it has been recently shown, in Cottone et al. (2008) that a damage detection strategy based on a proper functional calculated on the analytical signal of the structural dynamical response consents to identify very low damage levels. Moreover, it has been shown that if one applies a system of filters to the experimental signal in order to isolate the single modal component of the dynamical response, the method can be effective also in presence of instrumental noise and for damages occurring in more points of the structure. The necessity of the filtering operation relies on the fact that, in the above mentioned case, the functional to be minimized may have many local minima and most of the optimization procedures do not succeed to find a correct value of the damage level. It has been shown that a way to circumvent the excess of local minima problem is by isolating the structural modal components, through filtering operations on the signals (refer to Cottone et al. (2008) and Barone et al. (2008)). On the other hand, in some practical case, for example dealing with structures whose principal frequencies are too narrow, the procedure based on the filtering is not suitable.

In this paper, we apply the Ant Colony Optimization for continuous domains ($ACO_{\mathbb{R}}$) already proposed in Socha & Dorigo (2008), introducing a perturbation operator with α-stable Lévy distribution. The choice of the α-stable Lévy distribution, with its particular inverse power-law tails, is justified because it has been observed that the wider exploration potential of such a distribution, makes the algorithm to attain better performance than $ACO_{\mathbb{R}}$ based on Gaussian distribution for function showing many local minima. In this way, we show that it is possible to perform the minimization of the functional without extrapolating the modal component by filtering the structural response. Applications to structures with narrow principal frequencies show the good performance in detecting an incipient damage.

The paper is organized as follow. Firstly, the Hilbert transform of the structural response is introduced in order to represent the analytical signal, and then a proper functional in terms of the phase of the analytical signal whose minimum corresponds to the damage level, is constructed. Then, a modified $ACO_{\mathbb{R}}$ with Lévy perturbation will be presented following the papers Candela et al. (2009) and Cottone et al. (2010). Lastly, numerical examples are given, considering a 3 degrees of freedom shear type structure.

2 DAMAGE IDENTIFICATION BY ANALYTICAL SIGNAL

2.1 *Hilbert transform and analytical signals*

In this section we briefly recall some features of the analytical signal in order to set the appropriate symbols. Let $x(t)$ denote the response of a linear system, where t indicates the time. Its Hilbert transform is defined as

$$\hat{x}(t) = \frac{1}{\pi}\int_{-\infty}^{\infty} \frac{x(t)}{t-\tau} \mathrm{d}\tau \tag{1}$$

where the integral is meant in the Cauchy principal value sense. The so-called *analytical signal* (refer to Oppenheim & Schafer, (1989)) is the complex signal defined as $z(t) = x(t) + \mathrm{i}\,\hat{x}(t)$, with $\mathrm{i} = \sqrt{-1}$. The function $z(t)$ can also be expressed in terms of the amplitude function $A(t)$ and the phase function $\theta(t)$ as

$$z(t) = A(t)\exp(\mathrm{i}\theta(t)) \tag{2}$$

with

$$A(t) = \sqrt{x(t)^2 + \hat{x}(t)^2} \qquad \theta(t) = tan^{-1}\left(\frac{\hat{x}(t)}{x(t)}\right) \tag{3a,b}$$

It is useful to recall the very remarkable property of the Hilbert transform, useful in applications. Let us denote with $X(\omega) = \mathcal{F}\{x(t);\omega\}$ the Fourier transform of the function $x(t)$, then it is possible to show that the property

$$\mathcal{F}\{\hat{x}(t);\omega\} = -\mathrm{i}\,\mathrm{sgn}(\omega)\,X(\omega) \tag{4}$$

where $\mathrm{sgn}(\omega)$ is the signum function, holds true. Practically, one can compute $X(\omega)$, apply the latter and finally perform an inverse Fourier transform to obtain $\hat{x}(t)$. Moreover, it is easy to prove from the definition and the latter equation that the analytical signal can be computed making the inverse Fourier transform of the function $2X(\omega)U(\omega)$, being $U(\omega)$ the unit step function.

2.2 *Incipient damage identification procedure*

In this paper we restrict our attention to identify incipient damage in structural elements by analyzing a recorded dynamical response. To this aim, let us indicate with $\gamma^{\mathrm{eff}} = (\gamma_1^{\mathrm{eff}}, \ldots, \gamma_n^{\mathrm{eff}})$ the unknown damage indexes. They are a set of n real coefficients defined in the interval [0,1] characterizing the damage magnitude on n different structural elements, where the zero bound corresponds to structural integrity. For example, one can consider $(1 - \gamma_j^{\mathrm{eff}})$ being the j-th coefficient such that, multiplied by the j-th structural element's flexural stiffness, it characterize the decrement of stiffness. The apex eff means that the γ^{eff} indexes are effective and our aim is to develop a damage identification procedure capable to identify the value of each coefficient associated to the respective structural element. It has to be noted that the present formulation it is not suited to establish location inside the element where damage occurred.

Let us indicate by $\eta^{ex}(\gamma^{eff}; t)$ a general feature extracted from the real structure, where the apex ex denotes that it is a recorded data, i.e. coming from an experimental test. Further, indicate by η^{th} (γ; t) the same feature calculated from a structural model, depending on the γ variables.

The objective functional

$$J_\eta(\boldsymbol{\gamma}) = \frac{1}{T_f - T_i}\int_{T_i}^{T_f} \left(\eta^{th}(\boldsymbol{\gamma}; t) - \eta^{ex}(\boldsymbol{\gamma}^{eff}; t)\right)^2 dt \tag{5}$$

where $[T_i, T_f]$ is the observation temporal window, consents to detect the set of damage indexes γ^{eff}. It has been shown in Cottone et al. (2008), Barone et al. (2008) that the most sensitive feature that consents to identify small damage also in presence of measuring noise affecting the experimental data is the phase of the analytical signal. Two examples will clarify the approach.

2.2.1 *Example: Incipient damage identification in a cantilever bar*

Consider a mass-less cantilever bar with a lumped mass at the free end subjected to axial displacement. The bar stiffness is EA/l, being E the Young's modulus, A the section area and l the bar length. An occurring damage will be modeled by a reduction of the axial stiffness $EA(1-\gamma)/l$ where $\gamma \in [0, 1]$ is the previously defined damage index. The equation governing the vibration of the given system under an impulsive force in horizontal direction modeled as a Dirac's delta function $\delta(t)$ is given in the form

$$\ddot{x}(t) + 2\Lambda\dot{x}(t) + \omega_0^2\,(1-\gamma)x(t) = \delta(t)/m \tag{6}$$

where $\Lambda = \zeta_0\omega_0\sqrt{1-\gamma}$ denotes the damping term, ζ_0 the damping ratio and $\omega_0 = \sqrt{EA/ml}$ is the undamaged bar circular frequency. Assume that the experimental system is characterized by a damaged index $\gamma_{eff} = 0.01$, that is a very low stiffness reduction. Performing an hammer test on the experimental set-up, the acceleration data $\ddot{x}^{ex}(\gamma^{eff}; t)$ is recorded and the phase of the analytical signal is calculated and indicated by $\theta^{ex}(\gamma^{eff}; t)$. Then, the functional in eq.(5) is constructed by selecting a value of the unknown damage index γ and solving the theoretical model ruled by eq.(6) and finding the signal $\ddot{x}^{th}(\gamma, t)$. From this acceleration response the $\theta^{th}(\gamma; t)$ to be introduced in eq.(5) is calculated and one value of the functional $J_\theta(\gamma)$ is obtained. The absolute minimum of the functional is located at $\gamma = \gamma^{eff}$ In this simple case, the functional is very well-behaved, there is only a global minimum and its minimization is very easy. Conversely, in multi degrees of freedom (DOF) structures or in recorded signal affected by measurements and/or ambient noise, the functional $J_\theta(\gamma)$ has many local minima due to the multi-component behavior of the Hilbert transform (see Cottone et al. (2008)). This aspect is stressed in the next example.

2.2.2 *Example: Incipient damage identification in a 3 DOF structure*

Let us consider a three-storey shear-beam type building. The mass, stiffness and viscous damping of each storey are assumed to be the same, with $m_j = 1000$ kg, $k_j = 980$ kN/m, $c_j = 2814$ Ns/m, respectively, for j = 1, 2, 3. Damage is simulated by decreasing the stiffness at first and second floors by a damage indexes $\gamma_1^{eff} = 0.010$ and $\gamma_2^{eff} = 0.015$, while $\gamma_3^{eff} = 0$. The acceleration of the third floor is recorded and the phase component from the analytical signal is extracted and indicated as $\ddot{\theta}_3^{ex}(\gamma_1^{eff}, \gamma_2^{eff}, \gamma_3^{eff}; t)$.

For each combination of $\gamma = (\gamma_1, \gamma_2, \gamma_3)$ it is possible to calculate the model theoretical response, i.e. the acceleration of the third storey $\ddot{x}_3^{th}(\gamma_1, \gamma_2, \gamma_3; t)$ and the functional in terms of phase is computed. Both the noise and the multimodality of the response negatively affect $J_\theta(\gamma)$ and many local minima render the minimization to be a difficult task. At this point, two strategy of attack to the problem are possible. The first is to process the signal in order to restore a well-behaved functional as like as in the single DOF previously outlined. In Cottone et. al (2008), Barone et al. (2008) this has been achieved by filtering the single modal components both from the experimental and the theoretical signals and calculating the objective function. This strategy proved to perform very well also in experimental tests in Navarra & Pirrotta (submitted). Limits of this approach rely on the ability of the filter in isolating the modal responses. In case of structures with narrow principal frequencies such procedure might not be effective.

The second strategy to solve the problem is to use an optimization algorithm able to deal with strong presence of local minima. This is investigated in this paper, exploiting the Ant Colony Optimization with Lévy perturbation already proposed by some of the authors in Candela et al. (2009) and Candela et al. (2010).

3 LÉVY ANT COLONY OPTIMIZATION

3.1 *α-stable Lévy distribution in searching and optimization*

α-stable Lévy distributions are characterized by heavy-tailed probability density function that causes infinite variance and are defined by four coefficients. A random variable X is said to have a α-stable distribution if there are parameters $0 < \alpha \leq 2$, $\sigma > 0$, $1 \leq \beta \leq 1$, $\mu \in \mathbb{R}$ such that its characteristic function $\phi_X(\theta)$ has the form:

$$\phi_X(\theta) = \begin{cases} \exp\left[-\sigma^\alpha|\theta|^\alpha\left(1 - i\beta\text{sgn}(\theta)\tan\left(\frac{\pi\alpha}{2}\right)\right) + i\mu\theta\right], & \alpha \neq 1 \\ \exp\left[-\sigma\,|\theta|\left(1 + \frac{2}{\pi}i\beta\text{sgn}(\theta)\ln|\theta|\right) + i\mu\theta\right], & \alpha \neq 1 \end{cases} \tag{7}$$

The four parameters affect the shape of the distribution in an essential way and it is common to introduce an appropriate notation to take them into account. We denote with the symbol $X \sim S_\alpha(\sigma, \beta, \mu)$ an α-stable random variable with assigned parameters characterizing eq.(7). The coefficient α is called stability index and characterizes the tails of the stable distribution, decreasing as $x^{-1-\alpha}$. Lower values of α correspond to higher probability of having realization on the tails, while for $\alpha = 2$ the distribution $S_2(\sigma, \beta, \mu)$ coincides with the Gaussian distribution $N(\mu, \sqrt{2}\sigma)$. The scale σ, the skewness β and the location μ are parameters defining the shape of the distribution. Such a great variability makes Lévy distribution to be very suited in searching algorithms (see Gutowski (2001), Lee & Yao (2004), Lomholt et al. (2007)). Moreover, heavy-tailed distributions consent to have less localized space of search. In this paper, these features are introduced in the Ant Colony Optimization in continuous domain to minimize the objective function in eq.(5).

3.2 *The metaphor of ants foraging in combinatorial and continuous optimization and its Lévy based version*

Ant Colony Optimization (ACO), originally proposed for combinatorial optimization problems (Dorigo (1992); Dorigo & Gambardella (1997)) has been extended to continuous domains in Socha & Dorigo (2008), main reference of this section. It is use to denote the continuous domain formulation as $ACO_\mathbb{R}$ to make a distinction from ACO. The foraging behavior of real ants has motivated the ACO: i) ants firstly leave the nest for exploring the ambient in random way; ii) in case one ant finds the food, returns to the nest depositing a pheromone trail proportional to the food quality and quantity; iii) this trail is the only indirect possible communication with other ants; iv) pheromone does evaporate, i.e. it the shorter the path the higher the appeal of the pheromone trail. Although the algorithmic translation of this concepts into ACO are out of the scope of this paper, nevertheless it is worth to briefly outline how $ACO_\mathbb{R}$ works in order to better explain the role of the previously outlined Lévy distribution.

Let S be the search space and $X_1, X_2, \ldots, X_n$ the so-called decision variables, where n characterizes the dimension of the functional f: $S \to \mathbb{R}$ to be minimized. Construction of a solution $s_l \in S$ means to assign values $x^i \in \mathbb{R}$ to each X_i, for $i = 1, \ldots, n$, such that $s = (x^1, x^2, \ldots, x^n)$. A solution $s*$ is a global minimum if and only if $f(s*) \leq f(s)$ for every $s \in S$. The idea that is central to the way $ACO_\mathbb{R}$ works is the incremental construction of solutions based on the biased probabilistic choice of solution components. The bias is constituted by the pheromone trail guiding the searching process and is taken into account by means of n Gaussian kernel probability densities, $G^i(x)$ defined below. Practically we construct an archive T of k solutions $T = \{s_1, \ldots, s_l, \ldots, s_k\}$ where the generic solution s_l has n components, i.e. $s_l = (x_1^1, x_l^2, \ldots, x_l^n)$. Then, the solutions of the archive T are ordered according to their objective function values $f(s_l)$. For each $l = 1, \ldots, k$ we calculate the weights

$$\omega_l = \left(qk\sqrt{2\pi}\right)^{-1}\exp(-(l-1)^2/2(qk)^2) \tag{8}$$

that are dependent on the rank of the solution and are needed to calculate the Gaussian kernel

$$G^i(x) = \frac{1}{\omega_{tot}} \sum_{l=1}^{k} \omega_l g_l^i(x) = \frac{1}{\omega_{tot}} \sum_{l=1}^{k} \omega_l \left(\sigma_l^i \sqrt{2\pi}\right)^{-1} \exp\left(-\left(x - \mu_l^i\right)^2 / 2\left(\sigma_l^i\right)^2\right) \quad (9)$$

where $\mu_l^i = s_l^i$ and $\omega_{tot} = \sum_{l=1}^{k} \omega_l$ is a normalization factor. It must be noted that each density $g_l^i(x)$ is centered in s_l^i. The standard deviations σ_l^i are defined as average distance between s_l^i and the other solutions in T as

$$\sigma_l^i = \xi \sum_{r=1}^{k} \left|s_l^i - s_r^i\right| / (k-1) \quad (10)$$

The size of the archive k, ξ and q are user defined parameters of the algorithm whose correspondence to the ant metaphor is described in Socha & Dorigo (2008). Following this approach, Lévy distributed $g_l^i(x)$ could not be introduced, because Lévy densities are not known in analytical form. Yet, the following consideration is fundamental to introduce Lévy perturbations. Indeed, at each construction step i an ant selects a value x^i of X^i making a probabilistic choice $G^i(x)$, i.e. producing a sample whose density is given in eq.(9) and determined by the sole knowledge of the components $s_1^i, \ldots, s_k^i$. This corresponds to choose one distribution $g_l^i(x)$ with probability $p_l = \omega_l / \omega_{tot}$ and then sampling the chosen density by generating a random sample following the density $g_l^i(x)$. Therefore, the $ACO_{\mathbb{R}}$ with Lévy densities $g_l^i(x)$ can be implemented, simply generating samples of $S_\alpha(\sigma_l^i, 0, \mu_l^i)$ as described in Samorodnitsky & Taqqu (1994).

In the following the $ACO_{\mathbb{R}}$ with Lévy perturbations will be denoted as $ACO_{\mathbb{R}L}$, whereas it is plain that for $\alpha = 2$, $ACO_{\mathbb{R}L}$ coincides with $ACO_{\mathbb{R}}$. In Candela et al. (2009) some of the authors already attested the efficiency and the performances of the $ACO_{\mathbb{R}L}$ with respect to a set of test function. Briefly the results can be summarized as: The $ACO_{\mathbb{R}L}$ performs better or no worse than $ACO_{\mathbb{R}}$ for functions with many local minima for $\alpha = 1.6$ and $\alpha = 1.8$; lower values of the stability index give rise to higher degree of diversification in the searching process due to the tendency of the Lévy flights to escape from a region. In applying the $ACO_{\mathbb{R}L}$ to the minimization of eq.(5) we will select the value of $\alpha = 1.8$.

4 NUMERICAL ANALYSIS

4.1 *3-DOF structure with well spaced natural frequencies*

Let us consider the 3-DOF structure with parameters already specified in paragraph 2.2.2. The structure principal frequencies are $\omega_1 = 13.6$ rad/s, $\omega_2 = 38.5$ rad/s and $\omega_3 = 56.2$ rad/s, respectively. This example has been already studied in Cottone et al. (2008) by application of Butterworth filter to both the experimental and the theoretical signals, in order to extract the modal component. In that work, this was necessary because a well behaved functional without local minima was searched. We examine the same example with the perspective presented in this paper, without using the pre-filtering of the dynamical response. Instead, $ACO_{\mathbb{R}L}$ is used to minimize $J_\eta(\gamma)$ and we show that the algorithm is suited to estimate the assigned level of damage $\gamma^{\text{eff}} = (0.010, 0.015, 0.00)$. Just to visualize the problem, in Figure 1 the logarithm of the functional calculated on the phase of the analytical signal is depicted: It can be noted that many local minima are present, all along a line dependent on the structure parameters, and that a unique global minimum corresponds to the assigned damage parameters γ^{eff} In this simple example, the value of the functional in γ^{eff} is zero, i.e. in Figure 1 the effective value of the functional at the minimum is negative infinity.

$ACO_{\mathbb{R}L}$ has been implemented with $q = 0.1$, $\xi = 0.85$, $k = 50$ for 5000 ant's generation steps. 30 independent runs have been used to assess the performance of the algorithm. The search space has been confined just to the damage index γ_1 and γ_2 in the domain $[0, 0.2] \times [0, 0.2]$. Table 1 resumes the mean and the standard deviation of the damage indexes obtained and of the absolute percentage error $\epsilon_j = 100(\gamma_j^{\text{eff}} - \gamma_j)/\gamma_j^{\text{eff}}$. Figure 2 shows the variation of the best, the worst, the mean and the median in a single run. In this simple case, $ACO_{\mathbb{R}L}$ succeeds to identify the incipient damage and gives a very good quantitative estimate of the extent of the damage.

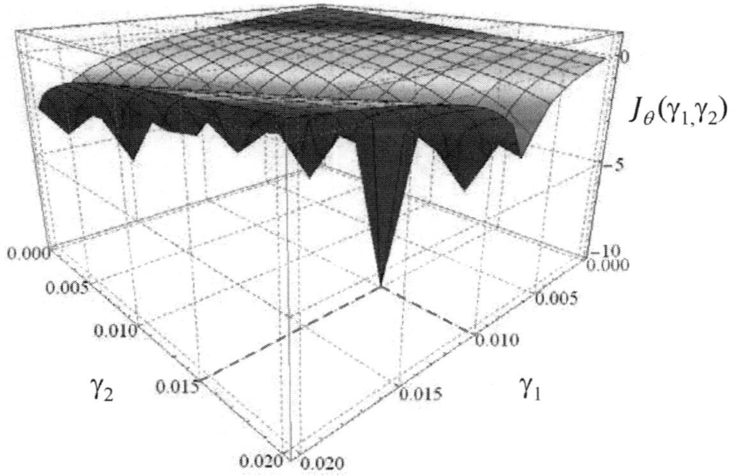

Figure 1. Logarithm of the functional $J_\theta(\gamma)$ calculated on the phase of the analytical signal. Thick dashed lines correspond to the global minimum to be identified. (right panel).

Table 1. Mean and standard deviation of the damage indexes.

Test	γ_1	γ_2	$\epsilon_1\%$	$\epsilon_2\%$
mean	0.010	0.015	0.37	0.34
st.dev	3.8E−5	6.0E−5	–	–

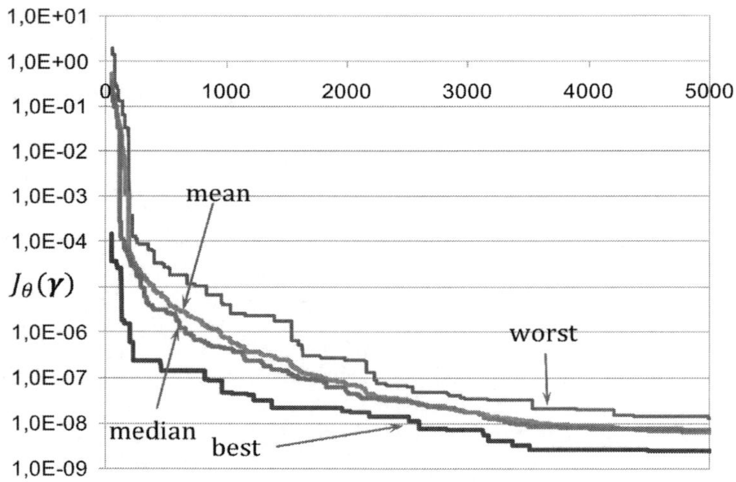

Figure 2. Evolution of the solution along 5000 ants generation steps.

4.2 *3-DOF structure with not well spaced natural frequencies*

Consider a 3-DOF with the parameters of the sound structure that are: $m_1 = 1000$ kg, $m_2 = 1300$ kg, $m_4 = 500$ kg, $k_1 = 980$ kN/m, $k_2 = 130$ kN/m, $k_3 = 140$ kN/m $c_j = 2814$ Ns/m, respectively, for j = 1, 2, 3. Such a structure presents not well spaced natural frequencies, in such a way that isolating the single modal component from the recorded signal by filtering might request very high order filters with consequent difficulties. It is assumed that the structure is characterized by the damage indexes $\gamma^{\text{eff}} = (0.010, 0.015, 0.00)$ as in the previous example. As outlined in section 2, the

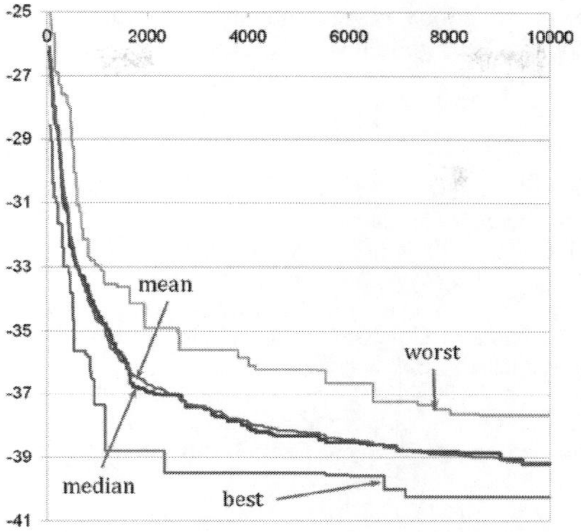

Figure 3. Evolution of the solution along 10000 ants generation steps.

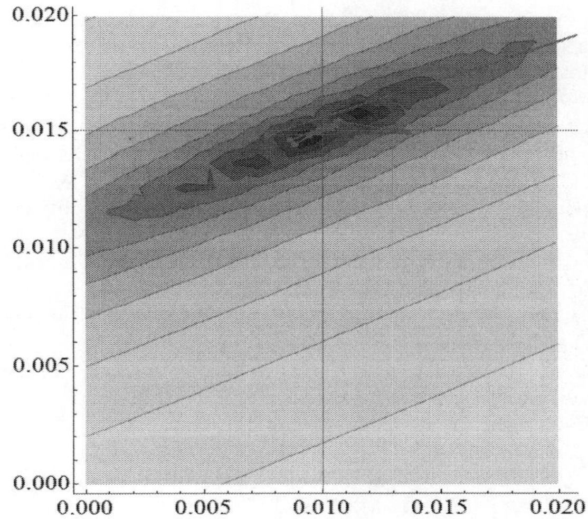

Figure 4. Trajectory (in red, thick line) of the best ant solution for one run, contrasted with the contour plot of the functional in the space of parameters γ_1, γ_2.

experimental analytical signal is calculated and the functional in eq.(5) is constructed. The search space in $ACO_{\mathbb{R}L}$ is the domain $[0, 0.2] \times [0, 0.2] \times [0, 0.2]$ in the variable parameters $\gamma_1, \gamma_2, \gamma_3$.

Also in this case, $ACO_{\mathbb{R}L}$ succeeds in locate the global minimum. Figure 3 shows the space of the objective function revealing that, in this case, the span between the best and the worst ant is greater than before. More interesting is Figure 4, where the trajectory of the best ant, that is the best ranked at each construction step, highlights the process of search captured into the groove where global and local minima are present.

5 CONCLUSIONS

In this paper we presented a method for the incipient damage identification based on the minimization of a functional calculated by some features of the analytical signal. Once the functional has been introduced, the minimization is performed by the meta-heuristic method of the Ant Colony Optimization with Lévy perturbation. Numerical results on 3 degree of freedom system confirm that, even in cases where the functional is very irregular and with many local minima, the damage index can be quantified with a very good level of accuracy.

REFERENCES

Adams, R.D., Cawley, P., Pye, C.J. & Stone, B.J. 1991. A vibration technique for non-destructively assessing the integrity of structures. *Journal of Mechanical Engineering Science* 20: 93–100.

Barone, G., Marino, F. & Pirrotta, A. 2008. Low stiffness variation in structural systems: Identification and localization. *Structural Control and Health Monitoring* 15:450–470.

Candela, R., Fileccia Scimemi, G., Cottone, G., Riva Sanseverino, E. 2009. Lévy flights for Ant Colony Optimization in continuous domains. CiE 2009. Computability in Europe 2009, Mathematical Theory and Computational Practice. Heidelberg (Germany). 19–24 July.

Candela, R., Fileccia Scimemi, G., Cottone, G., Riva Sanseverino, E. 2010. Composite laminates buckling optimization through Levy based Ant Colony Optimization. The Twenty Third International Conference on Industrial, Engineering & Other Applications of Applied Intelligent Systems. IEA-AIE 2010, Cordoba (Spain). 1–4 June.

Cerri, M.N. & Vestroni, F. 2000. Detection of damage in beams subjected to diffused cracking. *Journal of Sound and Vibration* 234: 259–276.

Cornwell, P., Doebling, S.W. & Farrar, C.R. 1999. Application of the strain energy damage detection method to plate-like structures. *Journal of Sound and Vibration* 224: 359–374.

Cottone, G., Pirrotta, A., Salamone, S. 2008. Incipient damage identification through characteristics of the analytical signal response. *Structural Control and Health Monitoring* 15(8): 1122–1142.

Davini, C., Gatti, F. & Morassi, A. 1992. A damage analysis of steel beams. *Meccanica* 22: 321–332.

Dorigo, M. 1992. Optimization, learning and natural algorithms (in Italian). PhD thesis, Dipartimento di Elettronica, Politecnico di Milano, Italy.

Dorigo, M. & Gambardella L. M. 1997. Ant colony System: A cooperative learning approach to the traveling salesman problem. *IEEE Trans. on Evol. Comp.* 1(1): 53–66.

Gounaris, G.D., Papadopoulus, C.A. & Dimarogonas, A.D. 1996. Crack identification in beams by coupled response measurements. *Computer & Structures* 58: 299–302.

Gutowski, M. 2001. Lévy flights as an underlying mechanism for global optimization algorithms. Available at: http://arxiv.org/pdf/math-ph/0106003.

Hearn, G. & Testa, R.B. 1991. Modal analysis for damage detection in structures. *Journal of Structural Engineering* 117: 1798–1803.

Lee, U. & Shin. J. 2002. A frequency response function-based structural damage identification method. *Computers & Structures* 80: 117–132.

Lee, C., Yao, X. 2004. Evolutionary programming using mutations based on Lévy probability distribution. *IEEE Trans. on Evol. Comp.* 8(1): 1–13.

Lomholt, M.A., Koren, T., Metzler, R. & Klafter, J. 2007. The advantage of Lévy strategies in intermittent search processes. Available at: http://arxiv.org/abs/0709.2352v1.

Luo, H. & Hanagud, S. 1997. An integral equation for changes in the structural dynamics characteristics of damaged structures. *International Journal of Solids and Structures* 34: 4557–4579.

Oppenheim, A.V. & Schafer, R.W. 1989. Discrete-Time Signal Processing, Prentice Hall, Englewood Cliffs, NJ.

Pandey, A.K., Biswas, M. & Samman, M.M. 1991. Damage detection from changes in curvature mode shapes. *Journal of Sound and Vibration* 145: 321–332.

Ratcliffe, C.P. 1999. A frequency and curvature based experimental method for locating damage in structure. *Journal of Vibration Acoustics* 226: 1029–1042.

Samorodnitsky, G. & Taqqu, M.S. 1994. Stable non-Gaussian random processes: Stochastic models with infinite variance. Chapman and Hall, New York.

Sampaio, R.P.C., Maia N.M.M. & Silva, J.M.M. 1999. Damage detection using frequency-response-function curvature method. *Journal of Sound and Vibration* 226: 1029–1042.

Shi, Z.Y., Law, S.S. & Zhang, L.M. 1998. Structural damage localization from modal strain energy changes. *Journal of Sound and Vibration* 218: 825–844.

Socha, K. & Dorigo, M. 2008. Ant colony optimization for continuous domains. *European Journal of Operational Research* 185: 1155–1173.

Thyagarajan, S.K., Schulz, M.J. & Pai, P.F. 1998. Detecting structural damage using frequency response functions. *Journal of Sound and Vibration* 210: 162–170.

Wang, Z., Lin, R.M. & Lim, M.K. 1997. Structural damage detection using measured FRF data. *Computer Methods in Applied Mechanics* 147: 187–197.

Reliability and Optimization of Structural Systems – Straub (ed)
© 2010 Taylor & Francis Group, London, ISBN 978-0-415-88179-1

Expected life-cycle costs of bolted connections for steel buildings subjected to seismic loading

David De León
School of Engineering, Autonomous University of Mexico State, Toluca, Mexico State, Mexico

ABSTRACT: The structural safety of bolted connections for buildings located on zones with high seismic hazard involves the consideration of uncertainties mainly related with the loading that demands the connection to a combination of stresses that may be critical for the building performance. Usually the bolts are subjected to strong shear forces trending to a fracture failure. Structural reliability provides a proper framework to appraise the connection safety and to associate it with the costs incurred at present and future times. These relationships allow for the determination of cost-effective measures and risk-based design recommendations looking for the optimization of long term performance.

1 INTRODUCTION

Steel buildings are a common design solution for seismic zones. However, the selection of the appropriate connection type is still an issue in Mexico (Miranda 1997a, 1997b, Miranda and Martínez 1999), especially after the amount of damages experienced due to the Northridge earthquake (Righiniotis and Imam 2004) occurred in California in 1994. The SAC Project (SAC Project 1994), developed in the US under FEMA's coordination (FEMA 273 1997), provides some insight to improve the understanding of the seismic behavior of steel connections.

Usually significant shear forces are developed in the bolts and the connection is stressed towards shear fracture failure (Bruneau et al. 1998). Uncertainties due to the magnitude of seismic loading and the connection response play an important role that needs to be quantified or adequately characterized (Ang and Tang 2007).

Frequently, the collapse limit state is emphasized to provide design recommendations (AISC 2005). But, given the character and extension of the damage produced by some earthquakes and the time the structure is off-service during repairs, the serviceability condition is also a concern. Structural reliability (Ang and Tang 1984) and life-cycle costing serve as the measuring tools to weight the cost/benefit relevance of the various connection alternatives (Wen and Fouth 1997) and to balance the trade-off between required safety and costs of the damage consequences.

A seismic hazard curve, previously developed for Mexico City (Esteva and Ruiz 1989) is used with scaling factors to assess the seismic vulnerability of the structures.

Throughout Monte Carlo simulation, statistics of the maximum acceleration demands are obtained at the connection location for typical buildings and, with these statistics and the connection model, statistics of the maximum responses are obtained. With these statistics and the state limit function, for a given connection type, probabilities of failure and damage are obtained for both demand levels: extreme and operational earthquakes. These probabilities are introduced into the life-cycle cost/benefit relationship for several connection types and the optimal type is obtained by comparing the expected life-cycle costs. The minimum value corresponds to the optimal connection type.

Damage costs include the repair cost and losses related to the potential fatalities, injuries and business interruption.

The results may also be used, after further refinements, to update the design specifications for seismic zones in Mexico.

2 FORMULATION OF THE DECISION CRITERIA

The expected life-cycle cost is usually calculated to assess the economic effectiveness of potential structural solutions and come up to optimal decisions under uncertain loading conditions.

Two alternative connection types are proposed and their performances are compared under the viewpoints of structural reliability and costs.

The expected life-cycle cost $E[C_L]$ is composed by the initial cost C_i and the expected damage costs $E[C_D]$:

$$E[C_L] = C_i + E[C_D] \tag{1}$$

The expected damage costs include the components of damage cost: expected repair $E[C_r]$, business interruption $E[C_e]$, injury $E[C_{in}]$ and fatality $E[C_f]$ costs and each one depends on the probabilities of damage and failure of the structure.

These component costs of damage are defined as:

$$E[C_r] = C_r(PVF)P_r \tag{2}$$

where:

C_r = average repair cost

PVF = present value function [Ang and De León, 1997].

$$PVF = \sum_{n=1}^{\infty}[\sum_{k=1}^{n}\Gamma(k,\gamma L)/\Gamma(k,\upsilon L)(\upsilon/\gamma)^k](\upsilon L)^n/n!\exp(-\upsilon L) \tag{3}$$

where ν = mean occurrence rate of earthquakes that may damage the structure, γ = net annual discount rate, and L = structure life.

And P_r = probability of repair, defined in a simplified way, as a factor of the failure probability.

Similarly, the business interruption expected cost, is expressed in terms of the loss of revenue due to the repairs or reconstruction works after the earthquake, assumed to last T years.

$$E[C_{DP}] = L_R(T)P_r \tag{4}$$

where:

L_R = loss of revenues per year in USD

The expected cost of injuries is proposed to be:

$$E[C_{in}] = C_{1I}(N_{in})P_f \tag{5}$$

where:

C_{1I} = average injury cost for an individual

N_{in} = average number of injuries on a typical steel building in Mexico given an earthquake with a mean occurrence rate ν.

For the expected cost related to loss of human lives, the cost corresponding to a life loss, C_{1L}, and the expected number of fatalities, N_D are considered. The cost associated with a life loss may be estimated in terms of the human capital approach, which consists in the calculation of the contribution lost, due to the death of an individual, to the Gross Domestic Product during his expected remaining life. The details of this calculation are explained in previous works (Ang and De León 2005). The expected number of fatalities is estimated from a curve previously developed for typical buildings in Mexico, in terms of their plan areas, given an earthquake with a mean occurrence rate ν.

$$E[C_L] = C_{1L}(N_D)P_f \tag{6}$$

In the next section, all the figures are estmated for typical costs in Mexico.

A typical geometry of a building, see Fig. 1, located on the soft soil of Mexico City is selected to analyze its critical frame under seismic loads. Statistics of its maximum response, at critical

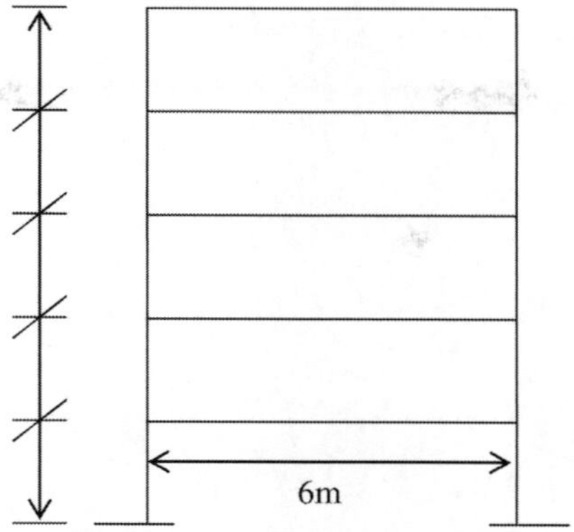

Figure 1. Typical frame for a steel building in Mexico.

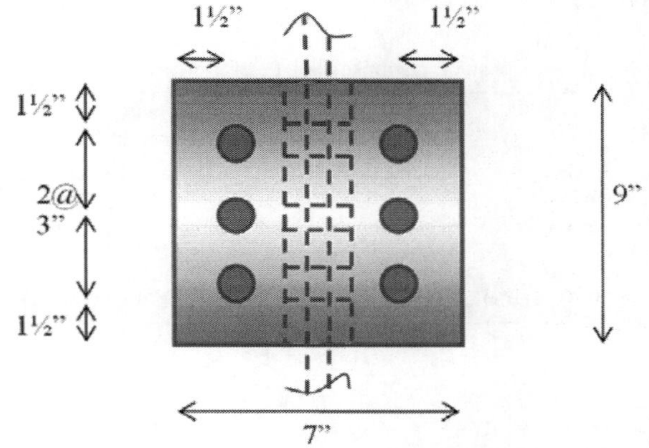

Figure 2. Front view of a sample of beam-column connection at column flange.

joint level, are obtained from the frame analyses subjected to Poissonian earthquakes (with mean occurrence rate ν) as scaled from the seismic hazard curve for Mexico City (Esteva and Ruiz 1989).

3 APPLICATION TO A STEEL BUILDING IN MEXICO

A medium height regular frames-based structural type steel building in Mexico is used to show the application of the procedure. One of these frames is shown in Fig. 1.

A series of bolted beam-column connections are designed (IMCA 1997) to transmit the forces at the critical joints, which were found to be at the corners of the first level, for the purpose of having a set of options with several safety margins.

Bolts are arranged to connect, through angles, the beam web to column flange and a general view of the proposed bolted connection is shown in Fig. 2.

The above described response statistics are used as an input to the FEM models of the alternative connections and a Monte Carlo simulation process is performed for each connection model in order to get the statistics of maximum shear force and moment. With these statistics and the limit state function of each connection, the corresponding failure probabilities are calculated. Potential failure modes, either bending moment or shear force being exceeded, are considered. The following limit states are used to calculate the failure probability.

$$g_1 = M_R - M_{Max} \tag{7}$$

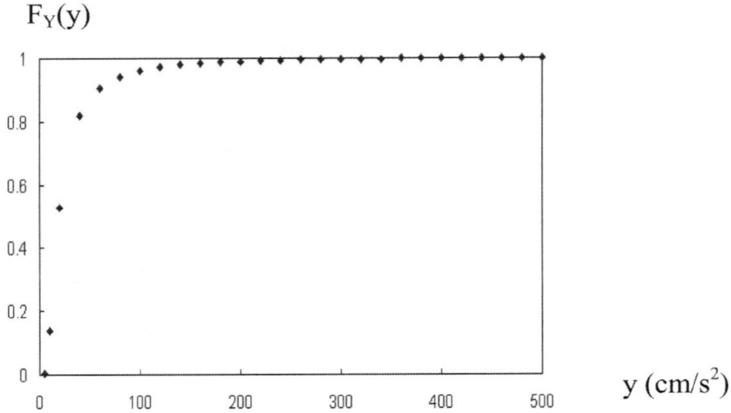

Figure 3. Annual probability of seismic intensity for Mexico City.

Table 1. Typical response statistics for critical joint.

	M_{max}(tn-m)	V_{max}(tn)
Mean =	0,344	0,050
St. Dev. =	0,362	0,052

Table 2. Sample of calculations for shear and moment for bolted connection.

# aleat. = P	y (cm/s^2)	$y/y_{spectralRDF}$	M_{max} (tn-m)	V_{max}(tn)
0,82422596	155,293872	0,35178134	0,43972667	0,06332064
0,07512694	55,0110198	0,12461438	0,15576798	0,02243059
0,03302182	53,5014565	0,12119483	0,15149353	0,02181507
0,51860788	82,7349802	0,18741642	0,23427053	0,03373496
0,4593356	76,9433636	0,17429689	0,21787112	0,03137344
0,34625381	68,3313961	0,15478853	0,19348566	0,02786194
0,76997367	131,262316	0,29734356	0,37167945	0,05352184
0,1227829	56,8602933	0,12880347	0,16100434	0,02318463
0,06217154	54,5348214	0,12353567	0,15441959	0,02223642
0,79798093	142,357645	0,32247739	0,40309674	0,05804593
0,18203045	59,4005297	0,13455777	0,16819722	0,0242204
0,93242496	282,250406	0,63937118	0,79921397	0,11508681
0,55684109	87,1267168	0,19736486	0,24670607	0,03552567
0,94768352	331,20888	0,75027496	0,9378437	0,13504949
0,41121002	72,9500794	0,16525106	0,20656382	0,02974519
0,70491243	112,339298	0,25447797	0,31809746	0,04580603

$$g_2 = V_R - V_{Max} \tag{8}$$

Where M_R, V_R, M_{Max} and V_{Max} are the moment and shear resistances and the maximum acting moment and shear, respectively. With the failure probabilities, and Eqs. (1) to (6), the expected life-cycle cost of each connection is obtained. The connection type to be recommended is the one with the minimum life-cycle cost.

The calculation process described in the last section is performed to the frame shown in Fig. 1 and the cumulative annual probability of seismic intensities, ground accelerations y, is shown in Fig. 3.

The response statistics, obtained at the critical joint level, are as shown in Table 1.

A sample of random acting critical forces, at the joint, is shown in Table 2.

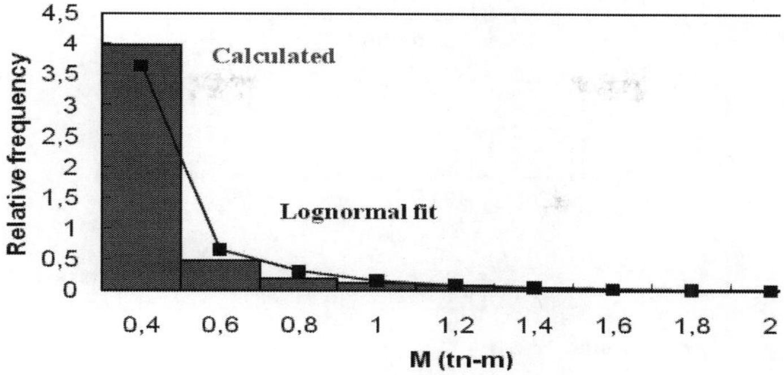

Figure 4. Probability distribution of maximum moments.

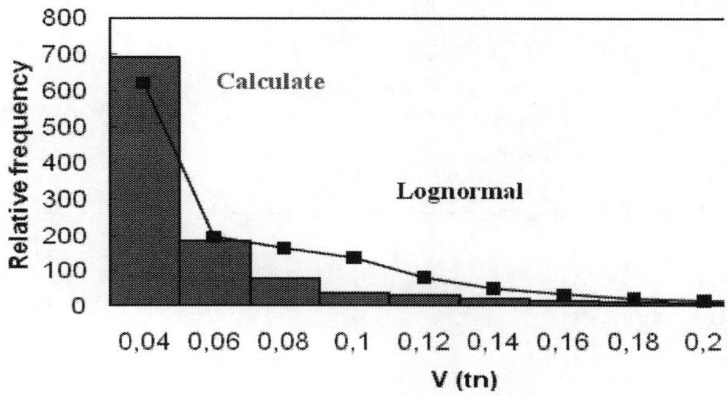

Figure 5. Probability distribution of maximum shear forces.

Table 3. Failure probabilities for alternative connections.

M_R (Ton-m) =	0,4	0,8	1,5	2	2,5	2,9	3,2
C_i (USD) =	500	1000	1500	2000	3500	8000	28000
P_f =	0,272	0,079	0,016	0,0067	0,0031	0,0018	0,0012

The ratio “y/yspectral RDF” is the ratio between the sampled intensity, from the seismic hazard curve, and the spectral intensity considered in the current code, for the purpose to use it as an amplification factor to get the moment and shear responses M and V, respectively.

From the obtained series of maximum moments and shear forces, the corresponding probability distributions are obtained and fitted to lognormal curves, as shown in Figs. 4 and 5.

The resisting moments, initial costs and failure probabilities, for the alternative connections, are shown in Table 3. In this Table, the initial cost is the individual cost of a connection. However, for the cost functions the cost and failure probability are related to the structural frame because the pertinent parameters for the cost functions are associated to the whole structure.

With the above failure probabilities, the expected life-cycle costs are calculated and the results are shown in Fig. 6.

From the series of alternative designs, their structural failure probabilities and the initial costs, the curve of initial cost is built. This curve is shown in Fig. 7.

The acceptable failure probability, as a function of consequences of failure, is obtained and shown in Fig. 8.

The cost of consequences is 63000 USD and the acceptable failure probability, from Fig. 8, is 0.0017, approximately the same as for the optimal design, from Table 3, for the design with $M_R = 2.9$ ton-m.

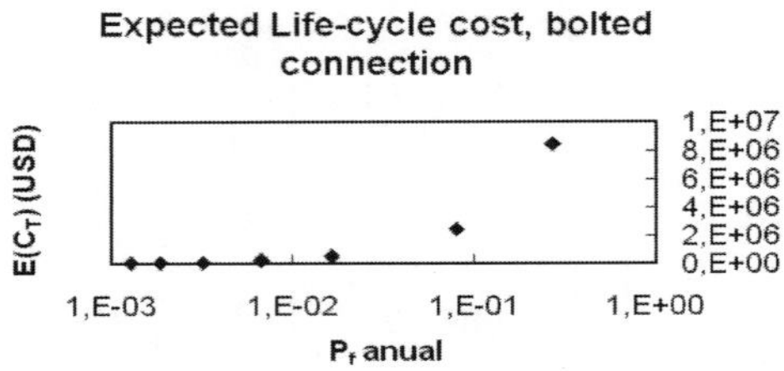

Figure 6. Expected life-cycle costs for alternative connections.

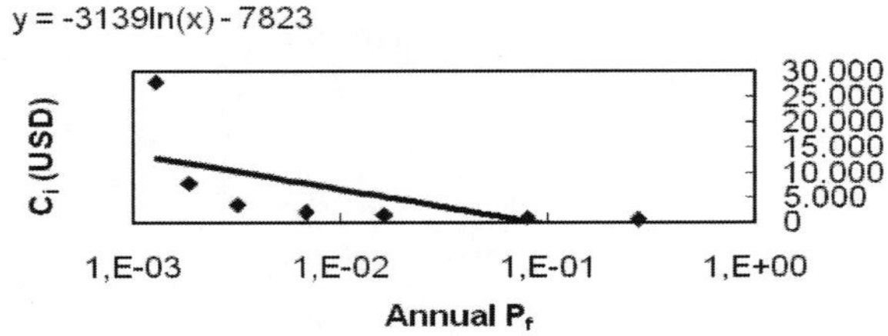

Figure 7. Initial costs for alternative connections.

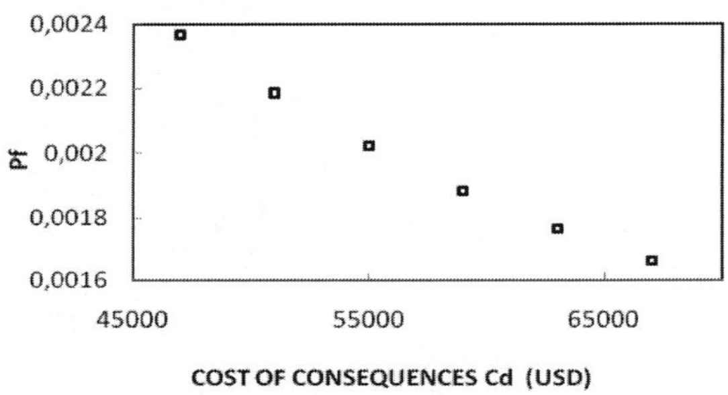

Figure 8. Acceptable failure probability for bolted connections for several costs of consequences.

4 DISCUSSION OF RESULTS

From the results obtained in the previous section, it is observed that the optimal connection type is the one with $M_R = 2.9$ ton-m.

The most important parameter in the calculations is the seismic hazard of the structure site because it is the main input for the mechanical demands M_{Max} and V_{Max} and, therefore, it represents a relevant issue for the level of failure probability and the expected life-cycle cost, which are critical for decision making.

The structure failure mechanism was a simplified one considering a combination of four connections failing simultaneously to produce a collapse mechanism consisting on the failure of the top and bottom sections of 2 columns located at the same floor. It is assumed that the failure probability of the four sections is the same. This simplification has to be improved in the next stage of the research work.

Other connection types need to be explored in order to complete and make more robust the optimization process. The limit states considered here may change according to the connection types and the location of critical joints may move according to the structural shape and regularity conditions of the structure.

The acceptable failure probability may be used to assess the viability of using the actual failure probability. The acceptable value depends on the level of failure consequences which, in turn, depends on the use of the structure and the seriousness of the consequences after failure, like life loss, contents loss and business interruption during the repair or reposition period. Critical functions, as in a hospital building, and damage extension on connections may increase the failure consequences.

5 CONCLUSIONS AND RECOMMENDATIONS

A reliability-cost based procedure has been implemented to derive criteria for the selection of bolted connections subjected to a seismic environment.

The procedure represents a step forward towards the development of performance-based techniques and tools that explicitly includes uncertainty and cost/benefit measures to back up engineering decisions about connections.

For some cases, as the one studied here, the critical joint is located in the first floor of the building and the bending moment is the critical failure mode.

The optimal annual failure probability is 0.0017.

Further research may lead to a wider range of applications in order to compare design, construction and retrofit alternative schemes.

Also, with additional work, the criteria may be used to update the Mexican code for design and retrofit specifications.

REFERENCES

AISC, (2005). *"Specification for Structural Steel Buildings"*. Chicago Illinois, E.U.

Ang, A. H-S., De León, D., (1997). "Determination of optimal reliabilities for design and upgrading of structures", *Structural Safety,* Vol. 19, Núm. 1:91–103.

Ang, A. H-S., De León, D., (2005). "Modeling and analysis of uncertainties for risk-informed decisions in infrastructures engineering", *Journal of Structure and Infrastructure Engineering*, Vol. 1, Núm. 1:19–31.

Ang, Alfredo H-S., Tang, W. H., (1984). "*Probability Concepts in Engineering Planning and Design*" Vol. II – *Risk, Reliability and Decisions*. John Wiley and Sons, New York.

Ang, Alfredo H-S., Tang, W. H., (2007). "*Probability Concepts in Engineering Planning and Design*" Vol. I – *Basic Principles*, 2nd. Edition. John Wiley and Sons, New York.

Bruneau, M., Whitakker, A., Uang, Ch. M. (1998). "*Ductile Design of Steel Structures*". McGraw Hill.

Esteva, L., Ruiz, S., (1989). "Seismic failure rates of multistory frames". *Journal of Structural Engineering, ASCE*, Vol. 115, Núm. 2:268–284.

FEMA 273 (1997). "*NEHRP Guidelines for the seismic rehabilitation of buildings*"

Gobierno del D. F., "*Normas Técnicas Complementarias para Diseño y Construcción de Estructuras Metálicas*", México DF, 2004.

IMCA (1997). "Memorias V *Simposio Internacional de Estructuras de Acero, IMCA*", Guadalajara, Jal.

Miranda, E. (1997a). "Plate-end connection for steel buildings in Mexico" Memorias V *Simposio Internacional de Estructuras de Acero, IMCA*, Guadalajara, Jal.

Miranda, E. (1997b). "Seismic design of beam-column connections". Memorias V *Simposio Internacional de Estructuras de Acero, IMCA*, Guadalajara, Jal.

Miranda, E., Martínez R., E. (1999). "Seismic design recommendations for steel structures with eccentrically-braced frames in México" Memorias VI *Simposio Internacional de Estructuras de Acero, IMCA*, Puebla.

Righiniotis, T. D., Imam, B. (2004). "Fracture reliability of a typical Northridge steel moment resisting connection" *Engineering Structures*, Vol. 26, Núm. 3:381–390, School of Engineering, University of Surrey, Guildford, Surrey GU2 7XH, UK.

SAC project (1994): *http://quiver.eerc.berkeley.edu:8080/library/index.html.*

Wen, Y. K., Fouth, D. A. (1997). "Proposed statistical and Reliability Frame-work for comparing and evaluating predictive models for Evaluation and Design and Critical Issues in developing such Framework". Report No. SAC/BD-97/03, *SAC Joint venture project*, Sacramento California.

Reliability and Optimization of Structural Systems – Straub (ed)
© 2010 Taylor & Francis Group, London, ISBN 978-0-415-88179-1

A hierarchical surrogate-based strategy for reliability-based design optimization

V. Dubourg
Clermont Université, IFMA, Laboratoire de Mécanique et Ingénieries, BP, Clermont-Ferrand, France
Phimeca Engineering, Centre d'Affaires du Zénith, Cournon d'Auvergne, France

J.-M. Bourinet
Clermont Université, IFMA, Laboratoire de Mécanique et Ingénieries, BP, Clermont-Ferrand, France

B. Sudret
Clermont Université, IFMA, Laboratoire de Mécanique et Ingénieries, BP, Clermont-Ferrand, France
Phimeca Engineering, Centre d'Affaires du Zénith, Cournon d'Auvergne, France

ABSTRACT: Reliability-based design optimization has received much attention in the last decade since it allows one to combine design optimization techniques with probabilistic performance criteria that take into account the unavoidable uncertainties inherent to engineering. From an industrial point of view though, this approach is still not applicable because the proposed resolution schemes rely upon simplifying assumptions (*e.g.* the first-order reliability method) that might not hold in practice; or in constrast, they require a large number of simulation runs (*e.g.* simulation-based reliability methods) which is not compatible with the time-consuming models used in the industry. This latter reason motivated the emergence of surrogate-based strategies. Surrogate models are analytical functions that are much faster to evaluate than the original model (*e.g.* a finite element model). They are usually built from a set of evaluations of the original model that is called a design of experiments. This paper makes use of a probabilistic surrogate known as kriging and proposes an adaptive refinement technique that gradually builds the design of experiments according to the required accuracy. This technique is then encapsulated in a nested reliability-based design optimization loop where the optimizer performs simulation-based reliability and reliability sensitivity analyses. For the sake of efficiency, the simulation technique that is used here is a variance reduction technique known as subset simulation.

1 INTRODUCTION

In structural mechanics, design optimization is the decision-making process that aims at finding the best set of design variables which minimizes some cost model while satisfying some performance requirements. Due to the inconsistency between these two objectives, the optimal solutions often lie on the boundaries of the admissible space. Thus these solutions are rather sensitive to uncertainty either in the parameters (aleatory) or in the models themselves (epistemic). Reliability-Based Design Optimization (RBDO) is a concept that accounts for this uncertainty all along the optimization process. Basically, the deterministic performance model is transformed into a more realistic probabilistic constraint. Despite its attractive formulation, the application field of RBDO is still mainly limited to academic examples. This is mostly due to the fact that it is either based on simplifying assumptions (*e.g.* FORM/SORM) that might not hold in practice; or in contrast, it requires computationally intensive stochastic simulations that are not affordable for real industrial problems. The present work attempts to propose an efficient strategy that would *in fine* bring the RBDO application field to more sophisticated examples, closer to real engineering cases.

The most basic formulation of a reliability-based design optimization problem reads as follows:

$$\boldsymbol{\theta}^* = \arg\min_{\boldsymbol{\theta}\in\mathcal{D}_{\boldsymbol{\theta}}} c(\boldsymbol{\theta}) \ : \ \begin{cases} f_i(\boldsymbol{\theta}) \leq 0, \ i=1,\ldots,n_c \\ \mathbb{P}\left(g_l\left(\boldsymbol{X}(\boldsymbol{\theta})\right)\leq 0\right) \leq P^0_{f\,l}, \ l=1,\ldots,n_p \end{cases} \tag{1}$$

where $\boldsymbol{X}$ is the random vector that describes the uncertainty in the parameters of the system to be designed, $\boldsymbol{\theta}$ is the vector of design variables, c is the cost function to be minimized, $\{f_i, \ i=1,\ldots,n_c\}$ is a set of deterministic constraints bounding the so-called admissible design space and $\{\mathbb{P}\left(g_l(\boldsymbol{X})\leq 0\right) \leq P^0_{f\,l}, \ l=1,\ldots,n_p\}$ are the probabilistic constraints specifying the minimum safety requirements. $P^0_{f\,l}$ should be carefully selected by the stake-holders according to the case-specific code of practice.

One should notice that the design vector $\boldsymbol{\theta}$ in Eq. (1) is a set of hyperparameters defining the random vector $\boldsymbol{X}$. In other words, in this work, design variables are exclusively considered as hyperparameters in the joint *probability density function* (PDF) of the random vector $\boldsymbol{X}$ because it simplifies the computation of reliability sensitivities (*i.e.* the gradients of the failure probability) as shown in Song et al. (2009). There is however no loss of generality since deterministic design variables might possibly be considered as artificially random (either normal or uniform) with small variance – *i.e.* asymptotically zero.

The recent book by Tsompanakis et al. (2008) provides an extensive review of the various tracks that have been explored by researchers in the structural reliability community in order to reduce the computational cost induced by this first basic formulation. The approach that is investigated here is the *surrogate-based* (or *response-surface-based*) approach. Starting with the premice that the performance function evaluation might involve a time-consuming computational task, the approach consists in replacing this function by a surrogate that is much faster to evaluate. This surrogate is usually built and possibly refined from a few evaluations of the real performance function which is referred to as a *design of experiments* (DOE). Various surrogates have been proposed in the structural reliability literature out of which: polynomial response surfaces, support vector machines, neural networks and kriging. This approach is argued here to be more flexible than the classical Taylor-expansion-based methods (*i.e.* approaches using the first-order reliability theory, FORM) that are extensively used in the common RBDO literature.

The remaining of this document is organized as follows. In section 2, the kriging surrogate is introduced. In Section 3, we make use of the genuine probabilistic nature of the kriging prediction in order to propose an adaptive refinement technique of the limit-state surfaces $\{g_i=0, \ i=1,\ldots,l\}$. Section 4 is devoted to the presentation of the proposed adaptive surrogate-based RBDO algorithm. It is then applied to a classical benchmark example in section 5. Discussion and perspectives are given in section 6.

2 THE KRIGING SURROGATE

In computer experiments, *surrogate modelling* addresses the problem of replacing a time-consuming mathematical model $\mathcal{M}$ called a *simulator* by an *emulator* that is much faster to evaluate. The emulator has the same input space $\mathcal{D}_{\boldsymbol{x}} \subseteq \mathbb{R}^n$ and output space $\mathcal{D}_y \subseteq \mathbb{R}$ as the simulator. For the sake of simplicity only scalar-output simulators are considered here. It is important to note that both $\boldsymbol{x}$ and y are deterministic in this context. The emulator is usually built from a set of observations named a *design of experiments* (DOE) $\{y_i = \mathcal{M}(\boldsymbol{x}_i), \ i=1,\ldots,m\}$.

Kriging (Santner et al. 2003) is one particular emulator that is able to give a probabilistic response $\widehat{Y}(\boldsymbol{x})$ whose variance (spread) depends on the quantity of available knowledge. In other words, the uncertainty in the prediction is purely epistemic and thus reducible as shown in the next sections.

In essence, kriging assumes that the output $y=\mathcal{M}(\boldsymbol{x})$ is a sample path of an underlying Gaussian stochastic process $Y(\boldsymbol{x})$ that would read as follows:

$$Y(\boldsymbol{x}) = \boldsymbol{f}(\boldsymbol{x})^{\mathsf{T}}\,\boldsymbol{\beta} + Z(\boldsymbol{x}) \tag{2}$$

where:

- $\boldsymbol{f}(\boldsymbol{x})^{\mathsf{T}}\,\boldsymbol{\beta}$ denotes the mean of the GP which corresponds to a classical linear regression model on a given functional basis $\{f_i, \ i=1,\ldots,p\} \in \mathcal{L}_2\left(\mathcal{D}_{\boldsymbol{x}}, \mathbb{R}\right)$;
- $Z(\boldsymbol{x})$ denotes the stochastic part of the GP which is modelled as a zero mean, constant variance σ_Y^2, stationary Gaussian process with a given symmetric positive definite autocorrelation model.

It is fully defined by its autocovariance function which reads:

$$C_{YY}\left(\boldsymbol{x},\,\boldsymbol{x}'\right)=\sigma_Y^2\,R\left(\left|\boldsymbol{x}-\boldsymbol{x}'\right|,\boldsymbol{\ell}\right),\quad \left(\boldsymbol{x},\,\boldsymbol{x}'\right)\in\mathcal{D}_{\boldsymbol{x}}\times\mathcal{D}_{\boldsymbol{x}} \tag{3}$$

where $\boldsymbol{\ell}$ is a vector of parameters defining R.

The most widely used class of autocorrelation functions is the anisotropic squared exponential model:

$$R\left(\left|\boldsymbol{x}-\boldsymbol{x}'\right|,\boldsymbol{\ell}\right)=\exp\left(\sum_{k=1}^{n}-\left|\frac{x_k-x_k'}{\ell_k}\right|^2\right) \tag{4}$$

The choice of the autocorrelation model should be made regarding the degree of regularity (derivability) of $\mathcal{M}$ (Rasmussen and Williams, 2006). Common regression models involves zero, first and second-order polynomials of x_k, $k=1,\ldots,n$. Note however that as in classical least-squares regression, the size p of the functional basis together with the input vector dimension n dramatically increases the number m of required observations. When attempting to model the output of a high-fidelity computer model, a good choice for the regression model would be an equivalent low-fidelity analytical model built on simplifying assumptions.

The best linear unbiased estimation of the GP Y at point $\boldsymbol{x}$ is shown (Santner et al., 2003) to be the Gaussian random variate $\widehat{Y}(\boldsymbol{x})\sim\mathcal{N}(\widehat{Y}(\boldsymbol{x}),\,\sigma_{\widehat{Y}}(\boldsymbol{x}))$ with moments:

$$\widehat{Y}\left(\boldsymbol{x}\right)=\boldsymbol{f}\left(\boldsymbol{x}\right)^{\mathsf{T}}\widehat{\boldsymbol{\beta}}+\boldsymbol{r}\left(\boldsymbol{x}\right)^{\mathsf{T}}\mathbf{R}^{-1}\left(\mathbf{Y}-\mathbf{F}\widehat{\boldsymbol{\beta}}\right) \tag{5}$$

$$\sigma_{\widehat{Y}}^2\left(\boldsymbol{x}\right)=\sigma_Y^2\left(1-\left\langle\begin{array}{cc}\boldsymbol{f}\left(\boldsymbol{x}\right)^{\mathsf{T}} & \boldsymbol{r}\left(\boldsymbol{x}\right)^{\mathsf{T}}\end{array}\right\rangle\left[\begin{array}{cc}\mathbf{0} & \mathbf{F}^{\mathsf{T}}\\ \mathbf{F} & \mathbf{R}\end{array}\right]^{-1}\left[\begin{array}{c}\boldsymbol{f}\left(\boldsymbol{x}\right)\\ \boldsymbol{r}\left(\boldsymbol{x}\right)\end{array}\right]\right) \tag{6}$$

where we have introduced r, $\mathbf{R}$ and $\mathbf{F}$ such that:

$$r_i=R\left(\left|\boldsymbol{x}-\boldsymbol{x}_i\right|,\boldsymbol{\ell}\right),\ i=1,\ldots,m \tag{7}$$

$$R_{ij}=R\left(\left|\boldsymbol{x}_i-\boldsymbol{x}_j\right|,\boldsymbol{\ell}\right),\ i=1,\ldots,m,\ j=1,\ldots,m \tag{8}$$

$$F_{ij}=f_i\left(\boldsymbol{x}_j\right),\ i=1,\ldots,p,\ j=1,\ldots,m \tag{9}$$

At this stage one can easily prove that $\widehat{Y}(\boldsymbol{x}_i)=\mathcal{M}(\boldsymbol{x}_i)$ and $\sigma_{\widehat{Y}}(\boldsymbol{x}_i)=0$ for $i=1,\ldots,m$, thus meaning the kriging surrogate is an *exact interpolator*.

Given a choice for the regression and correlation models, the optimal set of parameters $\boldsymbol{\beta}^*$, $\boldsymbol{\ell}^*$ and $\sigma_{\widehat{Y}}^{2*}$ can then be inferred using the *maximum likelihood principle* applied to the unique sparse observation of the GP sample path grouped in the vector $y=\langle\mathcal{M}(\boldsymbol{x}_i),\ i=1,\ldots,m\rangle$. This optimization problem can be solved analytically for both $\boldsymbol{\beta}^*$ and $\sigma_{\widehat{Y}}^{2*}$ assuming $\boldsymbol{\ell}^*$ is known. Thus, the problem is solved in two steps: the maximum likelihood estimation of $\boldsymbol{\ell}^*$ is first solved by a global optimization algorithm which in turns allows one to evaluate the optimal $\boldsymbol{\beta}^*$ and $\sigma_{\widehat{Y}}^{2*}$.

Figure 1 illustrates the kriging surrogate applied to the simple model $\mathcal{M}\ :\ x\mapsto x\ \sin x$ (red curve) in the input space $\mathcal{D}_x=[0,\ 20]$. The kriging surrogate uses a constant regression model and the squared exponential autocorrelation model given at Eq. (4). It is interesting to note that the further from an observation (red point) the prediction is made, the greater its associated uncertainty is. The uncertainty is represented here as the 95% confidence interval (blue shaded area) centered at the mean prediction (black curve).

3 THE PROPOSED ADAPTIVE REFINEMENT TECHNIQUE

Various refinement techniques have been proposed in the kriging-related literature, *e.g.* in global optimization (Jones et al., 1998) or in structural reliability (Bichon et al., 2008; Lee and Jung, 2008). They are all based on the genuine probabilistic nature of the kriging prediction. The refinement technique consists in finding the set of points that should be added to the DOE in order to improve the prediction and reduce its associated epistemic uncertainty in a specific region of interest. In global optimization, the region of interest is the vicinity of the optimum that can be sequentially achieved by means of the so-called efficient global optimization technique (Jones et al., 1998). In

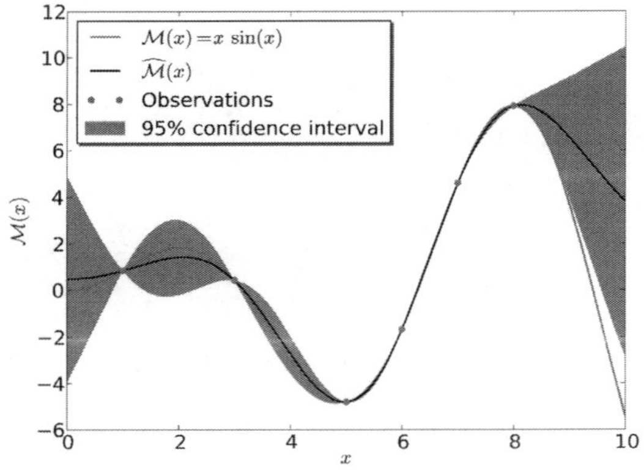

Figure 1. Illustration of the kriging surrogate.

reliability and reliability-based design optimization, the region of interest in the space of random variables $\boldsymbol{X}$ is the limit-state surface $g = 0$ – we drop the subscript l in this section for the sake of clarity.

Provided an initial kriging prediction of the performance function g, the probable vicinity of the limit-state surface $g = 0$ can be defined in terms of the following *margin of uncertainty*:

$$\mathbb{M} = \left\{ \boldsymbol{x} \, : \, -k\,\sigma_{\widehat{G}}\left(\boldsymbol{x}\right) \leq \widehat{G}\left(\boldsymbol{x}\right) \leq +k\,\sigma_{\widehat{G}}\left(\boldsymbol{x}\right) \right\} \tag{10}$$

where k might be chosen as $k = \Phi^{-1}(97.5\%) = 1.96$ meaning that a 95% confidence interval onto the prediction of the limit-state surface is selected. Any point that belongs to this margin has an uncertain sign, and the spread of this margin should thus be reduced in order to make the reliability estimation accurate. To do so, one need to add some knowledge (new points in the DOE) that are located in this margin.

The probability that a point $\boldsymbol{x} \in \mathcal{D}_{\boldsymbol{x}}$ falls into the margin $\mathbb{M}$ (with respect to the epistemic uncertainty in the prediction) can be expressed in closed-form as follows:

$$\mathbb{P}\left(\boldsymbol{x} \in \mathbb{M}\right) = \Phi\left(\frac{k\,\sigma_{\widehat{G}}\left(\boldsymbol{x}\right) - \widehat{G}\left(\boldsymbol{x}\right)}{\sigma_{\widehat{G}}\left(\boldsymbol{x}\right)}\right) - \Phi\left(\frac{-k\,\sigma_{\widehat{G}}\left(\boldsymbol{x}\right) - \widehat{G}\left(\boldsymbol{x}\right)}{\sigma_{\widehat{G}}\left(\boldsymbol{x}\right)}\right) \tag{11}$$

Then, finding the point that maximizes this quantity on the support of the PDF of $\boldsymbol{X}$ will finally bring the best improvement point in the DOE. Starting with this statement, many authors in the kriging literature decide to use global optimization algorithm in order to find *the best* improvement point.

In this paper, we propose a slightly different strategy that allows one to add *the best set* of improvement points. Indeed, it is argued here that adding several points per refinement iteration can significantly improve the refinement strategy in the sense that it will make the final DOE much more homogeneous in the vicinity of the limit-state and that it can be faster, given the availability to perform several simulation runs at the same time (distributed computations).

To do so, the probability to fall in the margin of uncertainty is multiplied by a weighting PDF w so that the following criterion:

$$\mathcal{C}\left(\boldsymbol{x}\right) = \mathbb{P}\left(\boldsymbol{x} \in \mathbb{M}\right)\, w\left(\boldsymbol{x}\right) \tag{12}$$

can itself be considered as a PDF up to an unknown but finite normalizing constant. The weighting density w can either be the original PDF of $\boldsymbol{X}$, or, as it is proposed here, a uniform PDF on a sufficiently large confidence region of the original PDF. Such a confidence region might be difficult to define for any given PDF, but as it is usually done in structural reliability, the original random vector $\boldsymbol{X}$ can be transformed into a probabilistically equivalent standard Gaussian random vector $\boldsymbol{U}$ for which the confidence region is simply an hypersphere with radius β_0. The reader is referred to Lebrun and Dutfoy (2009) for a more detailed discussion on such mappings. In that given space, β_0 can be easily selected as *e.g.* $\beta_0 = 8$ which corresponds to the maximal reliability index that can be justified numerically.

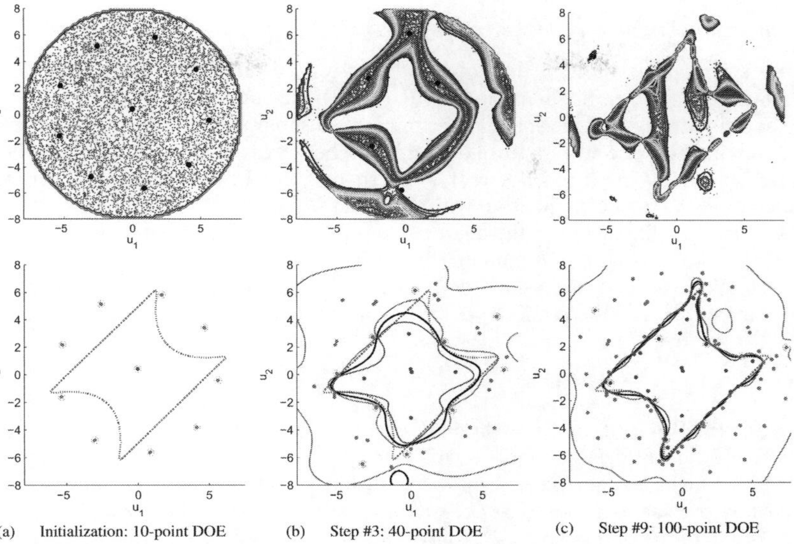

Figure 2. An illustration of the proposed limit-state surface refinement technique.

Given that pseudo-PDF, one can generate N (say $N = 10^4$) samples by means of any well suited *Markov chain Monte Carlo* (MCMC) simulation technique (Robert and Casella, 2004), such as the *slice sampling* technique (Neal, 2003). The N generated candidates are expected to be highly concentratred at the modes of $\mathcal{C}$ – and thus in $\mathbb{M}$. This population of candidates is then reduced to a smaller one that has essentially the same statistical properties (*i.e.* it uniformally spans $\mathbb{M}$) by means of the K-means clustering technique (Elkan, 2003).

Figure 2 illustrates the proposed refinement strategy applied to a nonlinear limit-state surface from Waarts (2000). The upper subfigures shows the contours of the refinement pseudo-PDF $\mathcal{C}$ at each refinement step together with the candidate population generated by slice sampling and its K cluster centers – $K = 10$ being given. In the lower subfigures one can see the real limit-state surface represented as the dashed black curve, its kriging prediction represented as the black line and its associated margin of uncertainty $\mathbb{M}$ which is bounded below by the red line and above by the blue line. Another interpretation of these figures is that any point within the blue bounded shape is positive with a 95% confidence and any point inside the red bounded shape is negative with the same confidence level.

We make a final remark considering the three approximate failure subsets $\widehat{F}^{-1}$, $\widehat{F}^{\,0}$, $\widehat{F}^{+1}$ defined as:

$$\widehat{F}^{i} = \left\{ \boldsymbol{x} \,:\, \widehat{G} + i\,k\,\sigma_{\widehat{G}}(\boldsymbol{x}) \leq 0 \right\}, \quad i = -1, 0, +1 \tag{13}$$

Due to the positiveness of standard deviation, the following statement holds:

$$\widehat{F}^{-1} \subseteq \widehat{F}^{\,0} \subseteq \widehat{F}^{+1} \Rightarrow \mathbb{P}\left(\boldsymbol{X} \in \widehat{F}^{-1}\right) \leq \mathbb{P}\left(\boldsymbol{X} \in \widehat{F}^{\,0}\right) \leq \mathbb{P}\left(\boldsymbol{X} \in \widehat{F}^{+1}\right) \tag{14}$$

Unfortunately, there is no proof that the real failure subset F satisfies the following relationship $\widehat{F}^{-1} \subseteq F \subseteq \widehat{F}^{+1}$ due to the empirical assumptions that were made in Section 2. However, the spread of the interval $[\mathbb{P}(\boldsymbol{X} \in \widehat{F}^{-1}),\ \mathbb{P}(\boldsymbol{X} \in \widehat{F}^{-1})]$ is a useful measure to check if the kriging surrogate is accurate enough for reliability analysis or not, and it is used in this paper as a stopping criteria for the proposed refinement procedure.

4 THE PROPOSED ADAPTIVE SURROGATE-BASED RBDO ALGORITHM

The proposed strategy to solve the RBDO problem in Eq. (1) consists in nesting the previously introduced kriging surrogate together with the proposed refinement strategy within a classical (but efficient) nested RBDO algorithm.

4.1 *The augmented reliability space*

The kriging surrogate is built in the so-called *augmented reliability space* such as defined in Taflanidis and Beck (2009). In Kharmanda et al. (2002), the augmented reliability space is defined as the tensor product between the space of the random variables and the design space: $\mathcal{D}_X \times \mathcal{D}_\theta$, but the dimension of this space suffers from both the number of random variables n and the number of design variables n_θ. It is also argued here that this space may cause some loss of information as the performance function is not in bijection with this space. In contrast, in Taflanidis and Beck (2009) and in this paper, the dimension of the augmented reliability space is kept constant by considering that the design vector $\boldsymbol{\theta}$ simply augments the uncertainty in the random vector $\boldsymbol{X}$. Indeed, the augmented random vector $\boldsymbol{X}(\boldsymbol{\theta})$ has a PDF h which sums the uncertainty in the design choices and the aleatory uncertainty in the random vector $\boldsymbol{X}$. Under such considerations, h reads as follows:

$$h(\boldsymbol{x}) = \int_{\mathcal{D}_\theta} f_{\boldsymbol{X}}(\boldsymbol{x} \,|\, \boldsymbol{\theta})\, \pi(\boldsymbol{\theta})\, \mathrm{d}\boldsymbol{\theta} \tag{15}$$

where f_X is the PDF of $\boldsymbol{X}(\boldsymbol{\theta})$ and π is the PDF of $\boldsymbol{\theta}$ that can be assumed uniform on $\mathcal{D}_\theta$.

The DOE should uniformally span a sufficiently large *confidence region* of this augmented PDF in order to make the surrogate limit-state surfaces accurate anywhere it can potentially be evaluated. More precisely, it should be accurate for extreme design choices $\boldsymbol{\theta}$ (*i.e.* located onto the boundaries of the optimization space $\mathcal{D}_\theta$) and extreme values of the marginal random vector $\boldsymbol{X}$ (to be able to compute reliability indices as large as *e.g.* $\beta_0 = 8$). A confidence region is essentially the multivariate extension of the univariate concept of *confidence interval*. Under the previous general assumptions, it is hard to give a mathematical form to the contour of this region. However, one may easily build an hyperrectangle region that bounds the sought confidence region.

Indeed, such an hyperrectangular region is defined as the tensor product of the margins' confidence intervals $\{X_i,\ i = 1, \ldots, n\}$. In order to compute the quantiles bounding these confidence intervals, one should additionally assume that the design parameters are exclusively involved in the definition of the margins – *i.e.* no parameters in the dependence structure (the *copula*) as it will never be the case in the RBDO applications. For each margin, the lower quantiles $\{q^-_{X_i},\ i = 1, \ldots, n\}$ (at the probability level $1 - \Phi(\beta_0) = \Phi(-\beta_0)$) and upper quantiles $\{q^+_{X_i},\ i = 1, \ldots, n\}$ (at the probability level $\Phi(+\beta_0)$) are respectively solutions of the following optimization problems:

$$q^-_{X_i} = \min_{\boldsymbol{\theta} \in \mathcal{D}_\theta} F^{-1}_{X_i}\left(\Phi(-\beta_0) \,|\, \boldsymbol{\theta}\right), \quad i = 1, \ldots, n \tag{16}$$

$$q^+_{X_i} = \max_{\boldsymbol{\theta} \in \mathcal{D}_\theta} F^{-1}_{X_i}\left(\Phi(+\beta_0) \,|\, \boldsymbol{\theta}\right), \quad i = 1, \ldots, n \tag{17}$$

where $\{F^{-1}_{X_i},\ i = 1, \ldots, n\}$ are the quantile functions of the margins. If the domain $\mathcal{D}_\theta$ is rectangular and if one is able to derive an analytical expression for the marginal quantile functions and there derivatives with respect to the parameters, then these optimization problems might be solved analytically. However assuming a more general setup where one has only numerical definition of these quantities, these problems can be efficiently solved by means of a simple gradient-based algorithm due to the convenient properties of the quantile functions – namely, the monotony with respect to the location and shape parameters. Finally, the sought hyperrectangle can be easily define by means of a simple indicator function being equal to one if $q^-_{X_i} \le x_i \le q^+_{X_i},\ i = 1, \ldots, n$ and zero otherwise. And this indicator function may then be used as the weighting PDF w in the refinement procedure proposed in Section 3.

4.2 *The adaptive surrogate-based nested RBDO algorithm*

The kriging surrogate together with its adaptive refinement procedure is finally plugged into a double-loop RBDO algorithm. The outer optimization loop is performed by means of the Polak-He optimization algorithm (Polak, 1997). Provided an initial design, the algorithm proceeds iteratively in two steps: *(i)* the direction is determined solving a quasi-SQP sub-optimization problem and *(ii)* the step size is approximated by the Goldstein-Armijo approximate line-search rule. The reliability and reliability sensitivity analyses are performed with the *subset simulation* variance reduction technique (Au and Beck, 2001) onto the kriging surrogate. The subset simulation technique for reliability sensitivity analysis is detailed in Song et al. (2009). The overall methodology was implemented within the FERUM toolbox (Bourinet et al., 2009). It makes use of the Matlab's toolboxes functions quadprog (for the SQP sub-optimization problem) and slicesample (to generate samples from the refinement pseudo-PDF $\mathcal{C}$).

5 NUMERICAL APPLICATION

The example treated in this section is taken from Royset et al. (2001). Consider a short column with a rectangular cross section of dimensions b and h, which is subjected to bi-axial bending moments M_1 and M_2 and axial force P. Assuming an elastic-perfectly plastic material, the reliability of the column is defined by the following performance function:

$$g(\boldsymbol{X}) = 1 - \frac{4\,M_1}{b\,h^2\,f_y} - \frac{4\,M_2}{b^2\,h\,f_y} - \left(\frac{P}{b\,h\,f_y}\right)^2 \tag{18}$$

where f_y denotes the yield strength of the material. In the present application, all the variables are considered as random with distributions given in table 1. The optimization problem consists in minimizing the construction cost which is assumed to be proportional to the cross section area considering an additional expected failure cost that is a hundred times greater than the construction cost, so that the objective function reads:

$$c(b,\,h) = c_0(b,\,h) + P_f\,c_f(b,\,h) = c_0(b,\,h)\,(1 + 100\,P_f) = b\,h\,(1 + 100\,P_f) \tag{19}$$

The search for the optimal design is constrained to the designs that satisfy the following limit aspect ratio rates: $1/2 \leq b/h \leq 2$ with $100 \leq b,\ h \leq 1000$. The minimum reliability level is chosen as $\beta = 3$. Note that in the original paper, the design variables are considered as deterministic. Since the present approach only deals with design variables that defines the joint PDF of the random vector $\boldsymbol{X}$, they are considered here as Gaussian with a small coefficient of variation and the optimization is performed with respect to their mean b and h.

The results are given in Table 2. The Deterministic Design Optimization (DDO) was performed using the mean values of all the variables in Table 1 without considering uncertainty and thus leads to a 50% failure probability. Note that the given optimal cost does not account for the expected failure cost. The other lines of Table 2 shows the results of the RBDO problem. The first row gives the reference results from Royset et al. (2001). The number of performance function calls was not given in the original paper, it has thus been estimated given the methodology the authors used and assuming they targeted a 5% C.o.V. on the failure probability estimated by Monte-Carlo simulation. The second row provides the results from a FORM-based nested RBDO algorithm. It can be seen that the first-order reliability theory assumptions are not conservative in this case. The third row gives the results obtained by the same nested RBDO algorithm, using however the subset simulation technique as the reliability (and reliability sensitivity) estimator. Finally, the last row gives the results obtained when using kriging as a surrogate for the limit-state surface. The kriging model used for this application used a constant regression model and a squared exponential covariance model. It was initialized with a 50-point DOE and sequentially refined with $K = 10$ points per refinement iteration.

6 DISCUSSION

The proposed surrogate-based approach allows to save a significant number of simulation runs without using the common first-order reliability theory assumptions. It is important to note that its

Table 1. Stochastic model for the short column case.

Variable	Distribution	Mean	C.o.V.
M_1 (N.mm)	Lognormal	250×10^6	30%
M_2 (N.mm)	Lognormal	125×10^6	30%
P (N)	Lognormal	2.5×10^6	20%
f_y (MPa)	Lognormal	40	10%
b (mm)	Gaussian	b	1%*
h (mm)	Gaussian	h	1%*

**Device uncertainty* to match the RBDO formulation in Eq. (1)

Table 2. Benchmark results for the short column case.

Method	Opt. design (mm)		Cost (mm^2)	# of func. calls	Reliability	(C.o.V)
DDO	$b = 258$	$h = 500$	1.29×10^5	50	$\beta_{sim} \approx 0.01$	(3‰)
RBDO ($\beta \geq 3$)						
Reference (DSA)	$b = 372$	$h = 559$	2.15×10^5	$\approx 4 \times 10^6$	$\beta_{sim} \approx 3.38$	(5%)
FORM-based (RIA)	$b = 399$	$h = 513$	2.12×10^5	9 472	$\beta_{HL} = 33.38$	(N/A)
Check of FORM assumptions			2.20×10^5	4×10^5	$\beta_{sim} \approx 3.19$	(3%)
Present w/o kriging	$b = 369$	$h = 560$	2.16×10^5	19×10^6	$\beta_{sim} \approx 3.35$	(3%)
Present w/ kriging	$b = 379$	$h = 547$	2.17×10^5	140	$\beta_{sim} \approx 3.32$	(3%)

β_{HL} is the Hasofer-Lind reliability index, β_{sim} is the equivalent reliability index estimated by *subset simulation*.

efficiency mainly relies upon the fact that the kriging surrogate is built in the augmented reliability space and that it can thus be used from one nested reliability analysis to the other. Paradoxically, it might also suffer from the potential high degree of nonlinearity of the performance function in that augmented space and this is the reason why the strategy still need to be validated and tuned abroad an extended set of benchmark examples.

REFERENCES

Au, S.-K. and J. L. Beck (2001). Estimation of small failure probabilities in high dimensions by subset simulation. *Prob. Eng. Mech. 16*(4), 263–277.

Bichon, B., M. Eldred, L. Swiler, S. Mahadevan, and J. McFarland (2008). Efficient Global Reliability Analysis for nonlinear implicit performance functions. *AIAA Journal 46*(10), 2459–2468.

Bourinet, J.-M., C. Mattrand, and V. Dubourg (2009). A review of recent features and improvements added to FERUM software. In *Proc. of the 10th Int. Conf. On Struct. Safe. And Rel. (ICOSSAR)*, Osaka, Japan.

Elkan, C. (2003). Using the triangle inequality to accelerate K-means. In *Proceedings of the Twentieth International Conference on Machine Learning (ICML-2003)*, Washington D.C.

Jones, D., M. Schonlau, and W. Welch (1998). Efficient Global Optimization of Expensive Black-Box Functions. *Journal of Global Optimization 13*(4), 455–492.

Kharmanda, G., A. Mohamed, and M. Lemaire (2002). Efficient reliability-based design optimization using a hybrid space with application to finite element analysis. *Struct. Multidisc. Opt. 24*(3), 233–245.

Lebrun, R. and A. Dutfoy (2009). A generalization of the nataf transformation to distributions with elliptical copula. *Prob. Eng. Mech. 24*(2), 172–178.

Lee, T. and J. Jung (2008). A sampling technique enhancing accuracy and efficiency of metamodel-based RBDO: Constraint boundary sampling. *Computers & Structures 86*(13-14), 1463–1476. Structural Optimization.

Neal, R. (2003). Slice sampling. *Annals of Statistics 31*, 705–767.

Polak, E. (1997). *Optimization Algorithms and Consistent Approximations*. Springer.

Rasmussen, C. and C. Williams (2006). *Gaussian Processes for Machine Learning* (Internet Edition ed.). Adaptive Computation and Machine Learning. Cambridge, Massachusetts: MIT Press.

Robert, C. and G. Casella (2004). *Monte Carlo statistical methods (Second Edition)*. Springer Series in Statistics. Springer Verlag.

Royset, J. O., A. Der Kiureghian, and E. Polak (2001). Reliability-based optimal structural design by the decoupling approach. *Reliability Engineering and System Safety 73*(3), 213–221.

Santner, T., B. Williams, and W. Notz (2003). *The design and analysis of computer experiments*. Springer series in Statistics. Springer.

Song, S., Z. Lu, and H. Qiao (2009). Subset simulation for structural reliability sensitivity analysis. *Reliability Engineering and System Safety 94*(2), 658–665.

Taflanidis, A. and J. Beck (2009). Stochastic subset optimization for reliability optimization and sensitivity analysis in system design. *Computers & Structures 87*(5–6), 318–331.

Tsompanakis, Y., N. D. Lagaros, and M. Papadrakakis (Eds.) (2008). *Structural Design Optimization Considering Uncertainties* (1 ed.). Taylor & Francis.

Waarts, P.-H. (2000). *Structural reliability using finite element methods: an appraisal of DARS: Directional Adaptive Response Surface Sampling*. Ph. D. thesis, Technical University of Delft, The Netherlands.

Reliability and Optimization of Structural Systems – Straub (ed)
© 2010 Taylor & Francis Group, London, ISBN 978-0-415-88179-1

Structural reliability assessment using Kriging metamodel and Monte Carlo simulation: AK-MCS method

Benjamin Echard, Nicolas Gayton & Maurice Lemaire
IFMA-LaMI, Clermont-Ferrand, France

ABSTRACT: This article describes a new method to perform structural reliability analysis. It is called AK-MCS for Active learning method combining Kriging and Monte Carlo Simulation. It proposes to perform a Monte Carlo simulation without evaluating the whole population on the expensive performance function. This is possible thanks to the predictions by Kriging metamodel which is based on a design of experiments of weak size. The method is performed on several examples and compared to other metamodels. It is proved to require less calls to the performance function and for a better accuracy on the probability of failure.

1 INTRODUCTION

Reliability aims at determining the probability of failure of a system by considering its inputs as random. The performance function $G(\mathbf{x}) = 0$ characterizes the response of the system: a negative value means that the failure occurs, a positive implies that the system is safe in this configuration. The border between the negative and positive domains is called the limit state ($G(\mathbf{x}) = 0$). To assess the probability of failure $P_f = \text{Prob}(G(\mathbf{x}) \leq 0)$, several methods exist. The most common of them remains the Monte Carlo simulation. The major problem of this method is the very numerous evaluations required. To avoid this, metamodels can be used. They aim at approximating the limit state with a strategic design of experiments (DoE) to obtain, for a less important computation time, a sufficiently accurate probability of failure. Quadratic response surfaces, Polynomial Chaos (Ghanem and Spanos 1991) or Support Vector Machine (Hurtado 2004) methods are possibilities. Aside from these metamodels, a stochastic approach has also been intensively investigated: Kriging.

Developed for geostatistics in the fifties and sixties by Krige and then theorised by Matheron (1973), it presents several interesting differences with the other metamodels. First, Kriging is an exact interpolation method, *i.e.* the prediction in a point of the design of experiments is the exact value of the performance function in this point. Furthermore, thanks to its stochastic propriety, Kriging provides not only a predicted value $\hat{G}(\mathbf{x})$ in a point but also an estimation of the local variance of the prediction (Kriging variance: $\sigma^2_{\hat{G}}(\mathbf{x})$). Thanks to these characteristics, Kriging has been used intensively in optimisation problems in the nineties with active learning methods such as the Efficient Global Optimisation EGO by Jones et al. (1998).

The first applications of Kriging to structural reliability problems are more recent. Romero et al. (2004) compare Kriging with polynomial regression and finite-element interpolation on progressive lattice-samplings with analytical functions. The results show that none of the methods is more efficient than the others. Following this, another method is proposed by Kaymaz (2005) to perform structural reliability analysis. It is compared to classic response surface methods. The method is based on the MATLAB toolbox DACE (Lophaven et al. 2002) and consists of finding the design point. The conclusions are that Kriging metamodel does not greatly improve the reliability results compared to quadratic response surface methods' results, unless Kriging parameters are well chosen. These methods are based on progressive design of experiments but they cannot be defined as active learning methodologies. Indeed, the design of experiments is not improved by learning from the data supplied by Kriging, *i.e.* Kriging prediction and variance. This was noticed by Bichon et al. (2008) who proposed a fully active learning method inspired by EGO and the Kriging contour estimation method by Ranjan et al. (2008). The method is found to be really efficient. However, high dimensions are a problem as the size of the initial design of experiments evolves in n^2 (n being the number of random variables). Furthermore, the method approximates the limit state in the whole design space and therefore, even in regions where configurations show very weak densities of probabilities and have negligible effects on the probability of failure. To avoid that, this article proposes to develop an active learning method which combines Kriging and Monte Carlo simulation to classify predicted negative and positive $\hat{G}$ values of a Monte Carlo population.

The proposed method called AK-MCS for Active learning method combining Kriging and Monte Carlo Simulation is detailed Section 2. Following this, the method is compared with other metamodels of literature Section 3 on two examples. The method AK-MCS is shown to be very efficient as the number of calls to the performance function is relatively small compared to the other solutions and this for a better accuracy on the estimated probability of failure.

2 PROPOSED METHOD: AK-MCS

The method that we propose here is named AK-MCS. It consists of an Active learning reliability method combining Kriging and Monte Carlo Simulation. The active learning principles for reliability with Kriging are proposed in the previous works (Bichon et al. 2008; Ranjan et al. 2008). However, the concept here is different. In fact, instead of approximating the limit state in the whole space and therefore making expensive evaluations of G for points with very weak densities of probability, it is preferred here to focus on a Monte Carlo population. AK-MCS classifies a Monte Carlo population of n_{MC} points without evaluating n_{MC} times the true performance function. To avoid these evaluations, Kriging is used for its exact interpolation characteristic and for its interesting proprieties in active learning methods. Indeed, its stochastic nature enables us to define a learning function to select the best next point to add to the design of experiments to improve the accuracy of the metamodel on the signs of the n_{MC} points. AK-MCS must be seen as a modification of crude Monte Carlo simulation.

The proposed method consists of 10 stages:

1. **Generation of a Monte Carlo population in the design space.** It is named S and is composed of n_{MC} points. At this stage, none of them is evaluated on the performance function. They represent candidate points to be evaluated if the active learning requires it. This population remains the same during the whole process of learning in AK-MCS unless Stage 9 is reached.
2. **Definition of the initial design of experiments (DoE).** To perform Kriging, a design of experiments is required. Here, it consists of a random selection of N_1 points among the population S. These points are evaluated on the performance function and are used as the initial design of experiments for the Kriging model. According to our experience, a dozen of points are enough. The initial design of experiments is preferred to be defined small and to add step by step only the point that is improving the most the metamodel. In the previous cited papers (Bichon et al. 2008; Ranjan et al. 2008), the initial design of experiments is much larger as they grow with the dimension of the problem.
3. **Computation of the Kriging model according to the design of experiments.** This stage is performed with the toolbox DACE, as this tool has been used in several Kriging reference articles (Kaymaz 2005; Marrel et al. 2008; Kleijnen 2009; Dellino et al. 2009). The correlation model is chosen as anisotropic Gaussian and the regression model is taken constant (Ordinary Kriging).
4. **Prediction by Kriging and estimation of the probability of failure.** First, the Kriging predictions $\hat{G}(\mathbf{x}^{(i)})$ for $i = 1, \ldots, n_{MC}$ are obtained thanks to DACE. Then, the probability of failure $\hat{P}_f$ is estimated with the signs of these predictions. It is obtained by the ratio between the number of points in the population S with a negative or null Kriging prediction and the total number of points in S.

$$\hat{P}_f = \frac{n_{\hat{G}\leq 0}}{n_{MC}} \tag{1}$$

5. **Identification of the best next point in S to evaluate on the performance function.** A learning function comes then to identify the point that improves the most the Kriging model. It is based on the Kriging prediction and its uncertainty, therefore the variances $\sigma^2_{\hat{G}}(\mathbf{x}^{(i)})$ for $i = 1, \ldots, n_{MC}$ are first computed with DACE. It is proposed by Bichon et al. (2008) to maximise a function called EFF for Expected Feasibility Function which provides an indication of how well the true

value of the performance function in a point can be expected to satisfy the constraint $G(\mathbf{x}) = 0$. It is shown to be efficient when a full approximation of the limit state is required. Here, accuracy on the sign is needed only among the points of the Monte Carlo population S and the limit state can be roughly approximated in configurations with very weak densities of probability. Therefore, it is preferred to define a more suitable learning function for Monte Carlo simulation. For that, it is first needed to understand which points are interesting to know, *i.e.* to evaluate on G. In Monte Carlo simulation, these points are the one who are likely to move from positive to negative (or negative to positive) predictions and therefore modify the probability of failure. This uncertainty on the sign of the prediction is directly related to the Kriging prediction and its variance, therefore these points can present three characteristics: to be close to the predicted limit state, to have an important uncertainty (high Kriging variance) or both at the same time. To identify them, the learning function U is proposed:

$$|\hat{G}(\mathbf{x})| - U(\mathbf{x})\sigma_{\hat{G}}(\mathbf{x}) = 0 \tag{2}$$

This function is an adaptation of the lower confidence bounding (*lcb*) function proposed in Cox and John (1997) for optimisation. It indicates the distance in Kriging standard deviations between the prediction and the predicted limit state. It corresponds to the Cornell's reliability index. The best next point to know becomes the one with the minimum $U(\mathbf{x})$ value among the n_{MC} candidates of S.

6. **Stopping condition on learning.** Once the best next point $\mathbf{x}^*$ in S to evaluate has been identified with the learning function, its learning function's value is compared to a stopping one taken as 2. This is referred as the stopping condition on learning. This value of 2 corresponds to a probability of 97.7% of having the right sign for the prediction. It is seen to guarantee great accuracy on the probability of failure on the examples and this value is proposed in Cox and John (1997) for their optimisation problems. In the rest of the article, AK-MCS using the learning function U is named AK-MCS+U. It must be noticed that any other learning functions can be used with AK-MCS, like AK-MCS+EFF for example.
7. **Update of the previous design of experiments with the best point.** If the stopping condition Stage 6 is not satisfied, the learning carries on and the best point $\mathbf{x}^*$ in S is evaluated on the performance function. Following this, it is added to the design of experiments: $N_{i+1} = N_i + 1$. The method, then, goes back to Stage 3 to compute the new Kriging model with the updated design of experiments composed of N_{i+1} points.
8. **Computation of the coefficient of variation of the probability of failure.** If the stopping condition Stage 6 is satisfied, the learning is stopped and the metamodel is said to be accurate enough on the performance function's signs of the n_{MC} points. The next stage consists of checking if the Monte Carlo population S is sufficiently large to obtain a low coefficient of variation on the Kriging prediction of the probability of failure (Stage 4). Here, a coefficient of variation below 5% is seen as acceptable. It is calculated by:

$$\text{C.O.V}_{\hat{P}_f} = \sqrt{\frac{1 - \hat{P}_f}{\hat{P}_f \, n_{MC}}} \tag{3}$$

9. **Update of the population.** If the coefficient of variation is found to be too high, S is increased with new points coming from another Monte Carlo population (generated like Stage 1). AK-MCS goes back to Stage 4 to predict the new population and the active learning method carries on until the stopping condition is met again. It is important to note that no information about previous performance function's evaluations is lost.
10. **End of AK-MCS.** If the coefficient of variation of $\hat{P}_f$ is low enough, AK-MCS stops and the last prediction of the probability of failure is considered as the result of the method.

3 AK-MCS APPLICATIONS

AK-MCS efficiency is illustrated on two examples in 2D selected for their complex limit states. Indeed, they present high non-linearity, non-convexity and non-connexity domains of failure. AK-MCS is compared to Monte Carlo simulation and results from literature.

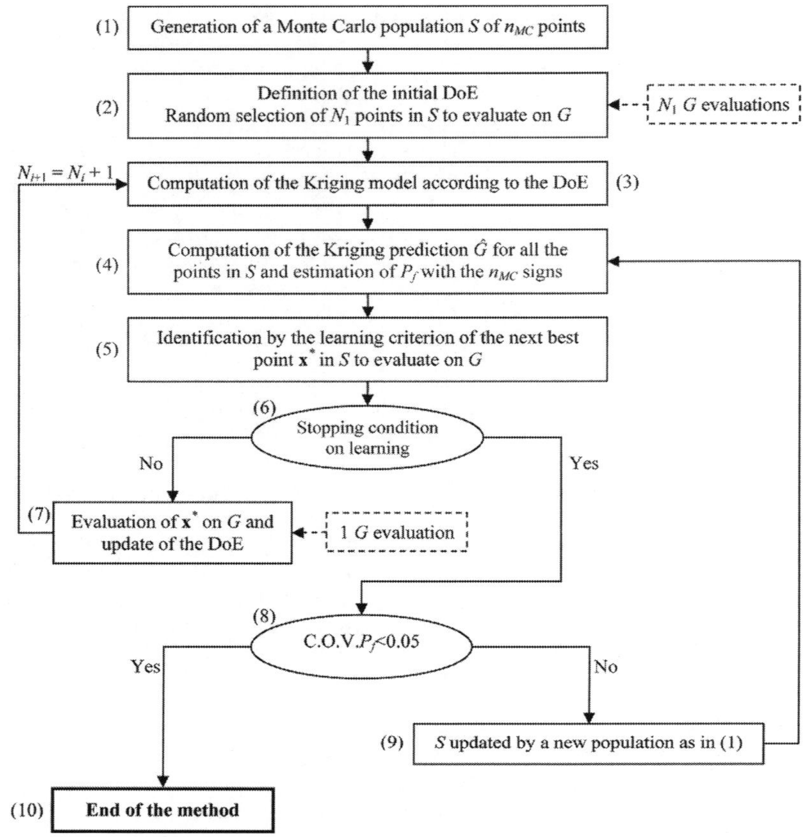

Figure 1. AK-MCS flowchart.

3.1 *Example 1: Series system with 4 branches*

The first example consists of a series system with 4 branches which has been studied by Waarts (2000) and Van Gemert (2005). The random variables x_i are standard normal distributed random variables. The performance function reads:

$$G(x_1, x_2) = \min \left\{ \begin{array}{l} 3 + 0.1\,(x_1 - x_2)^2 - (x_1 + x_2)/\sqrt{2}; \\ 3 + 0.1\,(x_1 - x_2)^2 + (x_1 + x_2)/\sqrt{2}; \\ (x_1 - x_2) + 6/\sqrt{2}; (x_2 - x_1) + 6/\sqrt{2} \end{array} \right\} \tag{4}$$

AK-MCS + U is compared with numerous metamodels proposed by Schueremans and Van Gemert (2005). The results are summarised in Table 1. The information is given as the number of calls to the performance function N_{call}, the number of points n_{MC} in the Monte Carlo population S, the predicted probability of failure $\hat{P}_f$, its corresponding reliability index β and the error percentage ϵ_β in comparison to the reference reliability index of the Monte Carlo simulation or of the reference value by Schueremans Van Gemert (2005). It is shown that AK-MCS + U is more efficient than the other metamodels found in literature. Indeed, the number of calls to the performance function is lower than most of the metamodels proposed by Van Gemert (2005). Directional Sampling (DS) and Importance Sampling + Neural Network (IS + NN) require less calls to the performance function. However, the nearly radial shape of the limit state function at hand is an adequate problem to solve with crude DS (Schueremans and Van Gemert 2005). This is the reason why N_{call} is extremely low for crude DS. About IS + NN, it must be seen that the reliability index is not accurate. An error of 3% is obtained in comparison to the reference reliability index. Therefore, AK-MCS + U shows

Table 1. Example 1, reliability results – Comparison of AK-MCS + U with a Monte Carlo simulation on the same population of size $n_{MC} = 10^6$ and with metamodels from literature.

Method	N_{call}	$\hat{P}_f$ (C.O.V$_{\hat{P}_f}$)	β	$\epsilon_\beta(\%)$
Monte Carlo	10^6	$4.416\ 10^{-3}$ (1.5%)	2.618	–
AK-MCS + U	117	$4.416\ 10^{-3}$	2.618	0
Directional Sampling (DS)	52	$4.5\ 10^{-3}$	2.61	0.04
DS + Response Surface	1745	$5.0\ 10^{-3}$	2.57	1.91
DS + Spline	145	$2.4\ 10^{-3}$	2.82	7.63
DS + Neural Network	165	$4.1\ 10^{-3}$	2.64	0.76
Importance Sampling (IS)	1469	$4.9\ 10^{-3}$	2.58	1.53
IS + Response Surface	1375	$4.5\ 10^{-3}$	2.62	0
IS + Spline	428	$4.5\ 10^{-3}$	2.62	0
IS + Neural Network	52	$5.7\ 10^{-3}$	2.53	3.44

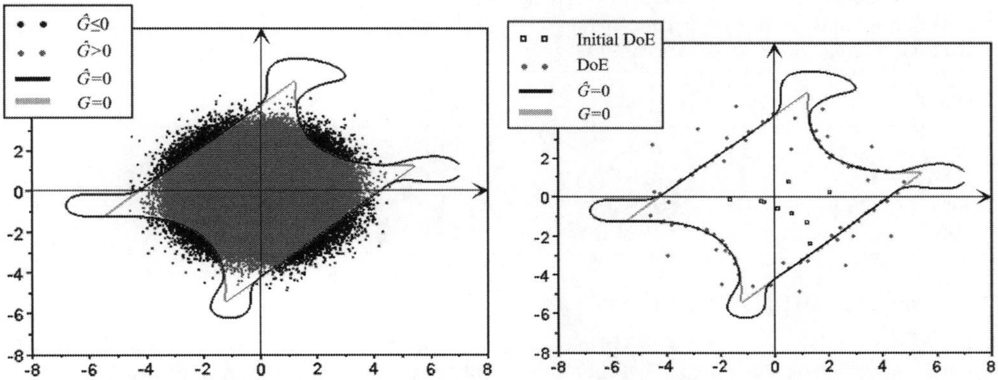

(a) Monte Carlo population predicted by AK-MCS+U.
(b) AK-MCS+U final DoE.

Figure 2. Example 1, AK-MCS + U results.

great effectiveness as the number of calls is low and the probability of failure is extremely accurate. In fact, the exact estimation of $\hat{P}_f$ by Monte Carlo is obtained.

Figures show the real interest of AK-MCS + U. Indeed, Figure 2(a) shows that an accurate separator is defined where the Monte Carlo points are. The locations with extremely low densities of probability are badly approximated as they do not present any interest in the evaluation of the probability of failure. Consequently, this method certifies a correct approximation of the probability of failure with a minimum number of calls to the performance function. Furthermore, the positions of the added points to the design of experiments show that the learning function U really focuses on the limit state (Figure 2(b)). The points are only added in the vicinity of the contour to obtain a very accurate sign for the Monte Carlo points near it. By plotting the evolution of the probability of failure in terms of the number of calls to the performance function (Figure 3), several steps can be observed in its evolution. These steps are coming from the way the learning function works. Indeed, as the limit state has 4 different branches, the minimum of the learning function U (*i.e.* the best next point to add) is first set on one of them. Once the metamodel is accurate enough for this branch, the minimum of the learning function goes to another branch and carries on. Furthermore, it can be seen that the probability of failure is well approximated with 70 calls to the performance function, the 50 following calls being only needed to satisfy the stopping condition. Depending on the aim of the analysis, it is possible to select a stopping condition like min $(U(\mathbf{x})) = 2$ or a more "visual" (but riskier) one by plotting the evolution of the probability of failure in terms of the number of calls to the performance function.

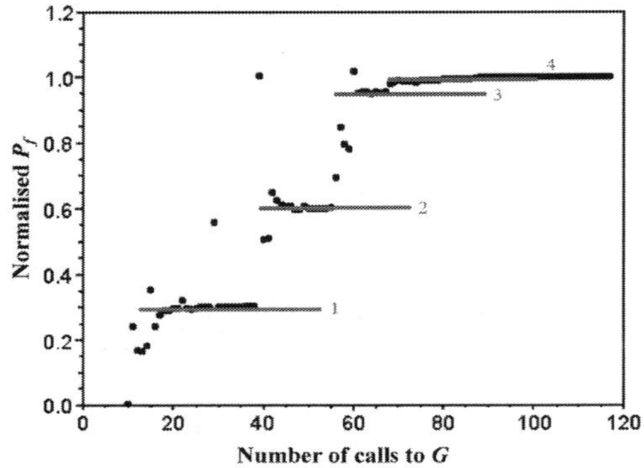

Figure 3. Example 1, Evolution of P_f normalised by P_f obtained by Monte Carlo simulation in terms of N_{call}.

Table 2. Example 2, reliability results – Comparison of AK-MCS + U (population size: $n_{MC} = 6\ 10^4$) with a Monte Carlo simulation on the same population, a passive Kriging combined with Latin Hypercube Sampling and a Subset Simulation using Phimeca-Soft.

Method	N_{call}	$\hat{P}_f$ (C.O.V$_{\hat{P}_f}$)	β	$\epsilon_\beta(\%)$
Monte Carlo	$6\ 10^4$	$7.34\ 10^{-2}$ (1.5%)	1.45	–
AK-MCS + U	416	$7.34\ 10^{-2}$	1.45	0
Passive Kriging + LHS	416	$6.73\ 10^{-2}$	1.50	3.44
Subset	5000	$7.65\ 10^{-2}$	1.43	1.45

3.2 *Example 2: Modified rastrigin function*

The next example consists of a modified version of the Rastrigin function (Törn and Zilinskas 1989). This modification is made to obtain a highly non-linear performance function involving non-convex and non-connex domains of failure (*i.e.* "scattered gaps of failure"). The random variables x_i are standard normal distributed random variables. The performance function reads:

$$G(x_1, x_2) = 10 - \sum_{i=1}^{2} \left(x_i^2 - 5\cos(2\pi x_i) \right) \tag{5}$$

The method is compared Table 2 with Monte Carlo simulation, Subset Simulations (Au et al. 2007) performed with Phimeca-Soft (Lemaire and Pendola 2006) and a passive Kriging metamodel, that is based on a fixed Latin Hypercube design of experiments which size is equal to the number of calls required by AK-MCS + U.

This example shows that AK-MCS + U can be used on highly non-linear limit states and on problems involving non-convex and non-connex domains of failure. Its performance is incomparable to Subset Simulation or even to passive Kriging combined with Latin Hypercube sampling. Indeed, the probability of failure estimated by AK-MCS + U with only 400 calls is found to be the same than the one obtained by Monte Carlo simulation for the same population (Table 2). Subset Simulations and Kriging based on a Latin Hypercube Sampling (LHS) give less accurate results. Furthermore, Figure 4, the evolution of the probability of failure shows several steps like in the previous example. It comes from the several "scattered gaps of failure" that can be observed Figure 5(a). This figure shows the design of experiments required by AK-MCS + U to satisfy the stopping condition and the similar probability of failure than the one estimated by a crude Monte Carlo simulation on the same population. It is seen that the points are concentrated in the vicinity of the limit state. Furthermore, Figure 5(b) shows that the limit state is well approximated where the Monte Carlo population is located. Indeed, the approximation coincides exactly with the limit state in the region covered by the Monte Carlo population. Then, the interpolation worsens where no Monte Carlo points are observed. Once again, this is not important as it corresponds to locations which present extremely weak densities of probability and therefore have a negligible effect on the probability of failure.

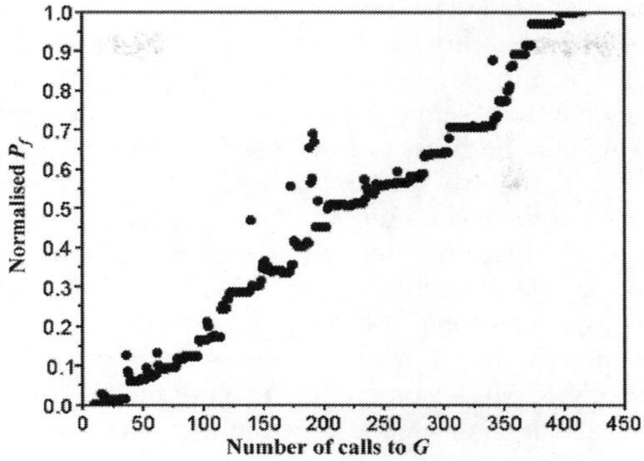

Figure 4. Example 2, Evolution of P_f normalised by P_f obtained by Monte Carlo simulation in terms of N_{call}.

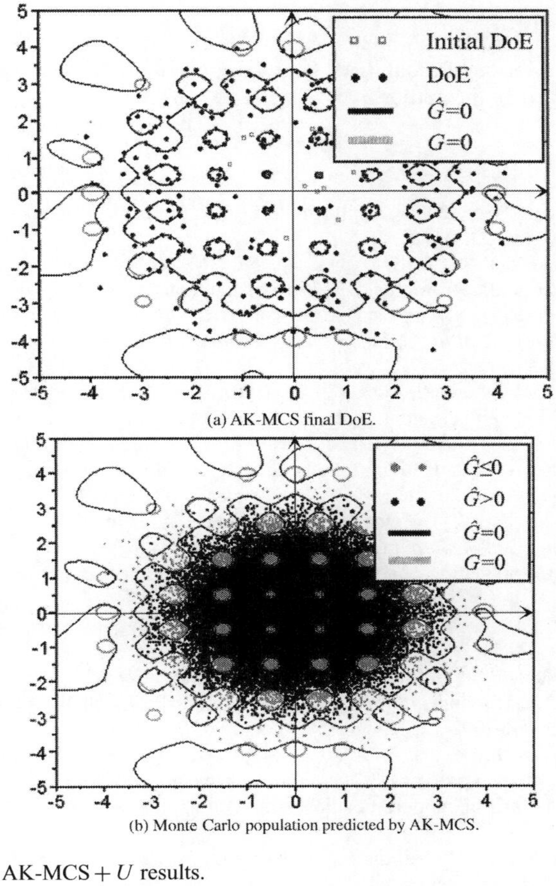

(a) AK-MCS final DoE.

(b) Monte Carlo population predicted by AK-MCS.

Figure 5. Example 2, AK-MCS + U results.

4 CONCLUSION

This article describes an Active learning method combining Kriging and Monte Carlo Simulation (AK-MCS) to perform structural reliability analysis. This approach combines the advantages of two methods: Kriging metamodel and Monte Carlo simulation. The idea is to perform a Monte Carlo simulation without evaluating the whole population on the performance function. Indeed, the population is predicted using a Kriging metamodel which is defined thanks to only a few points of the population that are evaluated. This strategy permits to focus only on configurations with sufficiently high densities of probability. Therefore, the limit state is not approximated accurately in the whole design space but only where Monte Carlo points are located. Moreover, the active learning is preferred to update the Kriging metamodel by adding a new point step by step to the design of experiments. This is done thanks to the Kriging variance and prediction which permit to define a learning function to identify the best next point to improve the design of experiments, that is to say the point which is the most likely to cross the predicted limit state and to disrupt the probability of failure. Here, the learning function U is proposed to quantify the risk of a change of the prediction sign. This article illustrates AK-MCS with U on two 2D examples with complex limit states' shapes. The proposed strategy is found to be economic in the number of calls to the expensive performance function and its results are very accurate for the probability of failure. To finish, on the concept of AK-MCS, some other Kriging methods such as AK-IS (Importance Sampling) or AK-SUBSET can be proposed to open the way to efficient calculations of very weak probabilities of failure.

ACKNOWLEDGEMENTS

This research was supported by the French Agence Nationale de la Recherche under the project APPRoFi (**App**roche Mécano **P**robabiliste Pour la conception **Ro**buste en **F**at**i**gue), which is gratefully acknowledged by the authors.

REFERENCES

Au, S. K., J. Ching, and J. L. Beck (2007). Application of subset simulation methods to reliability benchmark problems. *Structural Safety 29*(3), 183–193.

Bichon, B. J., M. S. Eldred, L. P. Swiler, S. Mahadevan, and J. M. McFarland (2008, October). Efcient Global Reliability Analysis for nonlinear implicit performance functions. *AIAA Journal 46*, 2459–2468.

Cox, D. D. and S. John (1997). SDO: A statistical method for global optimization. In M. N. Alexandrov and M. Y. Hussaini (Eds.), *Multidisciplinary Design Optimization: State-of-the-Art*, pp. 315–329. Siam, Philadelphia.

Dellino, G., P. Lino, C. Meloni, and A. Rizzo (2009, April). Kriging metamodel management in the design optimization of a CNG injection system. *Mathematics and Computers in Simulation 79*(8), 2345–2360.

Ghanem, R. G. and P. D. Spanos (1991). *Stochastic Finite Elements: A Spectral Approach*. Springer, Berlin.

Hurtado, J. E. (2004, July). An examination of methods for approximating implicit limit state functions from the viewpoint of statistical learning theory. *Structural Safety 26*(3), 271–293.

Jones, D. R., M. Schonlau, and W. J. Welch (1998, December). Efcient Global Optimization of expensive black-box functions. *Journal of Global Optimization 13*(4), 455–492.

Kaymaz, I. (2005, April). Application of Kriging method to structural reliability problems. *Structural Safety 27*(2), 133–151.

Kleijnen, J. P. C. (2009, February). Kriging metamodeling in simulation: A review. *European Journal of Operational Research 192*(3), 707–716.

Lemaire, M. and M. Pendola (2006). Phimeca-soft. *Structural Safety 28*(1–2), 130–149.

Lophaven, S. N., H. B. Nielsen, and J. Sondergaard (2002). DACE, a matlab Kriging toolbox, version 2.0. Technical Report IMM-TR-2002-12, Technical University of Denmark.

Marrel, A., B. Iooss, F. Van Dorpe, and E. Volkova (2008). An efcient methodology for modeling complex computer codes with gaussian processes. *Computational Statistics & Data Analysis 52*(10), 4731–4744.

Matheron, G. (1973). The intrinsic random functions and their applications. *Advances in Applied Probability 5*(3), 439–468.

Ranjan, P., D. Bingham, and G. Michailidis (2008, November). Sequential experiment design for contour estimation from complex computer codes. *Technometrics 50*(4), 527–541.

Romero, V. J., L. P. Swiler, and A. A. Giunta (2004, April). Construction of response surfaces based on progressive-lattice-sampling experimental designs with application to uncertainty propagation. *Structural Safety 26*(2), 201–219.

Schueremans, L. and D. Van Gemert (2005). Benet of splines and neural networks in simulation based structural reliability analysis. *Structural Safety 27*(3), 246–261.

Törn, A. and A. Zilinskas (1989). Global optimization. *Lecture Notes in Computer Science 350*.

Waarts, P. H. (2000). *Structural reliability using nite element methods: an appraisal of DARS (Directional Adaptive Response Surface Sampling)*. Ph. D. thesis, TUDelft.

Reliability and Optimization of Structural Systems – Straub (ed)
© 2010 Taylor & Francis Group, London, ISBN 978-0-415-88179-1

Life-cycle framework for maintenance, monitoring, and reliability of naval ship structures

Dan M. Frangopol & Nader M. Okasha
Lehigh University, Bethlehem, PA, USA

ABSTRACT: The proper life-cycle management of naval ship structures is best conducted according to an integrated and well coordinated life-cycle framework. This paper aims to present a framework by combining concepts for maintaining, monitoring, and reliability assessment of naval ship structures into an integrated process.

1 INTRODUCTION

A particular vessel can deteriorate in many ways over its lifetime. The purpose of a life assessment is to develop some global index that describes the condition of the aging vessel in terms of safety or serviceability (Bea and Mayoss 1995). Advances in modern day technology have brought into existence powerful tools that offer great potential in strengthening our ability to understand the performance of naval ship structures. This performance is deemed to deteriorate over the life-cycle of these structures. Hence, the safe use of these vessels is contingent upon their proper life-cycle management. The proper life-cycle management of naval ship structures is best conducted according to an integrated and well coordinated life-cycle framework. A life-cycle framework is an interdisciplinary process that is implemented over the lifespan of a structure in order to upkeep its safety and structural integrity against deterioration due to aging processes and expected abnormal loads (Frangopol 2009).

This paper aims to present a life-cycle management framework for application to naval ship structures. This framework combines concepts for maintaining, monitoring, and reliability assessment of naval ship structures into an integrated process for their proper life-cycle management.

2 LIFE-CYCLE RELIABILITY ASSESMENT

2.1 *Probabilistic strength analysis*

Being the most critical limit state in a ship structure, the longitudinal strength of the midship hull girder is of particular interest because the vertical bending moment is one of the highest loads that a ship structure has to withstand (Luís *et al.* 2009). Strength of the ship hull is traditionally conducted via section analysis at midship and mostly through incremental curvature procedures (Hughes 1983, IACS 2008, Lua and Hess 2003, 2006, Mansour 1997). However, Okasha and Frangopol (2010) made use of optimization search algorithms to significantly speed up the process of finding the ultimate failure moment rather than generating the full moment-curvature curve for obtaining this value. This facilitated the inclusion of this method in a sampling technique for generating in a cost-effective way the probabilistic distribution for the ultimate moment.

2.2 *Load prediction*

In the absence of Structural Health Monitoring (SHM) data, the load effects can be predicted using design code formulations. The loads obtained are lifetime maxima and are conservative. The IACS (2008) guidelines are used to determine the load effects. These guidelines provide formulas for predicting the still water bending moment and wave-induced (low frequency) bending moments. In monitoring ship structures, the wave-induced and dynamic loads are of particular interest, and mostly no information about the still water bending moment is provided. Therefore, the still water bending moments remain the same after monitoring. However, once SHM data become available, the wave-induced bending moment can be updated. Hence, this bending moment determined using the IACS (2008) guidelines constitutes the prior knowledge in the performance prediction process.

Unlike the case of the low frequency waves, prior knowledge may not be available for the high frequency waves. Hence, being the only source for this load effect, the extracted high frequency peaks are used with classical estimation techniques (such as the maximum likelihood method) to generate the probability distribution for the dynamic bending moments in hogging and sagging.

2.3 *Reliability computation*

Loads, material properties and resistance degradation are random functions of time (Ellingwood 2005). The instantaneous reliability index of the ship hull with respect to ultimate limit state is computed as $\beta(t) = \Phi^{-1}(P[g(t) > 0])$, where Φ is the CDF of the standard normal distribution, $P[E]$ is the probability of occurrence of the event E, and $g(t)$ is the value of the time-variant performance function computed at time t as

$$g(t) = x_R M(t) - x_{sw} M_{sw} - x_w (k_w M_w + k_d M_d) \tag{1}$$

where $g(t)$ is the time-dependent performance function, $M(t)$ is the time-dependent resisting bending moment, M_{sw} is the still water bending moment, M_w is the wave-induced bending moment, x_R is the model uncertainty associated to the resistance determination, x_{sw} is the model uncertainty related to the still water bending moment prediction, x_w is the model uncertainty associated to the wave-induced bending moment prediction, M_d is the dynamic bending moment, k_w is a correlation factor for wave-induced bending moment and k_d is the correlation factor between wave-induced and dynamic bending moments (Mansour *et al.* 1984).

3 INTEGRATION OF MONITORING

3.1 *Signal filtering and peak extraction*

The collected signal of the SHM data comprises the high frequency waves produced by the slamming effects superimposed over the low frequency waves produced by the regular sea wave-induced bending moments. For the reliability analysis, these two components need to be separated by filtering the signal so that they are statistically analyzed independently. In each monitoring wave cycle, the signal provides all the values recorded at the sampling rate, but only the peak sagging and hogging values are of interest in the reliability analysis. Therefore, the peak signals need to be extracted from the complete record for the sagging and hogging moments independently. A peak extraction algorithm was developed in Okasha *et al.* (2009) to obtain the datasets of maximum bending moments in hogging and sagging separately from the signal.

3.2 *Bayesian updating*

Prior knowledge is available from the IACS (2008) regarding low frequency wave loads. Therefore, the extracted low frequency peaks are used to update the prior loads using Bayesian updating. The literature linking the themes of Bayesian updating and extreme value modeling is sparse, in part due to computational difficulties. Some of these difficulties have recently been overcome by techniques such as Markov chain Monte Carlo (McMC) (Coles and Powell 1996). Bayesian-updating of extreme value distributions cannot lead to explicit posterior distributions (Robert 1994). The Metropolis-Hastings (MH) algorithm has been suggested for this purpose (Robert 1994, Robert and Cassella 1999). Okasha and Frangopol (2009) have used the slice sampler algorithm (Neal 2003) for this purpose.

4 INSPECTION AND MAINTENANCE

The need for effective marine maintenance is very obvious and also recognized by all involved parties, ship owners, classification societies, and authorities (Aage and Fog 1991). The notion

Table 1. Coating inspection criteria point scoring (adopted from the ABS 2009).

Coating Condition	Rating	Points	Condition Assessment	Risk Level	Color Code
Good	Excellent	0	Freshly coated	Low	Green
	Good	1	Minor spot rusting	Low	Green
	Good – Fair	2	Breakdown > 5%	Low-Medium	Green
Fair	Fair	3	Local Breakdown	Medium	Yellow
	Fair – Poor	4	Breakdown > 10%	Medium-High	Yellow
Poor	Poor	5	Breakdown > 20%	High	Red
	Worst	6	Breakdown > 30%	Very High	Red

that ship maintenance and repair is a relatively less expensive and less important undertaking than ship design or construction is untrue; as ships get older, the maintenance requirements and costs increase significantly (Bea *et al.* 1991). Maintenance is a key factor in the economic life cycle of ships and offshore installations, as well as in the safety and risk levels of their performance (Aage and Fog 1991). Ship maintenance can be preventive and corrective and it can be classified into periodical or condition based (Skaar 1991).

In order to allow higher speeds and increased cargo capacity, high-speed vessels often employ novel and aggressive structural designs using composite or aluminum alloys with innovative arrangements and fabrications to maximize lightship weight reduction (Salvino and Brady 2008). However, the reduced weight design philosophy leads to highly fracture prone, complex, hard to inspect and repair structures (Bea and Hutchison 1993).

Being alerted by reaching of the reliability levels to a predefined or critical threshold is one of the main motivations behind conducting a life-cycle assessment of ship structures (Bea and Mayoss 1995). The occurrence of such event may lead to initiating a visual inspection for determining the correct maintenance alternatives to pursue. Visual inspection is the most important method to obtain information of the condition of marine structures with respect to cracks and corrosion (Aage and Fog 1991). The American Bureau of Shipping (ABS) has established the Hull Inspection and Maintenance Program (HIMP) for providing guidance in inspecting and maintaining ship structures (ABS 2007b). Six different criteria are identified to be addressed during inspection (ABS 2009): (1) Coating condition, (2) Presence of general corrosion, (3) Presence of pitting or grooving or other localized linear corrosion, (4) Presence of deformation, (5) Presence of fractures, and (6) Compartment or space cleanliness. These criteria are to be documented based on a traffic light point rating system (0 through 6 scale) and an example of scaling is given in Table 1 for the coating condition (ABS 2009). Once the rating is obtained, alternatives for maintenance are identified based on the ABS (2007a) and SSC 386-III (Gallion 1995) guidelines. The impact of each maintenance alternative on the life-cycle performance is determined and used beside the cost of implementation in deciding the optimum maintenance alternative to be used.

The effects of maintenance on the reliability of the ship structure are accounted for by modifying certain variables in the mathematical formulation. For instance, without maintenance, the corrosion loss in a ship structure can be given as (Akpan *et al.* 2002)

$$r(t) = C_1(t - t_0)^{C_2} \tag{2}$$

where $r(t)$ is the thickness loss (*mm*), t_0 is the corrosion initiation time depending on coating life (*years*), C_1 is the annual corrosion rate (*mm/years*), C_2 is a constant usually set to unity, and t is the time expressed in years. If it is decided to apply maintenance to the ship by recoating its surface at year t_c, Eq. 2 is modified as

$$r(t) = \begin{cases} 0 & \text{if } t \le t_0 \\ C_1(t - t_0)^{C_2} & \text{if } t_0 < t \le t_c \\ C_1(t_c - t_0)^{C_2} & \text{if } t_c < t \le t_c + t_1 \\ C_1(t - t_0 - t_1)^{C_2} & \text{if } t_1 + t_c < t \end{cases} \tag{3}$$

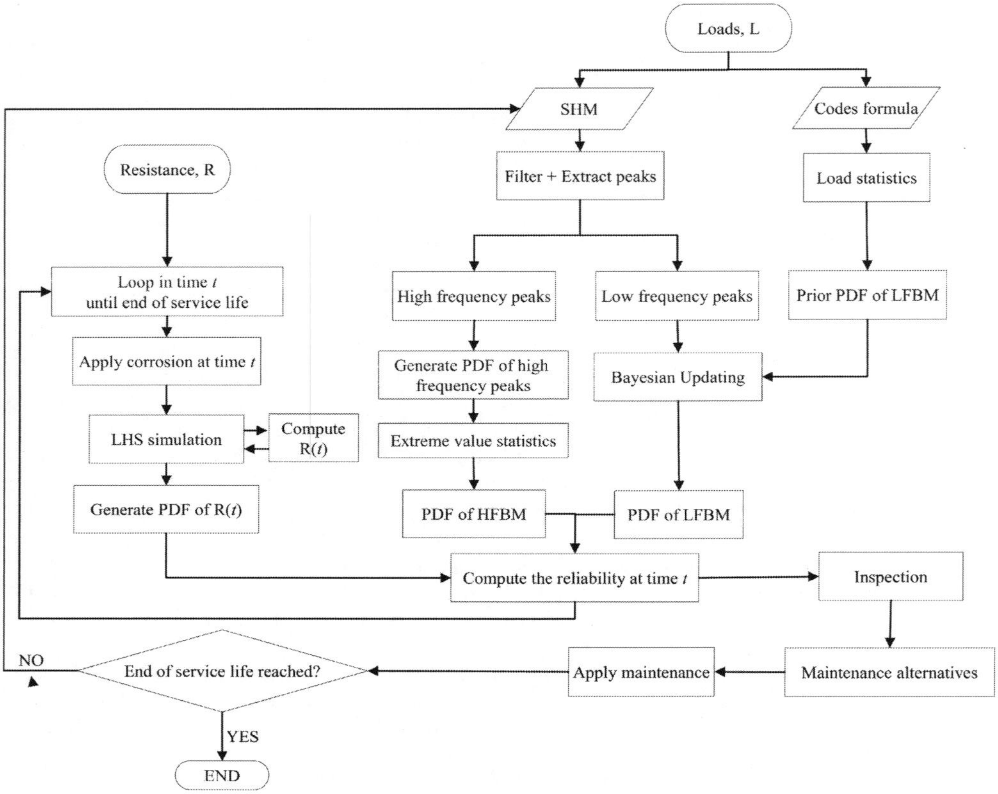

Figure 1. Framework for life-cycle management of ship structures.

where t_1 is the coating life of the new coat (*years*) because the new coat will halt corrosion for t_c years from when it is applied.

5 THE FRAMEWORK

The computational platform of the framework proposed in this paper is an integrated procedure composed of several sequential phases including: (a) life-cycle performance prediction; (b) maintenance optimization; and (c) updating of life-cycle performance using SHM data. This framework is designed for management of naval ships over their life-cycle, and thus serves as a reference for decision making from early stages until the end of the service life of the naval ships. A flowchart of the framework is shown in Figure 1.

The two pillars of the life-cycle performance are the resistance and load models. In this framework, the resistance is computed by generating the descriptors of the probabilistic resistance using a Latin Hypercube Sampling (LHS) simulation in which the strength is computed for each sample using an optimization search algorithm (Okasha and Frangopol 2010). This is done over the number of years specified for the service life, where at each year, or time increment, an LHS simulation is performed and the PDF of the resistance at that time R(t) is generated using classical estimation methods, such as the maximum likelihood method.

The loads can be estimated by two sources; the code formulations and SHM data. In fact, Bayesian updating can be used to combine both. Statistics for the code estimated loads need to be determined from established literature and used to generate the prior PDF of the Low Frequency Bending Moment (LFBM). The SHM data is filtered to separate the low and high frequency signals and their peaks are separately extracted. The low frequency peaks are used in Bayesian updating

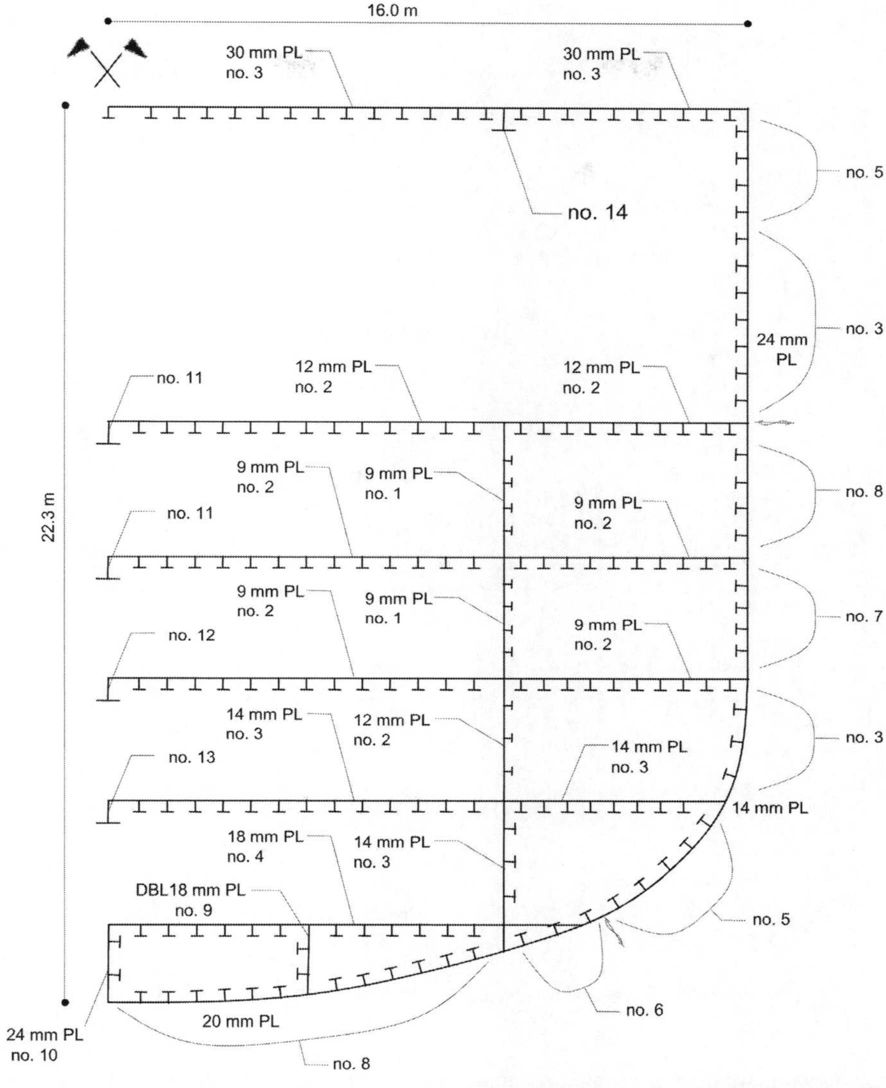

Figure 2. Cross-section geometry of a symmetric half hull at midship (adapted from Okasha *et al.* 2009).

with the prior PDF of LFBM to find the updated PDF of LFBM. Classical estimation techniques are used to generate the PDF of the high frequency peaks. This PDF is analyzed using extreme value statistics to find the PDF of the High Frequency Bending Moment (HFBM). Both the LFBM and HFBM are used in the reliability computations as shown in Equation 1.

Based on the obtained reliability profile, inspection is performed at times where the reliability is critical. According to the inspection results, maintenance alternatives are considered by analyzing their cost and effect on the reliability profile. Hence, a maintenance alternative is selected. At later stages of the service life, new SHM data may be obtained and the loads and hence the reliability profile are updated.

6 APPLICATION

These concepts are illustrated on the Joint High-Speed Sealift Ship (JHSS). A scaled structural segmented seakeeping model was constructed and successfully tested to obtain representative ship

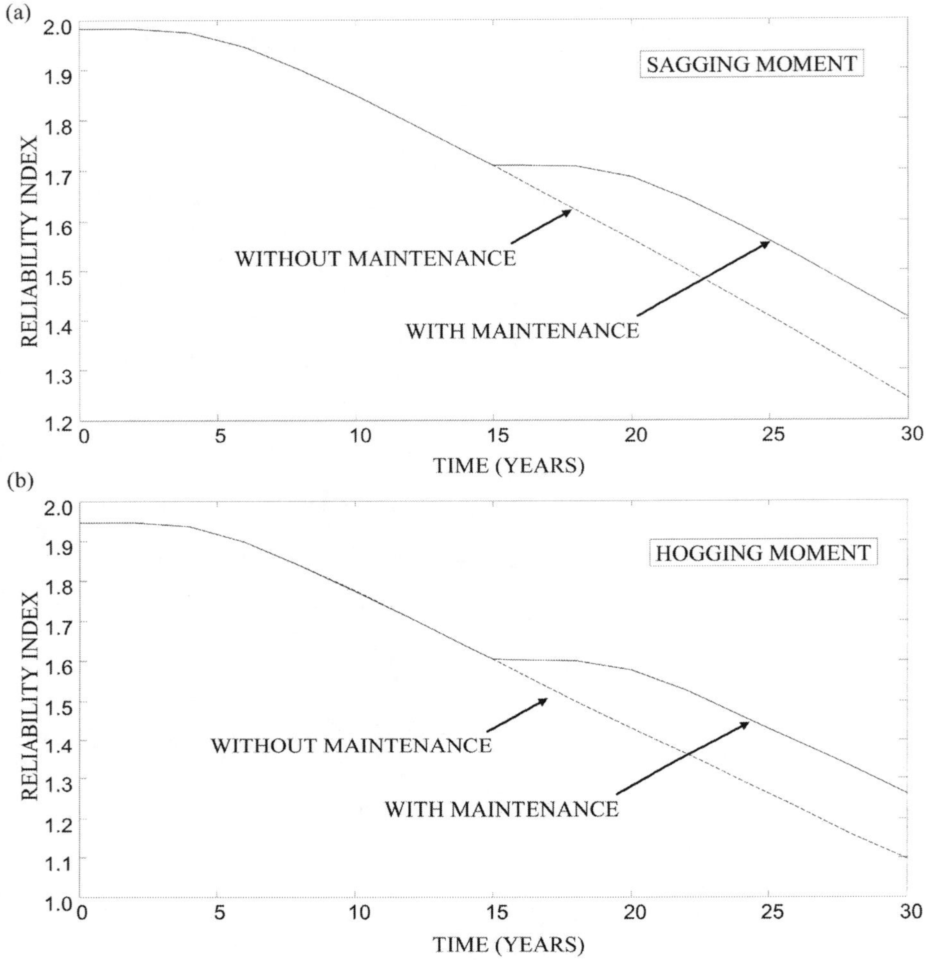

Figure 3. Reliability profile with and without maintenance in: (a) sagging moment; and (b) hogging moment.

motions and structural primary and secondary loading data; a schematic diagram of the midship design section is shown in Figure 2 (Devine 2009). Figure 3 shows the reliability index profile after integrating the SHM data for both the sagging and hogging bending moments. Also shown on the figure is the effect of maintenance on the reliability index profile, where the ship was recoated at the year 15 (in the middle of the service life) in order to slowdown the effect of the corrosion process for some time.

7 CONCLUSIONS

Efficiently ensuring the adequacy of the life-cycle performance of ship structures is fundamental for their safe use and can only be attained with integrating every valuable information gathered about the structural health into the assessment of this performance. The proper life-cycle management of naval ship structures is best conducted according to an integrated and well coordinated life-cycle framework.

This paper presented a life-cycle management framework for application to naval ship structures. This framework combines concepts for maintaining, monitoring, and reliability assessment of naval ship structures into an integrated process for their proper life-cycle management. The framework was implemented on the Joint High-Speed Sealift Ship and the SHM data obtained from testing a scaled model. The effect of maintenance on the reliability index profile was also studied.

ACKNOWLEDGMENTS

The support from the U.S. Office of Naval Research Contract Number N00014-08-1-0188 is gratefully acknowledged. The assistance of Ablerto Decò, a graduate research assistant at Lehigh University, in some of the computations and figures in this paper is also acknowledged. The opinions and conclusions presented in this paper are those of the authors and do not necessarily reflect the views of the sponsoring organization.

REFERENCES

Aage, C. & Fog, N.G. 1991. A Survey of Some Important Research Areas Related to Marine Maintenance. *Paper presented at the Marine, Structural Inspection, Maintenance and Monitoring Symposium*, Sheraton National Hotel, Arlington, Virginia, March 18–19, 1991.

ABS. 2007a. Guidance notes on the Inspection, Maintenance and Application of Marine Coating Systems. Third edition. American Bureau of Shipping, Houston, TX.

ABS. 2007b. Guide for Hull Inspection and Maintenance Program (HIMP). American Bureau of Shipping, Houston, TX.

ABS. 2009. Inspection Grading Criteria for the ABS Hull Inspection and Maintenance Program (HIMP). Third edition. American Bureau of Shipping, Houston, TX.

Akpan, U.O., Koko, T.S., Ayyub, B. & Dunbar, T.E. 2002. Risk Assessment of Aging Ship Hull Structures in the Presence of Corrosion and Fatigue. *Marine Structures* 15(3): 211–231.

Bea, R.G & Mayoss, R. 1995. Ship Maintenance Project, Volume 2, Corrosion Damage Evaluation. Ship Structure Committee: SSC-386 II.

Bea, R.G. & Hutchison, S.C. 1993. Maintenance of Marine Structures; A State of the Art Summary. Ship Structure Committee: SSC-372.

Bea, R.G., Pollard, R., Schulte-Strathaus, R. & Baker, R. 1991. Structural Maintenance for New and Existing Ship: Overview, Fatigue Cracking and Repairs. *Paper presented at the Marine, Structural Inspection, Maintenance and Monitoring Symposium*, Sheraton National Hotel, Arlington, Virginia, March 18–19, 1991.

Coles, S.G. & Powell, E.A. 1996. Bayesian methods in extreme value modelling: a review and new developments. *International Statistical Review* 64(1): 119–136.

Devine, E.A. 2009. An overview of the recently-completed JHSS Monohull and Trimaran structural seaways loads test program. Naval Surface Warfare Center, Carderock Division (NSWCCD) PowerPoint Briefing, 30 October, 2009.

Ellingwood, B.R. 2005. Risk-informed condition assessment of civil infrastructure: state of practice and research issues. *Structure and Infrastructure Engineering*, Taylor & Francis 1(1):7–18.

Frangopol, D.M. 2009. Life-Cycle Performance, Management, and Optimization of Structural Systems under Uncertainty: Accomplishments and Challenges. Keynote Paper in *Safety, Reliability and Risk of Structures, Infrastructures and Engineering Systems*, H. Furuta, D.M. Frangopol, and M. Shinozuka, eds., CRC Press, Taylor & Francis Group, Boca Raton, London, New York, Leiden, 2010, 13–14, and full (23 pages) paper on CD-ROM, Taylor and Francis Group, London, 2010, 38–60.

Gallion, K.A. 1995. Structural Maintenance Project, Volume 3, RMS – Repair Management System. Ship Structure Committee: SSC-386 III.

Hughes, O.F. 1983. Ship Structural Design – A Rationally-Based, Computed-Aided, Optimization Approach. *Wiley & Sons*, New York.

IACS. 2008. Common Structural Rules for Double Hull Oil Tankers. International Association of Classification Societies. Available at http://www.iacs.org.uk.

Lua, J. & Hess, P.E. 2003. Hybrid Reliability Predictions of Single and Advanced Double-Hull Ship Structures. *Journal of Ship Research*, SNAME, 47(2): 155–176.

Lua, J. & Hess, P.E. 2006. First-Failure-Based Reliability Assessment and Sensitivity Analysis of a Naval Vessel Under Hogging. *Journal of Ship Research*, SNAME 50(2): 158–170.

Luís, R.M., Teixeira, A.P. & Guedes Soares, C. 2009. Longitudinal Strength Reliability of a Tanker Hull Accidentally Grounded. *Structural Safety* 31: 224–233.

Mansour, A.E, Yan, H.Y. Zigelman, C.I. Chen, Y.N. & Harding S.J. 1984. Implementation of Reliability Methods of Marine Structures. *Trans. Soc. Naval Architects and Marine Engineers* 92: 11–20.

Mansour, A.E. 1997. Assessment of reliability of ship structures. Ship Structure Committee: SSC-398.

Neal, R.M. 2003. Slice Sampling. *The Annals of Statistics*, Institute of Mathematical Statistics 31(3): 705–741.

Okasha, N.M. & Frangopol, D.M. 2009. Integration of structural health monitoring in a system performance based life-cycle bridge management framework. (submitted).

Okasha, N.M. & Frangopol, D.M. 2010. Efficient Method Based on Optimization and Simulation for the Probabilistic Strength Computation of the Ship Hull. *Journal of Ship Research* (in press).

Okasha, N.M., Frangopol, D.M. & Deco, A. 2009. Integration of structural health monitoring in life-cycle performance assessment of ship structures under uncertainty. (submitted).

Robert, C.P. & Cassella, G. 1999. Monte Carlo Statistical Methods. *Springer Texts in Statistics*, Springer, Verlag, New York, USA.

Robert, C.P. 1994. The Bayesian Choice. *Springer Texts in Statistics*, Springer, Verlag, New York, USA.

Salvino, L.W. & Brady, T.F. 2008. Hull Monitoring System Development using a Hierarchical Framework for Data and Information Management. *Proceedings of 7th International Conference on Computer and IT Applications in the Maritime Industries*, (COMPIT'08), Liege, Belgium, April 2008.

Skaar, K.T. 1991. Future Class Requirements and Support Services for Inspection and Maintenance, Paper *presented at the Marine, Structural Inspection, Maintenance and Monitoring Symposium*, Sheraton National Hotel, Arlington, Virginia, March 18–19, 1991.

Reliability and Optimization of Structural Systems – Straub (ed)
© 2010 Taylor & Francis Group, London, ISBN 978-0-415-88179-1

Optimal maintenance planning considering deterioration uncertainty

Hiroshi Hattori
Kyoto University, Kyoto, Japan

Hitoshi Furuta, Koichiro Nakatsu & Nobuyoshi Tokioka
Kansai university, Osaka, Japan

Ken Ishibashi
Graduate school of Kansai university, Osaka, Japan

ABSTRACT: In order to establish a rational bridge maintenance program, it is inevitable to determine the deterioration curves of various components of bridges. However, it is difficult to identify appropriate deterioration curves because of various uncertainties involved in the prediction process and environment evaluation. Furthermore, it is necessary to take into account the effects of natural hazards. Under the situation, although the existing bridges should be kept in a safety condition through inspection and maintenance works, it is not easy to maintain them in a satisfactory level, due to the financial constraints. In this study, an attempt is made to develop an optimal maintenance planning method that can account for the effects of various uncertainties. In the proposed method, an optimal maintenance schedule is obtained with the aid of genetic algorithms, in which the deterioration curve can be treated as a random function. A numerical example is presented to demonstrate the applicability and efficiency of the proposed method.

1 INTRODUCTION

Recently, bridge maintenance has been becoming more and more important, because lots of existing bridges require repair and rehabilitation due to the aging and deterioration. In order to establish a rational maintenance program, it is inevitable to determine the deterioration curves of various components of bridges. However, it is quite difficult to identify appropriate deterioration curves because of various uncertainties involved in the prediction process and environment evaluation. Furthermore, it is necessary to take into account the effects of natural hazards such as earthquake, typhoon, flood, etc. Under the situation, although the existing bridges should be kept in a safety condition through inspection and maintenance activities, it is not easy to maintain them in a satisfactory level, because of the financial constraints.

A lot of research has been made on the maintenance scheduling so far, most of which are based on the prediction of the deterioration of components (Frangopol and Furuta 2001, Frangopol et al. 1997, Furuta et al. 1997). However, since the deterioration curve has various uncertainties, there is a problem in the accuracy of the prediction of the deterioration.

In this study, an attempt is made to develop an optimal maintenance planning method that can account for the effects of various uncertainties. In general, the repairing time is decided in a deterministic manner through optimization. There is no guarantee that the schedule updated with the aid of the difference between the prediction and observation may not be the optimal even for the small differences. By using the proposed method, it is possible to obtain a robust schedule for the case where the deterioration curve has uncertainties in the prediction process and environment evaluation (Furuta et al. 2008). Using such robust schedules, the bridge owners may not change the maintenance program unless the uncertainties and environmental changes are large.

In the proposed method, an optimal maintenance schedule is obtained by using Genetic Algorithm Considering Uncertainty (GACU). A numerical example is presented to demonstrate the applicability and efficiency of the proposed method.

2 PROBLEM OF MAINTENANCE SCHEDULING

The objective of bridge maintenance is to keep the bridge in a good condition and to extend its life. In order to establish an appropriate maintenance schedule for the bridge, it is very important to minimize the maintenance cost. However, there is a necessity to consider the risk of damage and failure caused by earthquakes. Then, since there is a trade-off relation between the maintenance cost and seismic risk, it is very difficult to satisfy both requirements in a manual way.

In the past researches, the trade-off has been treated by using the multi-objective optimization of the maintenance cost and seismic risk. However, the multi-objective optimization methods provide a lot of solutions so that it is not easy to choose an appropriate solution among the solution candidates. Furthermore, there is a problem in the prediction of deterioration curve. Since the maintenance schedule is optimized based on the prediction of deterioration, the accuracy of the deterioration prediction directly influences on the success of the optimization process. Namely, it is important to specify an appropriate deterioration curve. However, since the deterioration has uncertainties, it is difficult to achieve the high accuracy of the prediction. The prediction error of the deterioration curve may cause the increase of seismic risk or maintenance cost.

On the other hand, the annual budget of maintenance is one of dominant factors for the maintenance scheduling. By introducing a long term maintenance scheduling, it is possible to decrease the total amount of the maintenance cost if the annual budget is sufficient. However, the annual budget may fluctuate year by year. There is a possibility that the maintenance and renewal may not be performed according to the planned schedule because of the lack of annual budget. Then, it is necessary to equalize the annual budget as far as possible. However, the annual budget is not equally allocated by the optimization. If the annual budget must be equalized as the constraint the schedule obtained through the optimization has no guarantee of the optimality, namely, the obtained schedule is no longer optimal about the safety level and the seismic risk.

3 MAINTENANCE SCHEDULING CONSIDERING SEISMIC RISK

Life-Cycle Cost (LCC) of bridge structures consists of initial construction cost, maintenance cost, and failure cost. In usual, LCC analysis considers the damage and deterioration of materials and structures. However, in the region that often suffers from natural hazards such as typhoons and earthquakes, it is necessary to account for the effects of such natural hazards. Based on the seismic risk analysis, LCC is evaluated focusing on the effects of earthquakes that are major natural disasters in Japan. At first, LCC analysis is formulated to consider the social and economical effects due to the collapse of bridges occurred by the earthquake as well as the minimization of maintenance cost. The loss by the collapse of bridges due to the earthquake can be defined in terms of an expected cost and introduced into the calculation of LCC. A stochastic model of structural response was proposed, which accounts for the variation due to the uncertain characteristics of earthquake (Furuta et al. 2004a). Then, the probability of failure due to the earthquake excitation is calculated based on the reliability theory. Furthermore, LCC evaluations are performed not only for a single bridge but also many bridges forming road networks (Furuta et al. 2005a, 2006).

In the past researches (Furuta et al. 2004a, 2005b, 2005c), both the maintenance cost and seismic risk were treated by using the multi-objective optimization technique. However, it is not easy to select an appropriate maintenance plan among a lot of candidates that the multi-objective optimization provides as a set of Pareto solutions.

As mentioned before, there is such another problem that the prediction of the deterioration curve essentially involves various uncertainties. This implies that the accuracy of the prediction is low and insufficient, which may cause the increase of the seismic risk. Then, it is important to treat the uncertainty of the deterioration of the bridge for minimize the seismic risk.

4 THE PROPOSED METHOD

4.1 *Genetic Algorithm Considering Uncertainty (GACU)*

In order to obtain the maintenance schedule which has robustness to the uncertainty of deterioration, it is necessary to implement sampling many times. In GA Considering Uncertainty (GACU)

(Tamaki et al. 1999), objective function is defined as the expected value of $F'(x)$ to consider the search process as a sampling.

$$F'(x) = F(x) \quad with\ Uncertainty \tag{1}$$

$F(x)$ contains a variable element, that is, uncertainty, so that $F'(x)$ is changing according to the uncertainty. It is assumed that the number of sampling is age of individual. This sampling is performed by considering the evolution mechanism of inheritance, that is, gene of parents is resembled to that of children. The procedure of GACU is given as follows:

STEP 1. Generation of initial population: An individual in GA represents a solution candidate and forms a set of solutions called population. In general, the initial population is generated randomly.

STEP 2. Introduction of age structure: Here, age represents the numbers of survival during the evolution.

STEP 1. Crossover and mutation: New individuals are generated through the crossover and mutation. Crossover means the mating of two individuals. The rate of crossover controls the evolution process. The uniform crossover is employed here. Mutation is used to keep the diversity of the population so as to avoid the early convergence. The rate of mutation is prescribed to be small.

STEP 2. Evaluation: evaluation of new individuals and re-evaluation and adding age of surviving individuals

STEP 3. Natural selection: The selection of individuals is performed according to the fitness value of the individual to the envieonment.

STEP 2 to 5 are repeated until the convergence is achieved

4.1.1 *Fitness*

Fitness value of an individual is determined by Equation (2).

$$g(y_i) = \frac{(d_i - 1)g(p_i) + h(f(\hat{\omega}, x_i))}{d_i} \tag{2}$$

where $g(y_i)$ is the fitness of individual i. y_i is the genotype of individual i, x_i is the phenotype of y_i, p_i means the parent, $h(f)$ is the mapping function from the objective value to the fitness value, and w denotes the uncertainty involved in the evaluation of y_i, and is assumed to be normally distributed.

4.1.2 *Age structure*

The age of individual represents the numbers of surviving during the evolution. However, it is not a deterministic number but the mean value of surviving numbers through sampling. In this study, the following basic concept of age structure is considered:

a) A neighbor individual of the current individual is considered as identical individual.
b) The age of the neighbor individual is considered as the estimated number of sampling.
c) The fitness evaluation of the neighbor individual is calculated by using its age.

Then, the neighbor is identified as the distance between two individuals. In this study, the distance is defined as the hamming distance.

After measuring the distance between individuals, age of individual, which survives in the next generation, is increased.

$$d_p = d_p + 1 \tag{3}$$

On the other hand, age of child, which is born after crossover and mutation, P_α is defined as follows:

The probability that child takes over age of a parent, P_α, is given as Equation (4).

$$P_a(D) = \exp(-\frac{D^2}{\alpha}) \tag{4}$$

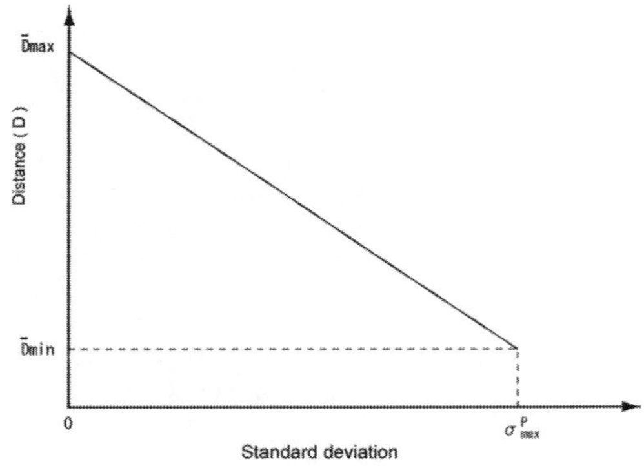

Figure 1. The relationship between standard deviation σ and bounded-distance.

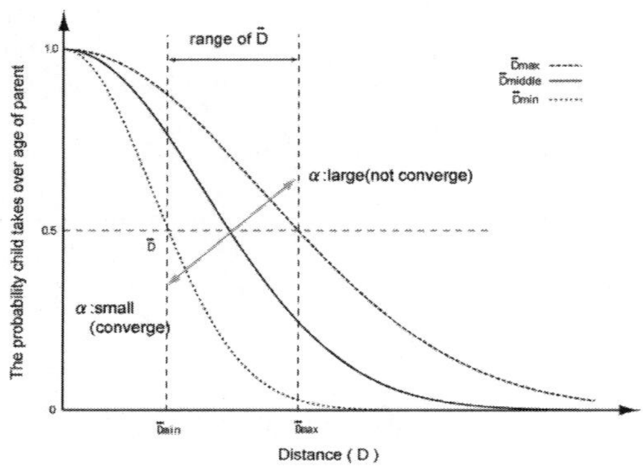

Figure 2. The relationship between critical-distance and age structure.

where D is the distance between a parent y_n, and a child y_c, the degree of increasing age is defined by the parameters α. The effects of age structure are obtained by changing this parameter depending on the convergence behavior.

$$\alpha = \frac{\overline{D}^2}{\log_e 2} \tag{5}$$

The bounded-distance which is D when $P_\alpha(D)$ is 0.5, $\overline{D}$, is given as Equation (6).

$$\overline{D} = \overline{D}_{\min} + (\overline{D}_{\max} - \overline{D}_{\min}) \frac{\sigma^P_{\max} - \sigma^P}{\sigma^P_{\max}} \tag{6}$$

where σ^P is the standard deviation of distance between arbitrary individuals of current generation, $\sigma^P_{\max}$ is the maximum value of σ^P before current generation. $\overline{D}_{\min}$ and $\overline{D}_{\max}$ are values of constant. The relationship between the standard deviation σ and the bounded-distance and the relationship between critical-distance and age structure are shown in Figure 1 and Figure 2.

5 BRIDGE MODEL

5.1 *Bridge pier model*

In this study, an RC bridge pier model is used for the numerical example of the maintenance scheduling. The failure type of the RC bridge pier is the flexure fracture. The front view, the side view and the cross sectional area of the bridge pier are shown in Figure 3.

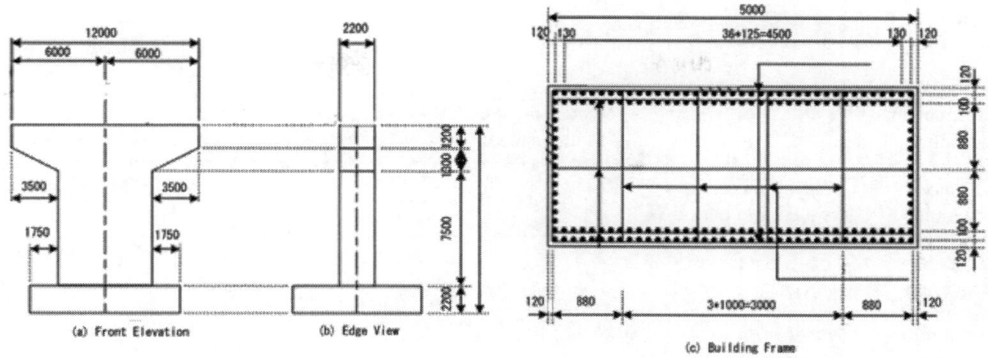

Figure 3. Bridge pier model.

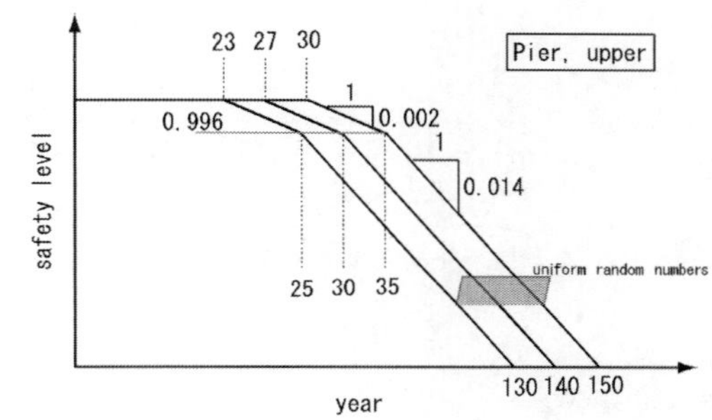

Figure 4. Deterioration model.

5.2 *Deterioration model of the bridge*

Figure 4 shows the deterioration curve of a member under the severe environment. The deterioration accelerates after 27 years. It is assumed that the preventive effects for corrosion lose after 27 years, the required safety level is prescribed as 0.8, and the service life is 100 years.

5.3 *Repair/reinforcement methods*

In this study, the repair/reinforcement methods shown in Table 1 are used for improving the performance of the bridge pier. The costs of those methods are also given in the same table according to the data provided by Ministry of Land, Infrastructure, and Transport in 2001.

6 MAINTENANCE PLANNING METHOD CONSIDERING SEISMIC RISK

In this study an attempt is made to develop a maintenance scheduling method that can consider the maintenance cost and seismic risk at the same time to achieve a rational maintenance schedule by using GA considering uncertainty.

6.1 *Objective function*

Here, the objective function is the total amount of the maintenance cost of all bridges.

6.2 *Sub-Objectives function*

There are a lot of efficient solutions when optimizing only the maintenance cost. Although the maintenance cost is almost equal, the characteristics of the maintenance schedule are different from

Table 1. Repair/reinforcement methods.

Maintenance Action	Effect	Effect on Reliability Index: Increasing Amount of Maintenance Effect	Effect on Reliability Index: Duration of Maintenance Effect (years)	Cost (1,000 Yen/m^2)
Surface Painting	Delays β decrease for a certain period.	—	5	20
Surface Covering	Delays β decrease for a certain period.	—	7	70
Section Restoring	Restores β to a percentage, and then allows it to deteriorate with the same slope as the initial deterioration curve.	β_{max} * 90%	—	530
Demi-neralization	Remove Chloride	β deteriorates with the same slope as the initial deterioration curve.		90
Desalination (Re-alkalize)	Alkalize concrete. Chloride	β deteriorates with the same slope as the initial deterioration curve.		90
Cathodic Protection	Delay the decrease of β for a certain period	—	25	100
Section Restoring with Surface Covering	Restore β to a percentage, delay the decrease of β for a certain period, and then β deteriorates with the same slope as the initial deterioration curve.	β_{max} * 90%	5	580

each other. Therefore, when the maintenance cost is almost equal, the seismic risk is considered as a sub-objective function. The seismic risk is defined in the following:

$$SeismicRisk = P_d \times P_h \times C_d \tag{7}$$

where P_d is the occurrence probability of seismic damage, P_h is the earthquake occurrence probability, and C_d is the seismic loss.

7 NUMERICAL SIMULATION

7.1 *Numerical simulation*

A numerical simulation is presented for the maintenance planning of bridge network. The uncertainty of deterioration curve is expressed by the normal distribution. In this example, the bridge network has 10 bridges and the maintenance plan of each bridge is established for 100 years. The result by the proposed method is compared with the ordinal GA method. The simulation is done 1,000 times and the maintenance costs obtained by the two methods are shown in Table 2.

From Table 2, it is confirmed that the maintenance cost can be decreased by using the proposed method. This can be done due to the robustness of the solution obtained by the proposed method. The ordinal method creates a front-loaded spending maintenance plan for adapting to the uncertainty

Table 2. Maintenance cost.

Maintenance cost	(Thousand Japanese Yen)
Ordinal method	Proposed method
238,620	214,840

Table 3. Seismic risk.

	Seismic risk	(Thousand Japanese Yen)
	Ordinal method	Proposed method
Best value	848,189	844,068
Worst value	856,458	848,919
Average value	849,335	845,000

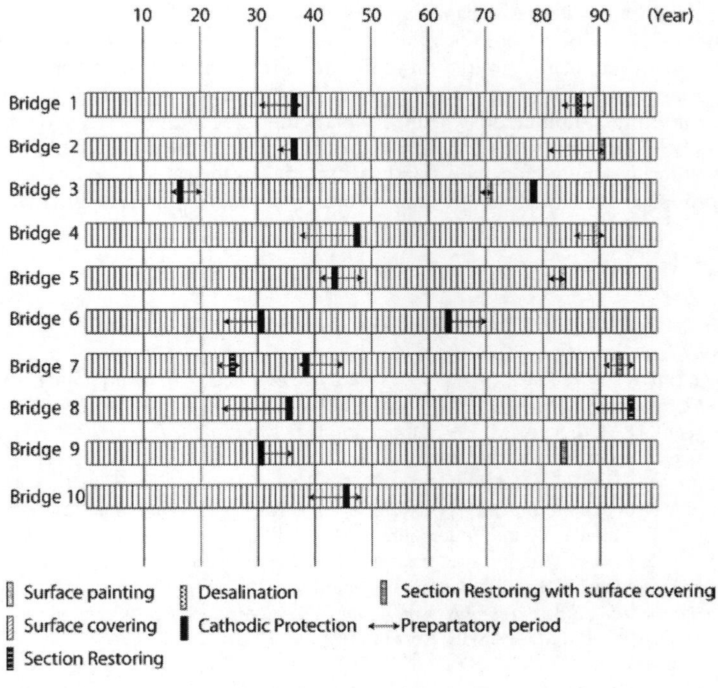

Figure 5. Maintenance plan by the proposed method.

of the deterioration and it gains the increase of the maintenance cost. On the other hand, the proposed method can correspond to the variation of deterioration, namely, hurry or postpone the maintenance actions because of the existence of leeway. This treatment can contribute to the reduction of maintenance cost by postponing the unnecessary maintenance action.

Table 3 shows that the proposed method can provide less seismic risk than the ordinary method. In this table, Best Value, Worst Value and Average value mean the maximum seismic risk, the minimum seismic risk and the average seismic risk in 1,000 times simulation, respectively. The above results imply that the proposed system can adapt to the uncertainties of the deterioration properly. Since the maintenance costs calculated by the ordinal method and the proposed method are almost the same, the effectiveness of the proposed system is verified.

In order to demonstrate the effectiveness of the proposed method, the maintenance plans obtained for all the bridges are shown in Figure 5. Here, the arrows mean the margin or leeway of the maintenance period. From Figure 5, it can be confirmed that the flexible preparatory periods performs the maintenance actions efficiently. Some repairing actions are pushed back because of the constraints of safety level or budget. Since the preparatory period of each action is optimized by the proposed method, the performance keeps good levels even when some maintenance action is pushed back. On the other hand, the ordinal method gives no preparatory period to each maintenance action; some unnecessary maintenance action must be allocated to acquire the prescribed safety level.

8 CONCLUSIONS

In this paper, an attempt was made to develop a new method of providing a solution for the optimal maintenance scheduling problem with uncertain situations. The bridge maintenance problem is formulated as an optimization problem with uncertain effects of repair and replacement. In order to establish a rational maintenance program, it is necessary to reduce the uncertainties. Thus, it is important to predict the deteriorating curve in a good accuracy. However, the prediction of deterioration unavoidably involves uncertainties. In this study, an attempt was made to develop a maintenance scheduling system that can adapt to the uncertainty of the deterioration of the bridge by introducing GACU. GACU combines the ordinal GA operation and Monte Carlo simulation. The application of Monte Carlo simulation to GA requires a great deal of computation time, the concept of age structure is introduced. The computation time is reduced by implementing the evolution and sampling simultaneously. Since this method does not use the probabilistic concept directly, it may be easily understood by inspectors and maintenance engineers. In addition, it is possible to consider the maintenance cost and seismic risk at the same time by using the proposed method. The proposed system can reduce the seismic risk without increase the maintenance cost.

REFERENCES

Frangopol, D. M. & Furuta, H. (eds.) 2001. *Life-Cycle Cost Analysis and Design of Civil Infrastructure Systems*, ASCE, Reston Virginia.

Frangopol, D. M., Lin, K.-Y., & Estes, A. C. 1997. Life-cycle cost design of deteriorating structures, *Journal of Structural Engineering*, ASCE, Vol. 123 (10), pp.1390–1401.

Furuta, H., Frangopol, D. M. & Saito, M. 1997. *Application of genetic algorithm to life-cycle cost maintenance of bridges, Proceedings of KKNN Symposium*, Seoul, Korea.

Furuta, H., Koyama, K., Dogaki, M. & Frangopol, D. M. 2004a. Seismic risk evaluation and life-cycle cost analysis of bridge structures in Japan, *Proceedings of 2nd ASRANeT Colloquium*, Barcelona, Spain.

Furuta, H., Kameda, T., Fukuda, Y., and Frangopol, D. M. 2004b, Life-Cycle Cost Analysis for Infrastructure Systems: Life Cycle Cost vs. Safety Level vs. Service Life, Proceedings of Joint International Workshops LCC03/IABMAS and fip/JCSS, EPFL, Lausanne, March 24–26,2003, (keynote lecture); in *Life-Cycle Performance of Deteriorating Structures: Assessment, Design and Management*, Frangopol, D. M., Bruhwiler, E., Faber, M. H. & Adey, B. (eds.), ASCE, Reston, Virginia, 19–25, 2004.

Furuta, H., Kataoka, H., Dogaki, M. & Frangopol, D. M. 2005a. Effects of seismic risk on life-cycle cost analysis for bridge maintenance, *Proceedings of 4th International Conference on Current and Future Trends in Bridge, Construction and Maintenance*, Kuala Lumpur, Malaysia.

Furuta, H., Koyama, K. & Oi, M. 2005b. Life-cycle analysis of bridge structures considering maintenance cost and seismic risk, *Proceedings of IFIP WG7.5 Working Conference on Structural Reliability and Optimization*, Aalborg, Denmark.

Furuta, H., Nakatsu, K. & Frangopol, D. M. 2005c. Optimal restoration scheduling for earthquake disaster using life-cycle cost, *Proceedings of 4th International Workshop on Life-Cycle Coast Analysis and Design of Civil Infrastructure Systems*, Coco Beach, USA.

Furuta, H., Koyama, K., Oi, M. & Sugimoto, H. 2006. Life-cycle cost evaluation of multiple bridges in road network considering seismic risk, *Proc. of 6th International Bridge Engineering Conference*, Boston, USA.

Furuta, H., Hattori, H., Ohama, T. & Yoshida, K. 2008. *Proceedings of the Twenty-First KKCNN Symposium on Civil Engineering*, Singapore.

Tamaki, H., Arai, T. & Abe, S. 1999. "A Genetic Algorithm Approach to Optimization Problems with Uncertainties", *Institute of Systems, Control and Information Engineers Journal*, vol.12, no.5, pp.297–303 (in Japanese).

Reliability and Optimization of Structural Systems – Straub (ed)
© 2010 Taylor & Francis Group, London, ISBN 978-0-415-88179-1

Bridge health monitoring using output-only vibration measurements excited by a single moving vehicle

C.W. Kim
Department of Civil and Earth Resource Engineering, Kyoto University, Kyoto, Japan

M. Kawatani
Department of Civil Engineering, Kobe University, Kobe, Japan

ABSTRACT: This paper investigates feasibility for bridge health monitoring from output-only vibration measurements under a single moving vehicle through a laboratory experiment. The parameters considering in this study are system frequency and damping constant rather than modal ones because signals obtained under a moving vehicle are not the responses of the observation bridge itself but those of the bridge-vehicle interactive system. A parameter from the AR coefficients is also adopted to detect abnormality of the bridge. Consideration also goes into diagnosis of the bridge from pattern change of identified system parameters due to damages. Observations demonstrate the possibility of making a decision for the current bridge condition from the identified system parameters of the vehicle-bridge interactive system.

1 INTRODUCTION

This paper covers feasibility investigations for Bridge Health Monitoring (BHM) from output-only vibration measurements under a single moving vehicle, so called traffic-induced vibrations, through a laboratory experiment. The background of using traffic-induced vibrations of bridges in BHM is that the traffic excitation is an important dynamic source especially for short span bridges.

This study focuses on the diagnosis of bridges from estimated dynamic parameters assuming the measured vibration responses as a kind of AutoRegressive (AR) process. Three parameters are considered: system frequencies; system damping constants; and an abnormality indicator from AR coefficients. First two parameters, frequency and damping constant, are common invibration-based BHM. Focusing on change of AR coefficients for BHM also may be an option because both frequency and damping constant depend on AR coefficients of the AR model (Nair et al. 2006). Therefore an alternative parameter from AR coefficients is considered for the vibration-based BHM in this study.

Since 1970s, the use of state-space models for modal parameter identification in time-domain has been increasing and also has yielded new approaches. Gersch et al. (1973), for example, used the time series of an AutoRegressive Moving Average (ARMA) process to describe the random response of a vibrating structure to a white noise excitation. Estimating the coefficients of ARMA model is a kind of nonlinear approach because both of the coefficients relating to AR and MA processes are unknown variables. Fortunately, the AR model with infinite order is equivalent to the ARMA model, which means that one can express the responses of a linear system subjected to white-noise input using the AR model with sufficient large order (Wang & Fang 1986, He & De Roeck 1997).

It is noteworthy that this study estimates system frequency and damping constant rather than modal ones because signals obtained under a moving vehicle are not the responses of bridge itself but those of the bridge-vehicle interactive system. Despite of the non-stationary property of the traffic-induced vibration, the idea of using the traffic-induced vibration data of short span bridges is that those parameters identified repeatedly under a given moving vehicle can provide a pattern or even a statistical one, which may give useful information to decide the bridge health condition.

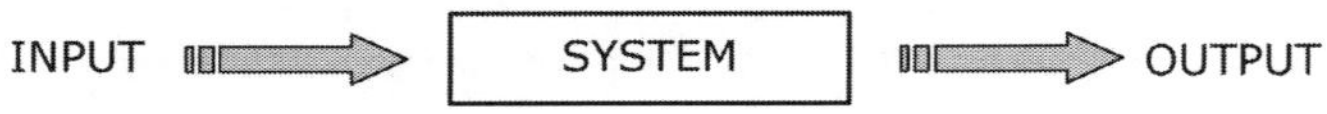

Figure 1. Three components for a system.

2 DYNAMIC PARAMETERS IDENTIFICATION

There are three important components in a system: the input; the output; and the system as shown in Figure 1. The idea for the system parameter identification by means of the output only vibration measurement is that the three components are linked with each other linearly and observing outputs makes it possible to identify the system (Ljung 1999). Of course the system especially in civil infrastructures is usually affected by many factors surrounding the structure. Those factors are unknown and are treated as noise.

If $y_c(t)$ denotes output of a bridge structure taken from an observation point, then the corresponding state equation of a continuous-time system is described as

$$\dot{\mathbf{x}}_c(t) = \mathbf{A}_c\mathbf{x}_c(t) + \mathbf{B}_c\mathbf{f}_c(t),\ \ \mathbf{y}_c(t) = \mathbf{C}_c\mathbf{x}_c(t) \tag{1}$$

where, $\mathbf{x}_c(t) = \begin{bmatrix} z(t) \\ \dot{z}(t) \end{bmatrix}$, $\mathbf{A}_c = \begin{bmatrix} 0 & 1 \\ -k_b/m_b & -c_b/m_b \end{bmatrix}$, $\mathbf{B}_c = \begin{bmatrix} 0 \\ -1/m_b \end{bmatrix}$ and $\mathbf{C}_c = \begin{bmatrix} 1 & 0 \end{bmatrix}$.

Therein, m_b, k_b, c_b, respectively indicate the mass, damping, and stiffness of the bridge. $z(t)$ denotes the displacement response, and $\mathbf{f}(t)$ represents external force vector. If the observability matrix is defined as $\mathbf{L}^T = [\mathbf{C}; \mathbf{CA}_c; \ldots; \mathbf{CA}_c^{n-1}]$, then the discrete state equation for the system from the equation of motion for a bridge through an observability transformation can be described as

$$\mathbf{x}(k+1) = \mathbf{Ax}(k) + \mathbf{Bf}(k),\ \ \mathbf{y}(k) = \mathbf{Cx}(k) \tag{2}$$

where $\mathbf{x}(k) = \mathbf{Lx}_d(k)$, and the subscript d indicates discretized one from the continuous system. The state transition matrix $\mathbf{A}$, input influence matrix $\mathbf{B}$ and output influence matrix $\mathbf{C}$ through an observability transformation are described as (Pi & Mickleborough 1989)

$$\begin{aligned} &\mathbf{A} = \mathbf{LA}_d\mathbf{L}^{-1} = [\{0\,1\,0 \cdots 0\}; \{0\,0\,1 \cdots 0\}; \cdots; -\{a_n\ a_{n-1} \cdots\ a_1\}],\ \ \mathbf{B} = \mathbf{LB}_d = [b_1; \cdots; b_n],\ \text{and} \\ &\mathbf{C} = \mathbf{C}_d\mathbf{L}^{-1} = [1\,0 \cdots 0] \end{aligned} \tag{3}$$

where $\mathbf{A}_d = e^{\mathbf{A}_c\Delta}$ and Δ represents the time interval of sampling. If the coefficient from a_1 to a_n of the system matrix $\mathbf{A}$ is known, then the eigenvalues of the matrix $\mathbf{A}$ provides the modal information of the system. Therefore, the next step is to derive the linear equation for the coefficient using the measurements (or observation) $\mathbf{y}$(k). The following two equations are obtainable from Equation 2

$$\mathbf{x}(k) = \mathbf{y}(k) - \mathbf{bf}(k) \tag{4}$$

$$\mathbf{y}(k+n) = -\mathbf{ax}(k) + \mathbf{Bf}(k) \tag{5}$$

where, $\mathbf{b}$ is the coefficient matrix for a white noise input described as

$$\mathbf{b} = [\{0\ b_1\ \cdots\ b_{n-1}\}; \{0\,0\,b_1\ 0\ \cdots 0\}; \{0\,0\,0\,0 \cdots\ b_1\}; \cdots; \{0\ \ \cdots\ \ 0\}],\ \text{and}\ \mathbf{a}=[a_1\ \cdots\ a_n]. \tag{6}$$

Substituting Equation 4 into Equation 5 and assuming $\mathbf{f}(k)$ as a white noise input $\mathbf{e}$(k) give the following AR model by using the fact that AR model with infinite order is equivalent to the ARMA model (Wang & Fang 1986, He & De Roeck 1997).

$$y(k) + \sum_{i=1}^{p} a_i y(k-i) = e(k) \tag{7}$$

To estimate the AR parameter, the autocorrelation function of $y(k)$ is used, which is obtainable by multiplying each term of Equation 7 with $y(k\text{-}s)$ and taking mathematical expectation. This process yields the following Yule-Walker equation

$$\mathbf{Ra} = -\mathbf{r} \tag{8}$$

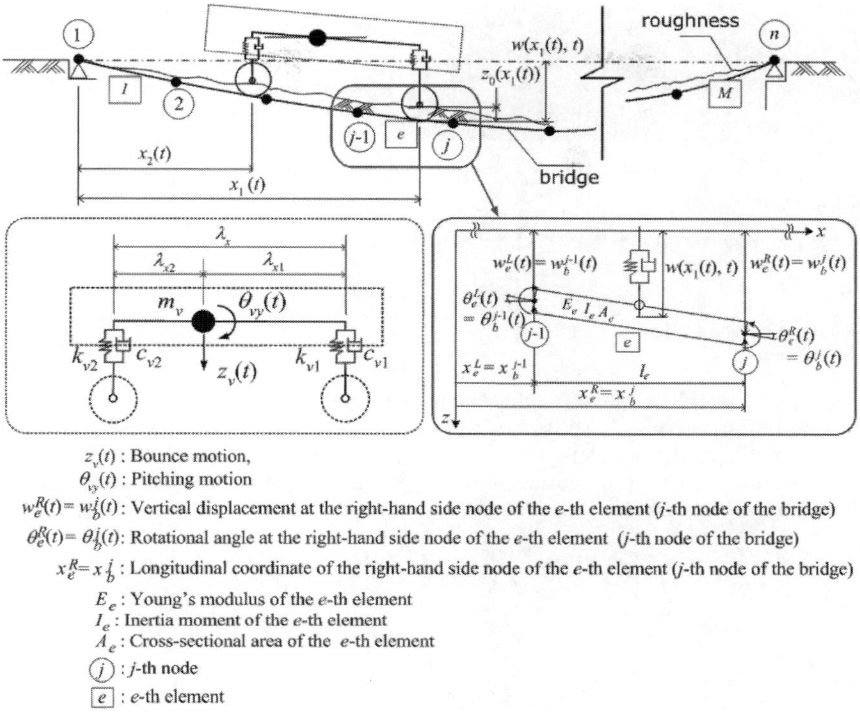

Figure 2. Scheme of a bridge-vehicle interactive system in moving vehicle laboratory experiment.

where $\mathbf{R}$ is a Toeplitz matrix about $R(p, s) = E[y(k-p)y(k-s)]$ which is the autocorrelation function of the signal. $\mathbf{r}^T = [R_1; \ldots; R_n]$. The Levinson-Durbin algorithm is adopted to solve Equation 8 for $\mathbf{a}$ (Ljung 1999).

The frequency and damping constant of each mode can be estimated from the eigenvalue of $\mathbf{A}$ as

$$z = \exp\left(-h_k\omega_k \pm j\omega_k\sqrt{1-h_k}\right) \tag{9}$$

where, h_k and ω_k are the damping constant and circular frequency of k-th mode of the system. j represents the imaginary unit.

The frequency and damping constant are estimated using the identified coefficient matrix $\mathbf{A}$ which comprises AR coefficients as shown in Equation 2. Therefore the AR coefficient itself represents the dynamic characteristics of the system, and is available as a parameter to decide health condition of bridges.

3 BRIDGE-VEHICLE INTERACTIVE SYSTEM

As previously stated, this study considers system parameters of the bridge-vehicle interactive system rather than the bridge system. It is meaningful to explain the mechanism of the interactive system. Figure 2 shows a schematic figure of the bridge-vehicle interactive system.

For the vehicle model shown in Figure 2, $z_v(t)$ and $\theta_v y(t)$ respectively represent vehicle's bounce and pitching motions. In that figure, m_v denotes the vehicle mass. Additionally, k_{vs} and c_{vs} respectively denote the spring constant and damping coefficient at the s-th axle of the vehicle. The subscript s indicates the position of an axle: that is, $s = 1$ and $s = 2$ respectively signify the first (or front) and second (or rear) axles. Distances from the vehicle's center of gravity to respective axles

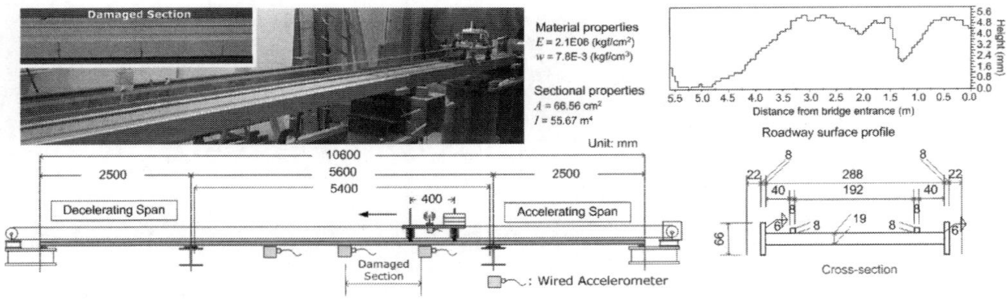

Figure 3. Experimental setup.

are denoted by λ_{x1} and λ_{x2}. $z_0(xs(t))$ indicates the roadway surface roughness at a position of $x_s(t)$ from the bridge entrance which is assumed as the reference position.

The combination of the interaction force at a contact point of a vehicle wheel with the bridge provides equations of motion for a bridge-vehicle interactive system. The dynamic equation of the bridge under a moving vehicle is writable as (Kim et al. 2005)

$$\begin{aligned}
&\mathbf{M}_{br}\ddot{\mathbf{q}}_r(t)+\left\{\mathbf{C}_{br}+\sum_{s=1}^{2}\boldsymbol{\psi}_s(t)c_{vs}\boldsymbol{\psi}_s^T(t)\right\}\dot{\mathbf{q}}_r(t)+\left\{\mathbf{K}_{br}+\sum_{s=1}^{2}\boldsymbol{\psi}_s(t)k_{vs}\boldsymbol{\psi}_s^T(t)\right\}\mathbf{q}_r(t)\\
&-\sum_{s=1}^{2}\boldsymbol{\psi}_s(t)c_{vs}\left\{\dot{z}_v(t)-(-1)^s\lambda_{xs}\dot{\theta}_{vy}(t)\right\}-\sum_{s=1}^{2}\boldsymbol{\psi}_s(t)k_{vs}\left\{z_v(t)-(-1)^s\lambda_{xs}\theta_{vy}(t)\right\}\\
&=\sum_{s=1}^{2}\boldsymbol{\psi}_s(t)\left[\left(1-\frac{\lambda_{xs}}{\lambda_x}\right)m_v g+\left\{c_{vs}\dot{z}_0(x_s(t))+k_{vs}z_0(x_s(t))\right\}\right]
\end{aligned} \tag{10}$$

where $\mathbf{M}_{br}$, $\mathbf{C}_{br}$ and $\mathbf{K}_{br}$ respectively represent the mass, damping, and stiffness matrices of the bridge. $\mathbf{q}_r(t)$ is the displacement vector of the bridge; overdots denote derivatives with respect to time. $\psi_s(t)$ is a load distribution vector to each node of the element on which a tire contacts.

Equation 10 shows a nonstationary system because the system damping and stiffness matrices consist of time-variant coefficients. Consequently, the parameters estimated by simplifying the interactive system as the AR process are those for the bridge-vehicle interactive system rather than the bridge system, and the effect of vehicle system results dispersion of the identified system parameters under random traffic condition. On the other hand, if the observations or outputs are taken repeatedly under the condition of a given single vehicle such as an inspection car the dispersion becomes weaker than under random traffics. Therefore this study focuses on the system parameters identified from the vibration data of short span bridge under a given single vehicle running.

4 LABORATORY EXPERIMENT

A moving vehicle laboratory experiment is performed to investigate feasibility of the proposed approach. The experiment setup is summarized in Figure 3. Roadway profiles are considered in the experiment. Regarding damage, three saw cuts are applied to both left and right sides of web plates between L/4 and L/2 of the bridge model. For the damaged section, the bending rigidity of the member decreases to around 89% of the intact state. Natural frequencies estimated from free vibrations for intact and damaged girders are 2.69 Hz and 2.59 Hz, respectively.

During the experiment, three different vehicle models of which the natural frequency for the bounce motion is changeable using a different set of mass and spring are considered in the experiment. Three vehicles called as VT-A, VT-B and VT-C are used in the experiment. Natural frequencies for the bounce motion of those vehicle models are respectively 2.93 Hz, 3.76 Hz and 3.03 Hz. Two different speeds of 0.93 m/s and 1.63 m/s are adopted to investigate the effect of the vehicle speed to damage identification results. Six traffic scenarios are considered: SCN1 of VT-A vehicle traveling with speed of 0.93 m/s; SCN2 of VT-A vehicle traveling with speed of 1.63 m/s; SCN3 of VT-B vehicle traveling with speed of 0.93 m/s; SCN4 of VT-B vehicle traveling with speed of 1.63 m/s;

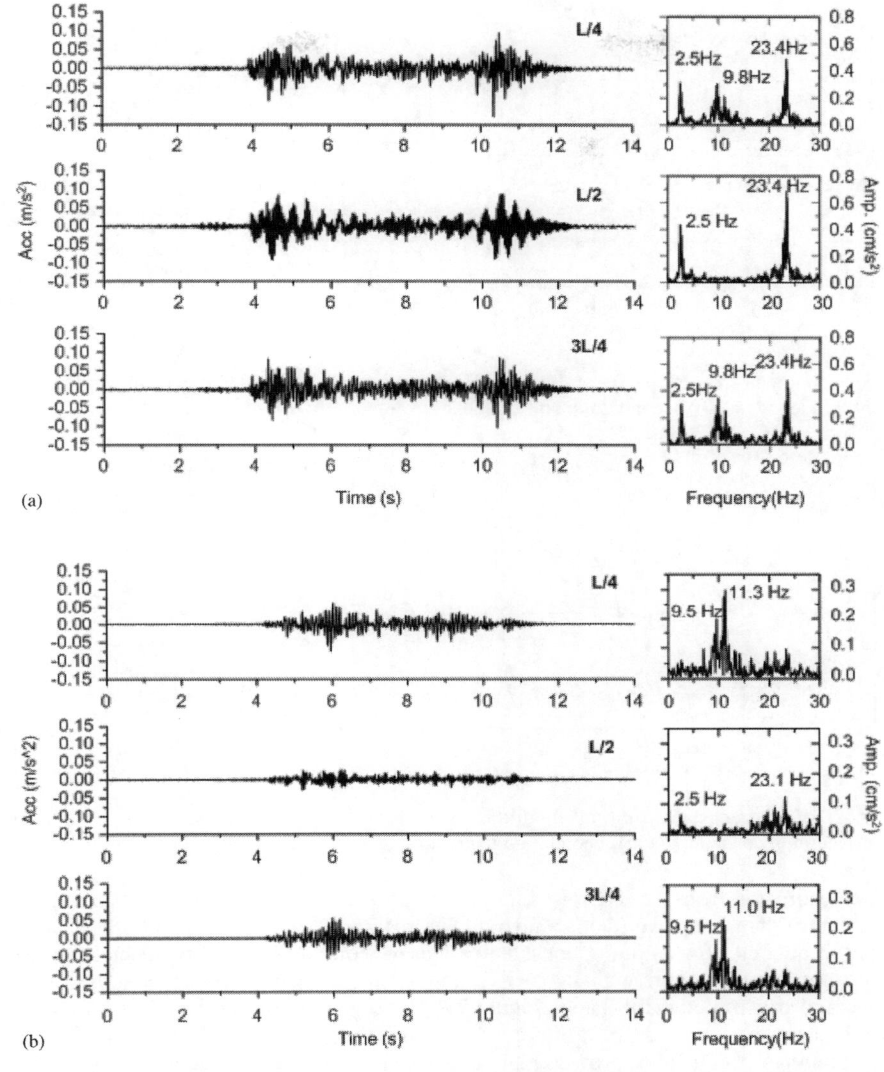

Figure 4. Accelerations of the bridge under traffic scenario SCN1: a) Intact; and b) Damaged.

SCN5 of VT-C vehicle traveling with speed of 0.93 m/s; and SCN6 of VT-C vehicle traveling with speed of 1.63 m/s.

Three points at 1/4, 1/2, and 3/4 of the span length are the observation points for acceleration responses of the bridge. The sampling rate of signals is 100 Hz. Examples of acceleration time histories of the bridge and their Fourier amplitude spectra before and after damage are shown in Figure 4. The appearance of additional dominant frequencies due to the damage in comparison with those of the intact bridge is also observed. Another interesting point is that dominant frequencies near 2.5 Hz and 23.4 Hz appearing in the intact bridge are weakened by the damage.

5 CONDITION SCREENING

Dominant frequencies and system damping constants estimated from the data taken from the six traffic scenarios are summarized in Figure 5 and Figure 6, respectively. In figures first 33 plots in the left part of the vertical dotted-line indicate identified dominant frequencies (or damping

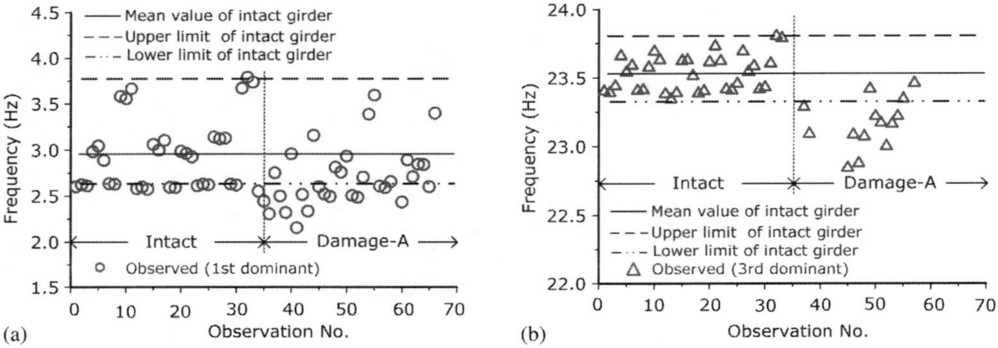

Figure 5. Dominant frequencies from all the traffic scenarios according to bridge health condition: a) first dominant frequency; and b) second dominant frequency.

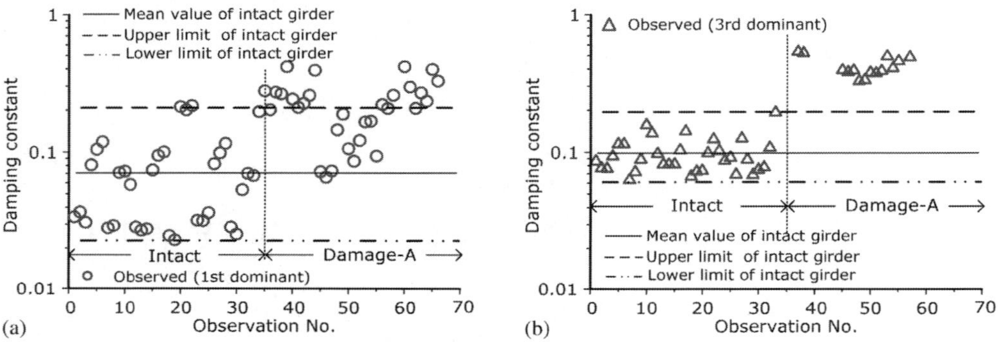

Figure 6. Damping constants from all the traffic scenarios according to bridge health condition: a) first dominant damping constant; and b) second dominant damping constant.

constants) estimated from the data of intact girder. On the contrary, the rest plots are those taken from the data of the damaged girder. a) and b) in Figures 5 and 6 respectively show the first and third dominant frequencies (or damping constants). Those second dominant frequencies and damping constants give less clear pattern change between intact and damage girders then those for first and third ones and are omitted in this paper. Figure 5 shows clear pattern change of identified dominant frequencies despite of their severe variations. An interesting point is, as can be read from Figure 5b), the third dominant frequencies provide more apparent pattern changes according to the damage.

Usually the damping constants derived from eigen-value of a system matrix may be subject to appreciable error (Pappa & Ibrahim 1981), and as a result larger coefficient of variance than that of the identified frequency is observed as shown in Figure 6. However, despite of their appreciable error the pattern change of identified damping constants due to the damage is very apparent comparing to that of the dominant frequency shown in Figure 5.

As another parameter, the Damage-Sensitive Feature (DSF) (Nair et al. 2006) which is defined in Equation 11 is adopted as another damage indicator.

$$DSF = \frac{a_1}{\sqrt{(a_1 + a_2 + a_3)}} \tag{11}$$

where G_1, G_2 and G_3 are the first three AR coefficients.

It is noteworthy that Nair et al. (2006) reported that the DSF is depending on the AR coefficients appeared to be the most promising because these coefficients are statistically the most significant among all the coefficients of the model.

Figure 7 shows those plots of DSF according to bridge's health condition such as intact and damage. It clearly demonstrates that the DSF also changes due to damage like frequency and

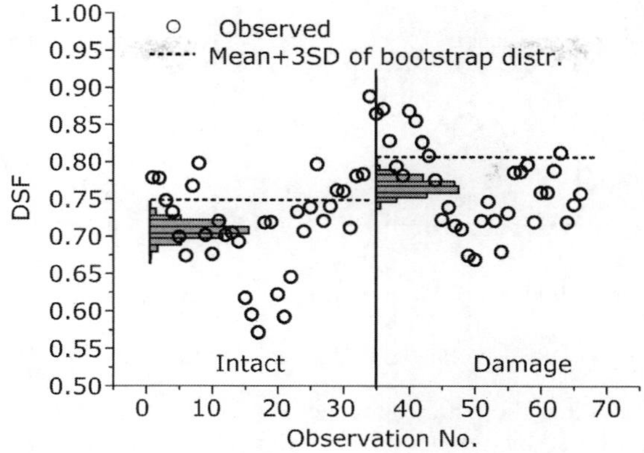

Figure 7. Damage indicator DSF according to bridge health condition.

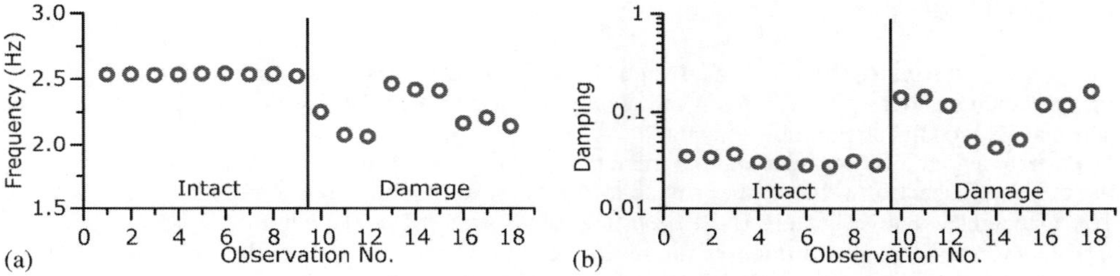

Figure 8. Dominant frequencies according to bridge health condition for SCN1: a) first dominant frequency; and b) first damping constant.

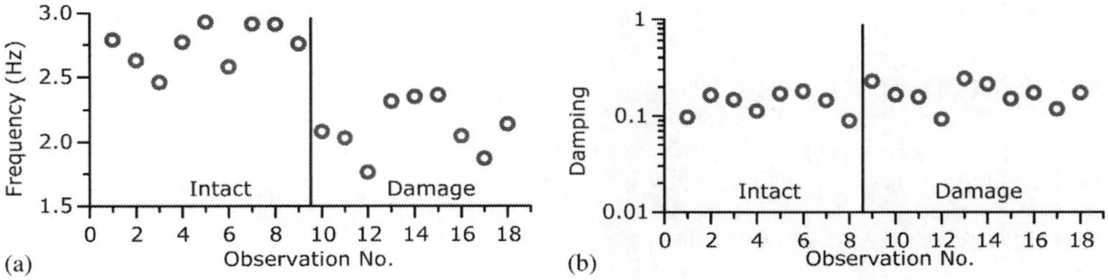

Figure 9. Dominant frequencies according to bridge health condition for SCN2: a) first dominant frequency; and b) first damping constant.

damping constants. In the figure the histograms show bootstrap distribution (Efron & Tibshirani 1993) of DSF values for 1000 resample. The reason for using bootstrap distribution is the confined number of observation data. The bootstrap distribution curve in Figure 7 gives more clear change of bridge conditions.

Observations also demonstrate that the lower speed scenarios give very clear pattern change due to the damage in comparing with the higher speed scenario. Figures 8 and 9 show identified modal parameters for low speed scenario SCN1 and higher speed scenario SCN2 respectively. The results demonstrate that the effect of the vehicle system to bridge vibrations, so called traffic-induced vibration of bridges or non-stationary vibration, increases with increasing speed, and as a result the identified results under higher vehicle speed yield more biased identification results than those of lower speed. Variations of DSF for SCN1 and SCN2 are summarized in Figure 10, which also shows lower speed traffic gives better chance to detect diagnosis in bridges.

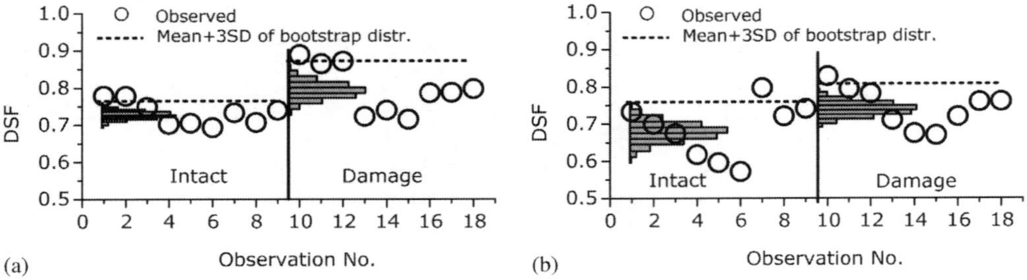

Figure 10. Damage indicator DSF according to bridge health condition: a) SCN1; and b) SCN2.

The point made here is that the pattern change of system modal parameters using the traffic-induced vibration data excited by a moving inspection vehicle can provide information for health condition of short span bridges.

6 CONCLUSIONS

This paper investigates feasibility of Bridge Health Monitoring (BHM) from output-only vibration measurements under a single moving vehicle through a laboratory moving vehicle experiment. The summarized results are as follows:

Pattern change of system modal parameters using the traffic-induced vibration data excited by the moving inspection vehicle can provide information for health condition of short span bridges. The Damage-Sensitive Feature (DSF) estimated by using AR coefficients may be an option for BHM. The bootstrap distribution of the resample means from the confined number of identified data provides a clear difference of the distribution between with and without damages.

The study also demonstrates that the effect of the vehicle system to bridge vibrations, so called traffic-induced vibration of bridges or non-stationary vibration, increases with increasing speed, and as a result the identified results under higher vehicle speed yield more biased identification results than those of lower speed.

ACKNOWLEDGMENT

A part of this study is supported by the Japan Society for the Promotion of Science (Grant-in-Aid for Exploratory Research under project No.19656112 and Grant-in-Aid for Scientific Research (C) under project No. 20560443) are Rent All Scholarship Foundation (Grant-in-Aid for Scientific Research), which are gratefully acknowledged.

REFERENCES

Gersch, W., Nielsen, N.N. & Akaike, H. 1973. Maximum likelihood estimation of structural parameters from random vibration data. J. of Sound and Vibration, 31(3), 295–308.

He, X. & De Roeck, G. 1997. System identification of mechanical structures by a high-order multivariate autoregressive model. Computers and Structures, 64(1–4), 341–351.

Kim, C.W., Kawatani, M., & Kim, K.B. 2005. Three-dimensional dynamic analysis for bridge-vehicle interaction with roadway roughness, Computers and Structures, 83(19–20), 1627–1645.

Ljung, L. 1999. *System identification – Theory for the user*, 2nd ed, PTR Prentice Hall, Upper Saddle River, N.J.

Nair, K.K., Kiremidjian, A.S. & Law, K.H. 2006. Time series-based damage detection and localization algorithm with application to the ASCE benchmark structure. J. of Sound and Vibration, 291, 349–368.

Pappa, R.S. & Ibrahim, S.R. 1981. A parametric study of the Ibrahim time domain modal identification algorithm. Shock and Vibration Bulletin, No. 51, Part 3, pp. 43–72.

Pi, Y.L. & Mickleborough, N.C. 1989. Modal identification of vibrating structures using ARMA model, J. of Eng. Mech., ASCE, 115(10), 2232–2250.

Wang, Z. & Fang, T. 1986. A time-domain method for identifying modal parameters. J. Appl. Mech., ASME, 53(3), 28–32.

Reliability and Optimization of Structural Systems – Straub (ed)
© 2010 Taylor & Francis Group, London, ISBN 978-0-415-88179-1

Risk analysis of fatigue-induced sequential failures by Branch-and-Bound method employing system reliability bounds

Young-Joo Lee & Junho Song
University of Illinois at Urbana-Champaign, Urbana, IL, USA

ABSTRACT: Various types of structural systems are often subjected to the risk of fatigue-induced failures. If such structural systems do not have adequate level of redundancy, local failures may initiate sequential failures toward exceedingly large damage. For risk-informed structural design and maintenance of such structural systems, it is thus essential to quantify the risk of fatigue-induced sequential failures. This paper introduces a newly developed Branch-and-Bound method employing system reliability Bounds method (termed as B^3 method). Using a disjoint cut-set formulation, the B^3 method identifies critical sequences of fatigue-induced failures in the decreasing order of their likelihood as it systematically updates both lower and upper bounds on the system failure probability. The updated bounds provide reasonable criteria for terminating the Branch-and-Bound search without missing critical sequences or estimating the system-level risk inaccurately. This paper presents the proposed method with numerical examples along with ongoing and future research work.

1 INTRODUCTION

Various structural systems are often subjected to the risk of fatigue-induced failures caused by repeated loading during their service (Karamchandani *et al.* 1992). If such structural systems do not have adequate level of redundancy, local failures may initiate sequential failures or exceedingly large damage such as system collapse. In risk-based decision-making for structural design, maintenance and retrofit, it is thus essential to quantify the risk of such structural failure and to identify critical sequences of local failures leading to a system failure.

Monte Carlo simulation is the most straightforward and widely used method; however, when structural analysis demands time-consuming computational simulations or when the failure probabilities are low, the computational and time costs required for converged results can be exceedingly large. A variety of branch-and-bound (B&B) methods were developed to identify critical failure sequences with significant likelihood in an efficient manner. Despite many research efforts, risk analysis methods based on the B&B approach are still either time-consuming or prone to miss critical failure sequences.

This paper presents a new Branch-and-Bound method employing system reliability Bounds method (termed as B^3 method) that identifies critical fatigue-induced failure sequences effectively and estimates the system risk accurately. The B^3 method identifies critical sequences of fatigue-induced failures in the decreasing order of their likelihood as it systematically updates both lower and upper bounds on the system failure probability. The updated bounds provide reasonable criteria for terminating the B&B search without missing critical sequences.

2 BRANCH-AND-BOUND METHOD EMPLOYING SYSTEM RELIABILITY BOUNDS

Despite many research efforts to develop effective searching techniques (Karamchandani *et al.* 1992), many of them have a risk of missing critical failure sequences. The branch-and-bound (B&B) method (Murotsu 1984) is considered one of the most effective searching methods, and has been widely used. To identify "globally" most-likely-to-occur system failure sequences, the method compares the probabilities of all the failure sequences that have been investigated and assumes further damage for the most likely sequence to continue the search. When the system failure is observed, the particular sequence is identified as a "system failure sequence." Although this selective search approach based on the probabilities of sequences helps reduce the number of sequences to explore, one still might need to explore a large number of sequences to obtain a reliable estimate on the structural system risk. This is due to the lack of reasonable criteria that would help terminate the search without underestimating the system-risk. The structural system failure event is often described as the union of the identified system failure sequences. During a search process, one can continuously update the lower bound of the system failure probability by the system reliability analyses employing the identified failure sequences. With the upper bound usually unknown, the trend of the lower bound updates alone cannot provide accurate termination criteria. The size of the updates on the lower bound does not decrease monotonically due to the statistical dependence between identified failure sequences.

By contrast, the search process of the B^3 method identifies *disjoint* failure sequences in order to (1) obtain both the lower and upper bounds of the system failure probability; (2) achieve monotonic decrease in the updates of the lower bounds as the search process proceeds; and (3) update the bounds of the system risk without performing additional system reliability analyses.

2.1 *Limit-state function formulations for disjoint failure sequences*

First, consider the following crack growth model by Paris & Erdogan (1963):

$$\frac{da}{dN} = C(\Delta K)^m \tag{1}$$

where a denotes the crack length, N is the number of load cycles, C and m are the material parameters, and ΔK denotes the range of the stress intensity factor. By Newman's approximation (Newman & Raju 1981), the range of stress intensity factor can be determined as

$$\Delta K = S \cdot Y(a) \cdot \sqrt{\pi a} \tag{2}$$

where S denotes the far-field stress range and $Y(a)$ is the "geometry" function. By substituting Equation 2 into Equation 1 and integrating the equation from the initial condition to the current time point, the relationship between the time duration (T) and the corresponding crack length (a) is derived as

$$\int_{a_0}^{a} \frac{1}{\left[Y(a)\sqrt{\pi a}\right]^m} da = \int_{a_0}^{a} C \cdot S^m dN = C \cdot N \cdot S^m = C \cdot \nu_0 \cdot T \cdot S^m \tag{3}$$

where a_0 is the initial crack length, and ν_0 is the loading frequency. A crack failure is assumed to occur when the crack exceeds a critical length a_{ci}. Then, the limit-state function for the failure of the i-th member within an inspection cycle $[0, T_s]$ is described as

$$g_i(\mathbf{X}) = T_i^0 - T_s \le 0, \text{where } T_i^0 = \frac{1}{C\nu_0 (S_i^0)^m} \int_{a_i^0}^{a_{c_i}} \frac{1}{\left[Y(a)\sqrt{\pi a}\right]^m} da \tag{4}$$

where a_i^0 is the initial crack length of the i-th component, T_i^0 denotes the time required for the crack grows from a_i^0 to a_{ci}, S_i^0 is the stress at the i-th member, and $\mathbf{X}$ denotes the vector of random variables.

In order to describe the failure sequences as *disjoint* events in the B&B search, we first formulate the case in which the i-th component fails before failures at any other components. The probability of this system event is described as

$$P_i = P\left\{\left[\bigcap_{\forall l \neq i} \left(T_i^0 < T_l^0\right)\right] \cap \left(T_i^0 < T_s\right)\right\} \tag{5}$$

The event in Equation 5 is a parallel system event consisting of n component events. Based on the results by component reliability analyses of n individual events, the probability can be computed by use of a system reliability analysis method.

Equation 5 can be generalized to failure sequences involving more than two component failures. For example, the probability of the progressive failure of a sequence $\{1 \rightarrow 2 \rightarrow \ldots \rightarrow (i-1) \rightarrow i\}$

is described as

$$P_i^{1,\ldots,(i-1)} = P\left\{\left[\bigcap_{\forall j \neq 1}\left(T_1^0 < T_j^0\right)\right] \cap \cdots \cap \left[\bigcap_{\forall l \neq 1,\ldots,i}\left(T_i^{1,\ldots,(i-1)} < T_l^{1,\ldots,(i-1)}\right)\right] \cap \left(T_1^0 + T_2^1 + \cdots + T_i^{1,\ldots,(i-1)} < T_s\right)\right\} \tag{6}$$

where $T_i^{1,\ldots,(i-1)}$ denotes the time required for the failure at the i-th component since the sequence of the failures in the subscript. Unlike T_i^0 in Equation 4, i.e., the time until the first failure for an undamaged structure, the time terms introduced for damaged structures (such as $T_i^{1,\ldots,(i-1)}$ in Equation 6) should be computed with the effects of load re-distributions considered. For convenient formulations of such time terms, a recursive formulation is derived for a general failure sequence $\{1 \rightarrow 2 \rightarrow \ldots \rightarrow (i-1) \rightarrow i\}$ as follows. The details of the derivation are found in Lee & Song (2010).

$$T_i^{1,\ldots,i-1} = \frac{1}{C\nu_0\left(S_i^{1,\ldots,i-1}\right)^m}\int_{a_i^0}^{a_{c_i}}\frac{da}{\left[Y(a)\sqrt{\pi a}\right]^m} - \sum_{k=1}^{i-1}\left(\frac{S_i^{1,\ldots,k-1}}{S_i^{1,\ldots,i-1}}\right)^m T_k^{1,\ldots,k-1} \tag{7}$$

As a result, the time terms in Equation 6 can be described in terms of deterministic and random parameters for any failure sequence during reliability analyses.

Let E_i, $i = 1, \ldots, N_{fs}$, denote the system failure sequences identified by a search process where N_{fs} denotes the number of all possible system failure sequences. If the failure sequences are described as disjoint events as in Equation 6, the system failure probability is obtained by summing up the probabilities of the identified system failure sequences, i.e.,

$$P(E_{sys}) = P\left(\bigcup_{i=1}^{N_{fs}} E_i\right) = \sum_{i=1}^{N_{fs}} P(E_i) \tag{8}$$

The probability of a system failure sequence, $P(E_i)$, is obtained by a system reliability analysis for the parallel system in Equation 6. This disjoint cut-set formulation enables us to update both the lower and upper bounds by the probabilities of the identified system-failure and non-failure cases, respectively.

2.2 *Systematic search scheme of the B^3 method*

The B^3 method uses a systematic search scheme that employs the aforementioned disjoint cut-set formulation and recursive limit-state functions. The search process is illustrated in Figure 1. The search starts by setting the lower and upper bounds of the system failure probability (denoted by P_{low} and P_{upp}) as zero and one, respectively. These bounds are updated continuously as the search process identifies cases of system failures and non-failures and computes their probabilities. The search process starts with a node that contains all the possible scenarios ("initial node", shown at the far left in Figure 1). This is considered one of the "mixed" nodes (gray nodes in Figure 1). Mixed nodes contain both system failure and non-failure cases. The stress distribution of the intact structure is obtained by a structural analysis.

Next, a "branching" process is performed to find the probabilities of individual component failures from the current structural condition. If the structural condition represented by a current node has n structural members (or locations) of potential crack failures, this branching process transfers the probability in the current "parent" node to the $(n+1)$ "child" nodes as shown in the second column of the nodes in Figure 1. As a result, the current "parent" node becomes "inactive," i.e., stops contributing to the system probability calculations. In the figure, dashed-line circles represent inactive nodes. Each of the first n child nodes, i.e., '1' through 'n', contains the probability that the corresponding member fails "first" during the given inspection cycle T_s. These are new "mixed" nodes and their probabilities are computed by system reliability analysis in Equation 5. By contrast, the last "white" node represents the case that no further failure occurs during the inspection cycle. Since these $(n+1)$ child nodes are mutually exclusive (or disjoint), the probability of the white node is computed by the probability of the parent node minus the sum of the probabilities of all the other child nodes. The upper bound is now decreased by the probability of the white node. This is the "bounding" process to update the upper bound.

The next step is to compare all the active mixed nodes in order to select the one with the highest probability. Then, a new structural analysis is performed using a model representing the damage scenario of the selected node. If the structural analysis reveals that the selected node represents a system failure case (e.g., "black" node in Figure 1), the lower bound is increased by the probability of the node (that has been already computed during the previous branching process). On the other hand, if a system failure is not observed by the structural analysis, the aforementioned "branching" process is performed to find the probabilities of the child nodes.

This repeated process of "branching" and "bounding" is continued until the following termination criteria are satisfied by the bounding process. If finding the most critical paths is the main interest, one can stop the search process when the most recently identified system failure sequence has a probability lower than a given threshold or the required number of failure sequences are identified. This is because the systematic search scheme guarantees finding the failure modes in the decreasing order of their likelihoods. On the other hand, if quantifying the risk of system failure is of interest, the process can be terminated when the width of the bounds is negligible compared to the overall risk level identified by the bounds. The proposed search procedure is explained by a flow chart in Figure 1.

When a B&B method uses a non-disjoint cut-set formulation, for example, the probability of a sequence $\{1\to2\to\ldots\to i{-}1\to i\}$ would be computed as

$$P_i^{1,\ldots,(i-1)} = P\{(T_1^0 < T_s)\cap\cdots\cap(T_1^0 + T_2^1 + \cdots + T_i^{1,\ldots,(i-1)} < T_s)\} = P\{(T_1^0 + T_2^1 + \cdots + T_i^{1,\ldots,(i-1)} < T_s)\} \tag{9}$$

Compared to the corresponding formulation of the B^3 method in Equation 6, this probability is easier to obtain because one can compute this probability by a component reliability analysis rather than a system reliability analysis. However, the identified failure paths are non-disjoint and correlated, so any update on the lower bound on the system failure probability requires a system reliability analysis for the union of all identified failure paths. By contrast, the B^3 method updates the lower bound just by adding the probability of newly identified system failure sequences to the current bound. Moreover, the B^3 method can provide the upper bound as well, which helps provide reasonable termination criteria for the search process. Even though a B&B method can identify critical failure sequences in the decreasing order of their probabilities (unless heuristics or problem-dependent truncation rules are introduced), the changes in the lower bound caused by newly identified system failure sequences are not decreasing monotonically due to the statistical dependence between the failure sequences. However, in the B^3 method, the increments of the lower bound are the same as the probabilities of the newly identified system failure cases, which are found in the decreasing order. Therefore, the increments on the lower bound diminish monotonically, which helps terminate the search process without performing more searches than necessary.

To achieve the aforementioned merits of the disjoint-cut-set formulation, the B^3 method needs to perform more component and system reliability analyses than conventional B&B based methods. Considering the efficiency in the search process and accuracy in system failure probability calculations, this additional task is worthwhile especially when the computational cost of the structural or finite element analysis is dominant. The component reliability analysis method used in the B^3 method should be able to identify the statistical dependence between the component events such that a system reliability analysis can be later performed with the dependence fully

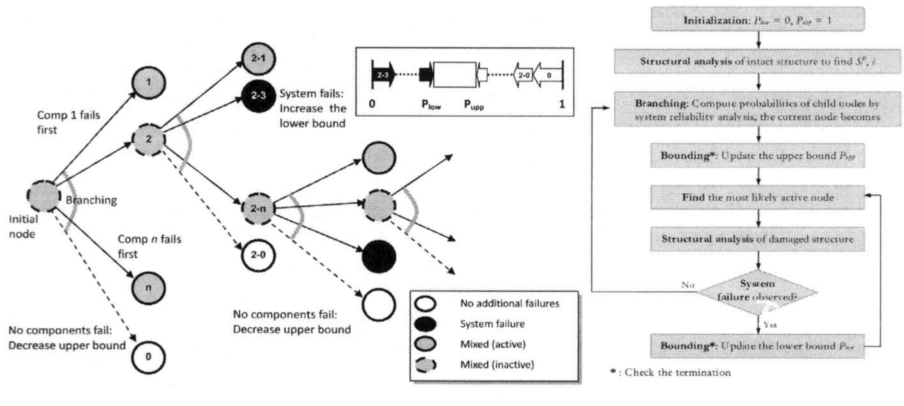

Figure 1. Search procedure (left); and flowchart of B^3 method (right).

considered. The system reliability analysis method should be able to compute the probability of parallel systems accurately. In particular, the method should be able to handle parallel systems with a large number of component events because the number of component events quickly increases as the search process proceeds. In this paper, we use the First-Order Reliability Method (FORM) and the Second-order Reliability Method (SORM). A comprehensive review on these methods can be found in Der Kiureghian (2005). For the system reliability analysis of parallel system events based on the first-order concept of system reliability (Hohenbichler & Rackwitz 1983), we use a multivariate normal integral method by Genz (1992) that can provide accurate estimation on the probabilities of large-size parallel systems in an efficient manner.

3 ILLUSTRATIVE EXAMPLE: GENERALIZED DANIELS SYSTEM

The B^3 method is illustrated by an example of a generalized "Daniels system" (Daniels 1945, Song & Der Kiureghian 2003, Straub & Der Kiureghian 2007). The system fails only when all the members/wires in a layer fail. The system failure event is described as a complex system event because of the load re-distributions caused by member failures. As shown in Figure 2 (left), the structure consists of six bars that are assumed to be perfectly brittle and to have identical, deterministic elastic moduli. The cross sectional areas of the bars are assumed to be $A_1 = 0.03\ \text{m}^2$, $A_2 = A_3 = 0.015\ \text{m}^2$, and $A_4 = A_5 = A_6 = 0.01\ \text{m}^2$. In this example, the uncertainties of C in Paris equation, initial crack length a_i^0, and external load are considered as random variables with the mean values 1.36×10^{-13} (mm/cycle/(MPa·mm)^m), 0.11 (mm) and 1,200 (kN), respectively. For the sake of simplicity, all of the random variables are assumed to follow lognormal distributions with the coefficient of variation (c.o.v.) of 0.1 and to be statistically independent of each other. The following deterministic parameters are used: the loading frequency (ν_0): 500,000 (cycle/year), the inspection cycle (T_s): 4 years, $m = 3$, $Y(a) = 3$, and the critical crack lengths: $a_{c1} = 30$ mm, $a_{c2} = a_{c3} = 15$ mm, and $a_{c4} = a_{c5} = a_{c6} = 10$ mm. Assuming the load is equally distributed by remaining bars on each story, the stress in each bar is calculated by dividing the external load 1,200 (kN) by the sum of the cross sectional areas of the remaining bars.

Figure 2 (right) shows the outcomes of the first branching, i.e., the probabilities of the seven child nodes. Due to the given conditions and assumptions, the node numbers corresponding to the bars in the same layer show the same probabilities. The probability of the last white node is computed by the probability of the parent node (i.e., 1.0) minus the sum of the probabilities of the other six child nodes. The upper bound is decreased by this probability, i.e., 0.1605. It is noted that only one structural analysis is required per branching regardless of the number of the child nodes. For the second branching, the active mixed nodes with the highest probabilities are identified. In this example, the nodes '4', '5', and '6' have the highest probability. Any of them can be selected for the next branching and this does not affect the final result. Suppose the node '6' is selected. The

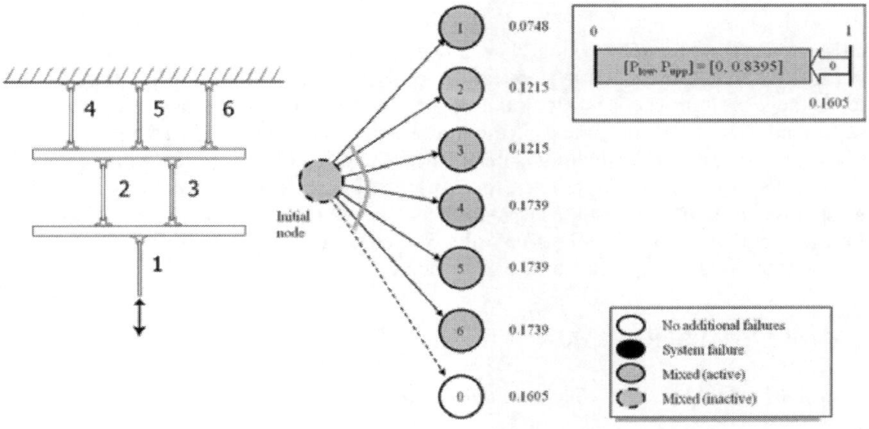

Figure 2. A generalized Daniels System (left); and the result of the first branching (right).

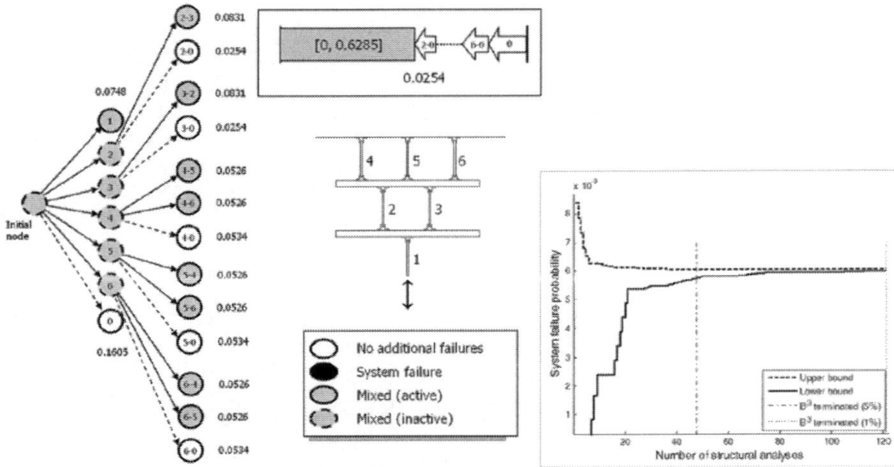

Figure 3. Event tree after the sixth branching (left); and the updated bounds during B^3 analysis (right).

corresponding structural member (i.e., the bar '6') is removed from the system. Since the structure still survives, a structural analysis is performed to find the stresses after the load re-distribution caused by the failure of the bar. The second branching is performed based on the results of the structural analysis. Figure 3 shows some of the active mixed nodes with relatively higher probability after four more branchings. The nodes '2–3' and '3–2' have the highest probabilities among all active mixed nodes, so the next branching starts from one of these nodes. Suppose the node '2–3' is selected for the next branching. However, this case causes system failure because all the bars in the second layer are disconnected. Therefore, the node '2–3' is identified as a system failure node and the lower bound is increased by its probability.

It is noteworthy that each of structural analyses leads to either another branching or a system failure case identification, which decreases the upper bound or increases the lower bound, respectively. In other words, every structural analysis performed during the B^3-based search contributes to narrowing the bounds of the system failure probability. Figure 3 also shows the updates of the bounds with the number of "structural analyses" during the search. When the width of the bounds reaches 1% of the upper bound, the lower and upper bounds of the system failure probability are estimated as 5.995×10^{-3} and 6.058×10^{-3} with 121 structural analyses. To verify this result, crude Monte Carlo simulations are performed. For each set of generated random variables, the system failure/non-failure is checked by performing structural analyses while following an event tree of load re-distributions. After 10^7 simulations, the system failure probability is estimated as 6.050×10^{-3} with a coefficient of variation (c.o.v.) of 4.053×10^{-3}, which belongs to the interval by the B^3 analysis.

It is also noted that the upper bound curve becomes almost flat relatively early in the search, at a level fairly close to the exact system failure probability by Monte Carlo simulations. This is because, as the search proceeds, structural analyses are performed for systems with more damage and less redundancy. Therefore, most of the updates are made on the lower bound. Even if non-failure is observed and thus branching is performed, the probabilities of the white nodes decrease as the system loses its redundancy. Therefore, in most cases, it is not necessary to wait until the bounds converge to each other too closely. Another important merit of the B^3 analysis is to identifiy critical failure sequences in the decreasing order of their probabilities, which allows us to terminate the search without missing critical failure sequences.

4 ONGOING/FUTURE RESEARCH

4.1 *Numerical example: Three-dimensional tripod jacket structure*

In order to test the applicability of the proposed method to large-size complex structural systems, we considered a three-dimensional tripod jacket structure consisting of 66 members shown in

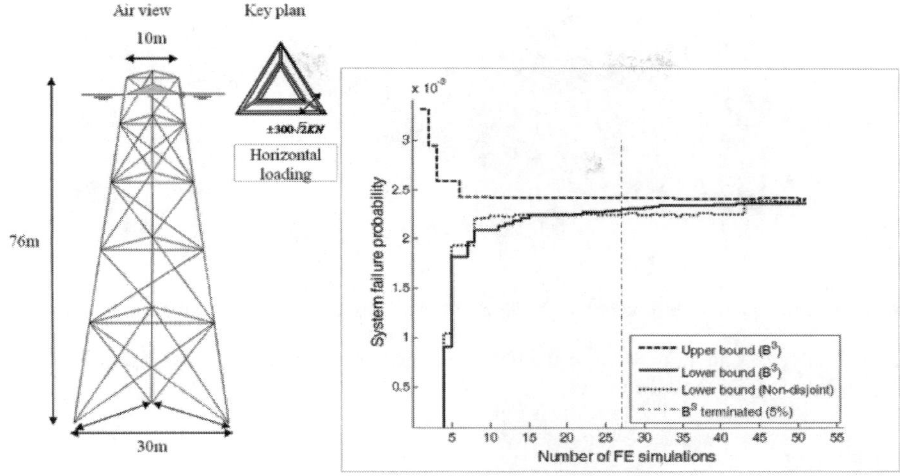

Figure 4. Three-dimensional tripod jacket structure (left); and the updates of the bounds (right).

Figure 4. The details of this example are described in Lee & Song (2010). The figure also shows the updates of the bounds by the B^3 method and those of the lower bound by the conventional B&B based approach (in Equation 9) with the number of FE simulations. The lower and upper bounds by the B^3 method at 5% gap termination point are 2.302×10^{-3} and 2.416×10^{-3} respectively with only 27 FE analyses. The system failure probability estimated by Monte Carlo simulation with one million samples is 2.392×10^{-3}, with a c.o.v. of 2.04×10^{-2}, which belongs to the interval by the B^3 method and is fairly close to the upper bound.

At the same termination point, the lower bound by the conventional B&B method is 2.270×10^{-3}, which is a seemingly good result. However, the approach has a couple of drawbacks in terms of search termination. First, it is not guaranteed that the updates in the lower bound diminish monotonically even though the system failure modes are identified in the decreasing order of their likelihoods because of the statistical dependence between the failure modes on the system failure probability. Second, the lower bound of the conventional approach requires more structural analyses for the convergence than the bounds by the B^3 method. By contrast, the lower bound by the B^3 method converges to the upper bound monotonically.

4.2 *Future work: application of B^3 method to aircraft longeron system*

In order to demonstrate the applicability to advanced FE simulations, the method is currently being applied to the risk analysis of an aircraft "longeron" that is longitudinally attached to the ribs of the fuselage to resist vertical forces. Longeron systems have the following characteristics. First, despite their structural complexity, it is easy to identify "hot spots" with relatively high stresses, which are generally located around fasteners. Second, local failure of a fastener can be described as the event that its crack length exceeds a critical level. Lastly, it is relatively simple to reflect the damage or failure of the identified component in the FE model by eliminating the failed fastener. The longeron example would further demonstrate the efficiency of the proposed method in risk analysis employing time-consuming computational simulations by use of a sophisticated FE model.

Figure 5 shows an initial FE model of longeron system, which has been recently developed by use of ABAQUS®. Assuming the longeron system is located in an upper fuselage under bending caused by vertical acceleration, the system is subjected to pure cyclic tension/compression loads. Due to the cyclic loads, fatigue cracks grow from any fastener holes in the splice system to the edges of the parts, and the FE model allows us to obtain the stress response changed due to the fatigue failures. Figure 5 (right) also displays an example of stress distribution computed by deterministic FE analysis with the initial FE model. It shows the stress distributions of the center section before (upper figure) and after the occurrence of a local cracking failure (bottom figure). This deterministic FE analysis will provide essential information to perform component and system

Figure 5. Initial longeron FE model (left); and the stress distribution before and after a crack failure (right).

reliability analyses. The model will be further modified to obtain a realistic example of risk analysis of fatigue-induced sequential failures.

5 CONCLUSIONS

This paper presents a new Branch-and-Bound (B&B) method employing system reliability Bounds (termed as B^3 method) that was developed in order to estimate the probability of system failure caused by fatigue-induced sequential failures by identifying critical system failure sequences efficiently and accurately. The disjoint-cut set formulation and the proposed search procedure allow us to identify critical sequences of fatigue-induced failures causing a system failure in the decreasing order of their likelihoods. Unlike other existing B&B based approaches, the proposed method provides both lower and upper bounds on the system failure probability while the size of the updates on the lower bound diminishes monotonically. The updated bounds provide reasonable criteria for terminating a B&B search without missing critical sequences. The method was demonstrated by numerical examples, and the merits of the proposed approach were demonstrated through comparison with the results by a conventional B&B based approach and Monte Carlo simulations. The method is currently being applied to an aircraft longeron system to demonstrate the applicability of the method to advanced FE simulations of structural systems.

REFERENCES

Daniels, H.E. 1945. The statistical theory of the strength of bundles of threads. *Proc. of the Royal Society of London. Series A, Mathematical and Physical Sciences*, 183(995): 405–435.

Der Kiureghian, A. 2005. First- and Second-order Reliability Methods, *Engineering Design Reliability Handbook*, edited by E. Nikolaidis, D.M. Ghiocel, and S. Singhal, CRC Press, Boca Raton, FL., Chap. 14.

Genz, A. 1992. Numerical computation of multivariate normal probabilities. *Journal of Computational and Graphical Stat.*: 141–149.

Hohenbichler, H., & Rackwitz, R. 1983. First-order concepts in system reliability. *Structural Safety*, 1: 177–188.

Karamchandani, A., Dalane, J.I., & Bjerager, P. 1992. Systems reliability approach to fatigue of structures. *Journal of Structural Engineering*, 118(3): 684–700.

Lee, Y.-J. & Song, J. 2010. Identification of critical sequences of fatigue-induced failures by branch-and-bound method employing system reliability bounds. *AIAA Conference*, April 12–15, Orlando, FL.

Murotsu, Y. 1984. Automatic generation of stochastically dominant modes of structural failure in frame. *Structural Safety*, 2(1): 17–25.

Newman, J.C. & Raju, I.S. 1981. An empirical stress intensity factor equation for the surface crack. *Engng. Fract. Mech.*, 15: 185–192.

Paris, P.C. & Erdogan, F. 1963. A critical analysis of crack propagation laws. *J. Basic Eng.*, Trans. ASME., 85: 528–534.

Song, J. & Der Kiureghian, A. 2003. Bounds on system reliability by linear programming. *Journal of Engineering Mechanics*, 129(6): 627–636.

Straub, D., & Der Kiureghian, A. 2007. Risk acceptance in deteriorating structural systems. *Proc. JCSS Workshop on Risk Acceptance and Risk Communication*, Stanford, CA.

Reliability and Optimization of Structural Systems – Straub (ed)
© 2010 Taylor & Francis Group, London, ISBN 978-0-415-88179-1

Probabilistic damage assessment by means of Bayesian FE model updating

Geert Lombaert
Structural Mechanics Division, K.U. Leuven, Leuven, Belgium

Babak Moaveni
Department of Civil & Environmental Engineering, Tufts University, Medford, MA, USA

Joel P. Conte
Department of Structural Engineering, University of California at San Diego, La Jolla, CA, USA

ABSTRACT: Localizing and quantifying potential damage in large and complex structures is one of the most challenging problems in structural health monitoring. Vibration-based Finite Element (FE) model updating has proven to be a powerful methodology to identify (i.e., detect, localize and quantify) structural damage. However, due to the presence of uncertainties on the identified modal parameters used in model updating, probabilistic damage identification methods are preferable. In this paper, Bayesian FE model updating is applied for the damage identification of a full-scale seven-story reinforced concrete building slice tested on the UCSD-NEES shake table. The modal parameters identified based on ambient vibration data are used to identify damage as a decrease of the effective stiffness in a number of predefined substructures of a FE model of the structure.

1 INTRODUCTION

In recent years, structural health monitoring has received increasing attention in the civil engineering research community with the objective to develop methods through which structural damage can be identified at the earliest possible stage and the remaining useful life of structures evaluated (damage prognosis). Vibration-based, non-destructive damage identification makes use of changes in dynamic characteristics (e.g., modal parameters) to identify structural damage. Extensive literature reviews on vibration-based damage identification were provided by Doebling et al. (Doebling, Farrar, Prime, and Shevitz 1996) and Sohn et al. (Sohn, Farrar, Hamez, Shunk, Stinemates, and Nadler 2003). Damage identification consists of: (1) detecting the occurrence of damage, (2) localizing the damage zones, and (3) estimating the extent of damage in the various damage zones. Numerous vibration-based methods have been proposed to achieve these goals.

Recently, unique tests on a full-scale seven-story reinforced concrete test structure have been performed on the UCSD-NEES shake table at the Englekirk Structural Engineering Center of the University of California, San Diego (UCSD) (Panagiotou, Restrepo, and Conte 2007). The objective of this test program was to verify the seismic performance of mid-rise reinforced concrete shear wall buildings designed for lateral forces obtained from a displacement-based design methodology, which are significantly smaller than those dictated by current force-based seismic design provisions in the United States (Panagiotou, Restrepo, and Conte 2007). The shake table tests were designed so as to damage the building progressively through several historical seismic motions reproduced on the shake table. At various levels of damage, several low amplitude white noise base excitations were applied through the shake table to the building which responded as a quasi-linear system with dynamic parameters depending on the level of structural damage. In addition to white noise base excitation tests, ambient vibration tests were also performed on the building specimen at different damage levels. Different state-of-the-art system identification methods were applied to acceleration response measurements in order to estimate modal parameters (natural frequencies, damping ratios and mode shapes) of the building in its undamaged (baseline) and various damage states (Moaveni 2007).

In this study, a Bayesian model updating strategy is applied to identify damage from the modal characteristics as determined from the ambient vibration tests at various damage states. Damage is identified as a decrease in the effective stiffness in the finite element (FE) model of the structure. In Bayesian inference, probability theory is used to represent epistemic uncertainty. Probability density functions represent the plausibility or the degree of belief that is attributed to uncertain parameters. The prior probability distribution represents our knowledge prior to the observations and is transformed into the posterior probability distribution that accounts for both the uncertain prior information and the uncertain experimental data. In this way, all conceivable solutions are considered, and characterized with probability density functions that reflect how likely the solutions are in view of the uncertain prior information and the uncertain experimental data. Beck and Katafygiotis (Beck and Katafygiotis 1998) were among the first to propose a Bayesian framework for model updating using vibration data in the time domain. A similar methodology based on modal data has been proposed by Vanik et al. (Vanik, Beck, and Au 2000) and Papadimitriou (Papadimitriou 2004).

In this study, a Monte Carlo simulation of the posterior probabilty density is used to generate an ensemble of models that are consistent with the prior information and the observed data. Mosegaard and Tarantola (Mosegaard and Tarantola 1995) apply such a strategy to an inverse problem of gravimetry. Beck and Au (Beck and Au 2002) use a similar strategy for finite element model updating in structural dynamics. Schevenels et al. (Schevenels, Lombaert, Degrande, and François 2008) have used this methodology for a probabilistic assessment of the resolution of the Spectral Analysis of Surface Waves (SASW) test. The outline of the paper is as follows. First, the test structure is presented, with an overview of the dynamic tests that have been performed. The modal characteristics of the first three longitudinal modes of the structure are given for four damage states of the structure. Next, the finite element model of the test structure is presented. The updating parameters in the Bayesian inference scheme consist of the stiffness in a number of substructures. Damage is therefore identified and localised as a decrease of the effective stiffness in these substructures. Finally, a choice is made for the prior probability density distribution of the updating parameters and the posterior probability density function is sampled using Markov Chain Monte Carlo techniques.

2 THE SEVEN-STORY TEST STRUCTURE

2.1 *The test structure and the dynamic tests*

The test structure (figure 1a) represents a slice of a full-scale reinforced concrete shear wall building consisting of a main shear wall (web wall), a back wall perpendicular to the main wall (flange wall) for transversal stability, a concrete slab at each floor level, an auxiliary post-tensioned column to provide torsional stability, and four gravity columns to transfer the weight of the slabs to the shake table. Pinned-pinned slotted slab connections capable of transferring in-plane diaphragm forces are placed between the web and flange walls at floor levels in order to minimize the moment transfer and coupling between the two walls. Figures 1a and 1b show the test structure mounted on the shake table and an elevation view with its general dimensions, respectively. Construction drawings and details about the material test data are available in Panagiotou et al. (Panagiotou, Restrepo, and Conte 2007).

A sequence of dynamic tests was applied to the test structure during the period October 2005 – January 2006 including ambient vibration, free vibration, and forced vibration tests (white noise and seismic base excitations) using the UCSD-NEES shake table. The seven-story test structure was damaged progressively through a sequence of four historical ground motion records used as table input motions and the modal parameters of the test structure were identified at different damage states (Moaveni 2007). The four earthquake records applied to the test structure consist of (http://peer.berkeley.edu/smcat): (1) longitudinal component of the 1971 San Fernando earthquake ($M = 6.6$) recorded at the Van Nuys station (EQ1), (2) transversal component of the 1971 San Fernando earthquake recorded at the Van Nuys station (EQ2), (3) longitudinal component of the 1994 Northridge earthquake ($M = 6.7$) recorded at the Oxnard Boulevard station in Woodland Hill (EQ3), and (4) 360 degree component of the 1994 Northridge earthquake recorded at the Sylmar station (EQ4). In the following, the damage state S0 is defined as the baseline state of the structure before its exposure to the first seismic excitation (EQ1), while damage states S1, S2, S3 and S4 correspond to the state of the structure after exposure to the first (EQ1), second (EQ2), third (EQ3), and fourth (EQ4) seismic excitations, respectively. Damage state S0 does not correspond to the uncracked state of the test structure, since the latter had already been subjected to low amplitude

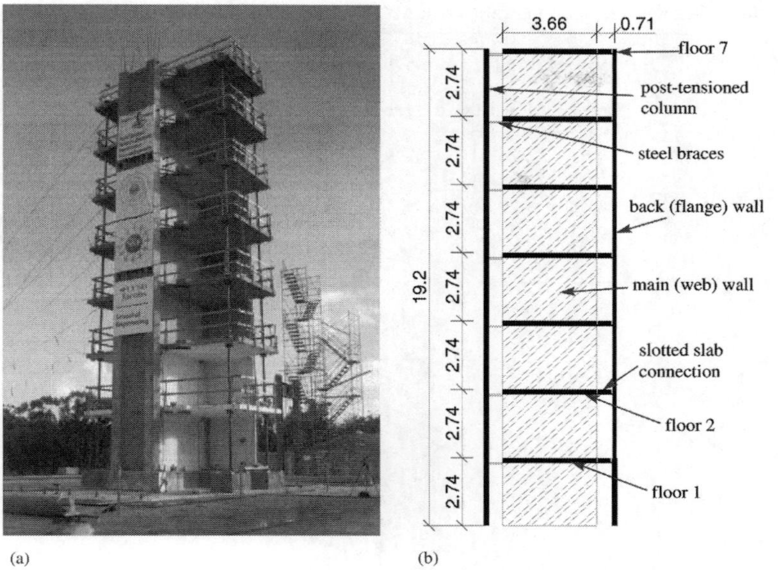

Figure 1. The seven-story test structure: (a) view of the structure mounted on the shake table and (b) elevation view of the test structure (units: m).

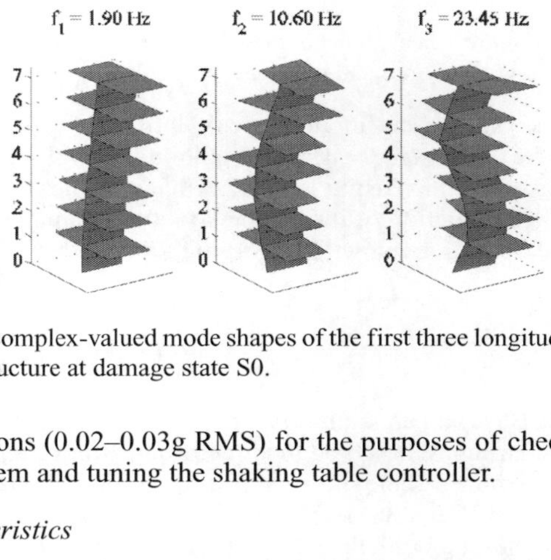

Figure 2. Real part of the complex-valued mode shapes of the first three longitudinal vibration modes (1st-L, 2nd-L, 3rd-L) of the test structure at damage state S0.

white noise base excitations (0.02–0.03g RMS) for the purposes of checking the instrumentation and data acquisition system and tuning the shaking table controller.

2.2 *The modal characteristics*

Modal parameters of the test structure were identified using state-of-the-art system identification methods based on measured data from low amplitude dynamic tests (i.e., ambient vibration tests and white noise base excitation tests) performed at various damage states (Moaveni 2007). In this study, the natural frequencies and mode shapes of the first three longitudinal vibration modes are used for damage identification of the test structure at damage states S0, S2, S3, and S4. These data have been determined from the measured ambient vibration data of 14 longitudinal accelerometers on the web wall (at each floor level and at mid-height of each story) by means of the Data-Driven Stochastic Subspace Identification (SSI-DATA) algorithm (Van Overshee and De Moor 1996). Figure 2 shows the real part of the complex-valued mode shapes of the first three longitudinal vibration modes (1st-L, 2nd-L, 3rd-L) of the test structure based on ambient vibration data from damage state S0. The real part has been determined by projecting all mode shape components onto their major principal axis in the complex plane.

Table 1 reports the natural frequencies, damping ratios, and modal assurance criterion (MAC) values of these three longitudinal vibration modes (1st-L, 2nd-L, 3rd- L) identified from ambient vibration excitation by means of the SSI-DATA algorithm at all damage states considered. The MAC

Table 1. Natural frequencies, damping ratios and MAC values identified from ambient vibration tests.

	Natural Frequency [Hz]			Damping Ratio [%]			MAC		
Damage state	1st-L	2nd-L	3rd-L	1st-L	2nd-L	3rd-L	1st-L	2nd-L	3rd-L
S0	1.91	10.51	24.51	2.3	2.4	0.5	1	1	1
S1	1.88	10.21	24.31	2.9	2.7	0.6	1.00	1.00	1.00
S2	1.67	10.16	22.60	1.3	1.4	0.9	1.00	0.97	0.96
S3	1.44	9.23	21.82	2.7	1.3	1.4	1.00	0.96	0.92
S4	1.02	5.67	15.09	1.0	1.7	1.0	0.99	0.87	0.81

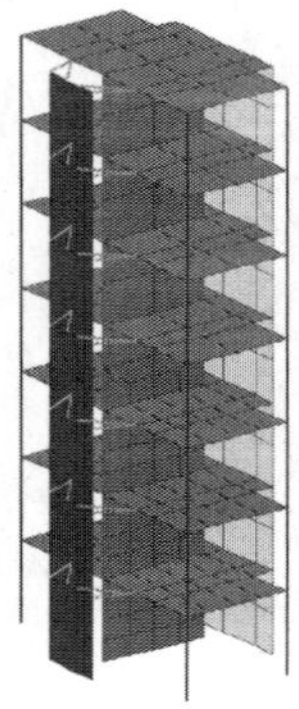

Figure 3. The FEDEASLab finite element model of the seven-story test structure: web wall (red), floor slabs (green), back wall (yellow), gravity columns and post-tensioned column (blue), and braces (cyan).

values are calculated between normalized mode shapes identified at various damage states with their counterparts identified at the undamaged state of the building. It is observed that: (1) the identified natural frequencies decrease with increasing level of damage, while the identified damping ratios do not show a clear trend as a function of increasing structural damage. (2) The calculated MAC values exhibit a general decreasing trend with increasing level of damage.

3 BAYESIAN FE MODEL UPDATING

3.1 *The FE model*

In this study, a probabilistic Bayesian updating strategy is used to identify (detect, localize and quantify) the damage in the test structure based on the identified natural frequencies and mode shapes for the first thee longitudinal modes of the shear wall building slice. Damage in the structure is identified as a change in material stiffness (effective modulus of elasticity) of the finite elements in the different substructures of the FE model. The FE model of the structure is a three-dimensional linear elastic model developed using the general-purpose FE structural analysis program FEDEASLab (Filippou and Constantinides 2004). The web wall at each of the first three stories is modeled using 16 shell elements, while the higher stories (5 to 7) are modeled using 4 shell elements each. The 4th story of the web wall is modeled using 8 shell elements. The floor slabs are modeled using 24 shell elements for each of the first three floors and 12 shell elements for each of the higher floors (4 to 7). The flange wall and the post-tensioned column are modeled using 8 and 10 shell elements per story, respectively. The gravity columns and braces connecting the post-tensioned column to the building slabs are modeled using truss elements. The inertia properties of the test structure are discretized into lumped translational masses at each node of the FE model. The resulting model consists of 322 linear elastic shell and truss elements, 340 nodes and 2538 degrees of freedom. An elaborate description of the model can be found in Moaveni (Moaveni 2007).

Initially, the material properties of the structure were based on concrete cylinder tests at various heights along the structure. These values have been used as the starting values for a deterministic updating of the finite element model by Moaveni (Moaveni 2007). This updated model is taken as the reference model in the following. In order to identify damage, the structure is subdivided into a number of substructures. As it can be anticipated that most of the damage will occur in the web wall, the substructures consist of every half of the part of the web wall for each of the first three stories, and the entire part of the web wall for each of the four remaining stories. This amounts to a total of 10 substructures that are considered in the damage identification. The Young's moduli of the 10 substructures in the reference model are scaled by the parameters θ_{Mk} that are collected in the vector $\boldsymbol{\theta}_M$. This defines a model class $\mathcal{M}_M$ of mechanical models that maps the parameter space $\Theta_M \subseteq \mathbb{R}^n$ to a set of models $\mathcal{M}_M^*$.

3.2 *Bayesian inference*

According to Bayes theorem, the posterior probability $p(\boldsymbol{\theta}|\mathbf{d})$ that accounts for the experimental data $\mathbf{d}$ is obtained from the prior distribution $p(\boldsymbol{\theta})$ as follows:

$$p(\boldsymbol{\theta}|\mathbf{d}) = cp(\mathbf{d}|\boldsymbol{\theta})p(\boldsymbol{\theta}) \quad (1)$$

where $p(\mathbf{d}|\boldsymbol{\theta})$ is the probability of observing the data $\mathbf{d}$ given the parameters $\boldsymbol{\theta}$ and the normalization constant c ensures the posterior PDF integrates to one. Once the observations $\mathbf{d}$ are made, and the corresponding numerical values $\tilde{\mathbf{d}}$ are inserted in the probability distribution $p(\tilde{\mathbf{d}}|\boldsymbol{\theta})$, this function no longer explicitly depends on the variable $\mathbf{d}$ and is defined as the likelihood function $L(\boldsymbol{\theta}|\tilde{\mathbf{d}})$.

The Beta distribution $p_{\text{Beta}}(x; \alpha, \beta)$, parametrized by α and β, and defined for $x \in [0, 1]$ is used to define the joint prior density $p(\boldsymbol{\theta}_M)$ of the scaling factors of the stiffness of the 10 substructures of the web wall. The Beta distribution is scaled to the range $[0, 2]$ for $\theta_{M,m}$. This implies that an upper limit of two is imposed for the scaling factor applied to the effective stiffness in the reference model of the structure. The parameters α_m and β_m are chosen as $\alpha_m = 6$ and $\beta_m = 6$ leading to a mean value of 1 and a coefficient of variation of 28%. The same prior joint PDF is used for every damage state.

The likelihood function $L(\boldsymbol{\theta}|\tilde{\mathbf{d}})$ is calculated from the *observed prediction error* equations for the natural frequency and mode shape of each identified mode m (Vanik, Beck, and Au 2000; Papadimitriou 2004):

$$\tilde{\omega}_m = \omega_m(\boldsymbol{\theta}_M) + e_{\omega m} \quad (2)$$

$$\gamma_m \tilde{\boldsymbol{\phi}}_m = \mathbf{L}\boldsymbol{\phi}_m(\boldsymbol{\theta}_M) + \mathbf{e}_{\phi m} \quad (3)$$

where $e_{\omega m}$ and $\mathbf{e}_{\phi m}$ are the observed prediction errors on the natural frequency and the mode shape of mode m, respectively. The matrix $\mathbf{L}$ defines a mapping from the N_{DOF} degrees of freedom (DOF) of the finite element model to the N_o observed modal displacements. The scalar γ_m is a scaling factor which is obtained through a least-squares fit (Papadimitriou 2004) of the experimental and computed mode shapes.

The *observed* prediction errors $e_{\omega m}$ and $\mathbf{e}_{\phi m}$ in equations (2) and (3) are due to the measurement error and the prediction error. The errors $e_{\omega m}$ and $\mathbf{e}_{\phi m}$ are assumed to be mutually independent and modelled as Gaussian distributed random variables:

$$p_{e_{\omega m}}(e_{\omega m}|\sigma_{\omega m}) = \left(2\pi\sigma_{\omega m}^2\right)^{-1/2} \exp\left(-\frac{1}{2}\frac{e_{\omega m}^2}{\sigma_{\omega m}^2}\right) \quad (4)$$

$$p_{\mathbf{e}_{\phi m}}(\mathbf{e}_{\phi m}|\sigma_{\phi m}) = \left((2\pi)^{N_o}\sigma_{\phi m}^{2N_o}\right)^{-1/2} \exp\left(-\frac{1}{2}\frac{\|\mathbf{e}_{\phi m}\|^2}{\sigma_{\phi m}^2}\right) \quad (5)$$

where $\sigma_{\omega m}$ is the standard deviation of the error in natural frequency, $\sigma_{\phi m}$ is the standard deviation of the error in the modal displacements and N_o is the number of observed modal displacements. It is now assumed that the ratio between the standard deviation $\sigma_{\omega m}$ of the error in natural frequency and the measured natural frequency $\tilde{\omega}_m$ has the same value σ_ω for all modes, so that $\sigma_{\omega m} = \sigma_\omega \tilde{\omega}_m$. The standard deviation $\sigma_{\phi m}$ of the error in the modal displacements is assumed to be proportional to the root mean square value $\|\gamma_m \tilde{\phi}_m\|/\sqrt{N_o}$ for each mode, so that $\sigma_{\phi m} = \sigma_\phi \|\gamma_m \tilde{\phi}_m\|/\sqrt{N_o}$. The unknown dimensionless standard deviations σ_ω and σ_ϕ are collected in the vector $\boldsymbol{\theta}_{\mathbf{e}} = \{\sigma_\omega, \sigma_\phi\}^{\text{T}}$, that is considered in the Bayesian inference as well, so that the vector $\boldsymbol{\theta} = \{\boldsymbol{\theta}_M, \boldsymbol{\theta}_{\mathbf{e}}\}^{\text{T}}$. The prior PDF $p(\boldsymbol{\theta}_{\mathbf{e}})$ of the parameters $\sigma_\omega, \sigma_\phi$ is chosen inversely proportional to each parameter, corresponding to mutually independent "noninformative" prior distributions as proposed by Jeffreys (Jaynes 2003).

The likelihood function $L(\boldsymbol{\theta}_M, \boldsymbol{\theta}_{\mathbf{e}}|\tilde{\mathbf{d}})$ is computed from equations (4) and (5) as:

$$L(\boldsymbol{\theta}_M, \boldsymbol{\theta}_{\mathbf{e}}|\tilde{\mathbf{d}}) \equiv p(\tilde{\mathbf{d}}|\boldsymbol{\theta}) = \Pi_m^N p_{e_{\omega m}}(e_{\omega m}|\sigma_{\omega m}) p_{\mathbf{e}_{\phi m}}(\mathbf{e}_{\phi m}|\sigma_{\phi m}) \quad (6)$$

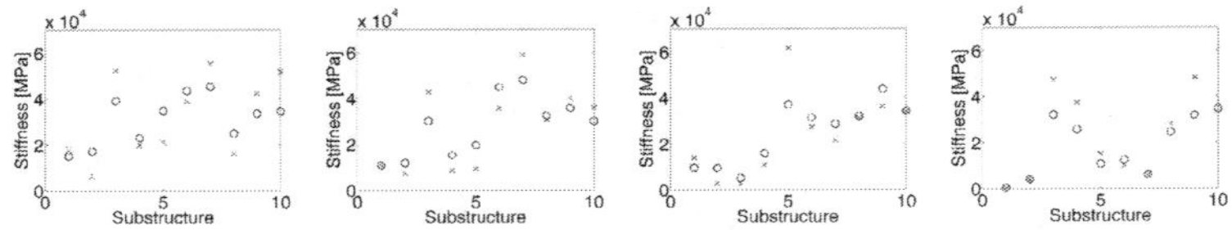

(a) S0 (b) S2 (c) S3 (d) S4

Figure 4. Maximum a posteriori (circles) and maximum likelihood estimate (crosses) of the effective Young's modulus in the damage states (a) S0, (b) S2, (c) S3, and (d) S4.

4 POSTERIOR PROBABILITY DISTRIBUTION

The posterior joint probability density function $p(\boldsymbol{\theta}_M, \boldsymbol{\theta}_\mathbf{e} | \tilde{\mathbf{d}})$ has been sampled by means of a Markov Chain Monte Carlo (MCMC) algorithm (Robert and Casella 2004). MCMC algorithms generate a sequence of states $\boldsymbol{\theta} = \{\boldsymbol{\theta}_M, \boldsymbol{\theta}_\mathbf{e}\}^\mathrm{T}$ such that the complete chain converges to a prescribed probability distribution, without requiring a closed-form expression of the latter. The transition from one state to the next state occurs by a random perturbation of the current state, that is determined by the so-called proposal distribution. The candidate state is either accepted or rejected, in which case the next state is the same as the previous one. The Metropolis-Hastings algorithm (Robert and Casella 2004; Hastings 1970) has been used to sample the posterior distribution.

Based on the samples of the MCMC procedure, the maximum a posteriori (MAP) and maximum likelihood estimate (MLE) have been determined for the four damages states considered in the identification. Figure 4 shows both estimates of the scaling factors $\boldsymbol{\theta}_M$, multiplied with the Young's moduli of the substructures in the reference model. The first six values in the figures therefore correspond to the effective Young's modulus of the lower and upper half of the web wall at each of the first three stories, while the four remaining values represent the effective Young's moduli of the web wall at the four top stories. When both estimates are compared for the four damage states, it is observed that the MLE of the stiffness in a substructure is generally situated more towards the lower or higher limit value of the range considered in the identification.

For the first two substructures, that correspond to the lower and upper half of the web wall at the fist story, a clear trend can be observed throughout the damage states. The MAP estimates of the effective Young's moduli gradually decrease from values around 20000 MPa in state S0 to much lower values below 5000 MPa in state S4. The low value in the reference state S0 could be due to the damage caused by the white noise base excitation that has been used to check the equipment prior to the seismic excitation. For the other substructures, no clear trend is observed. Within a single damage state, the values first decrease with increasing height and then decrease. In a similar way, the value for a single substructure can decrease from one damage state to another and increase subsequently. The average value of the stiffness in the substructures 3 to 10, however, seems to decrease as well from state S0 to S4.

The posterior marginal probability density of the scaling factor has been estimated based on a normal kernel function. The estimated marginal PDF of the scaling factors is transformed into a marginal PDF of the effective Young's moduli of the substructures, using the values in the initial FE model. Figure 5 compares the posterior and prior marginal probability density of the effective Young's modulus for the substructures 1 up to 4. This is where most of the damage has been observed during the tests.

For the two first substructures, it is observed that as the damage progresses (from S0 to S4), the posterior marginal PDF shifts progressively to lower values of the stiffness and becomes more narrow (i.e., with a smaller coefficient of variation). Given the fact that the posterior PDF for all damage states is calculated from the same prior PDF, this means that with increasing level of damage the data become more informative about the stiffness at the lower stories, where a very large decrease of the stiffness is observed. For the substructures 3 and 4, that correspond to the lower and upper half of the web wall at the second story, the posterior marginal PDF shifts to lower values from damage states S0 to S3. This trend is not continued to damage state S4, where the posterior marginal PDF is very close to the prior marginal PDF, so that it seems that the data in state S4 do not contain sufficient information to identify the (decreased) stiffness at these locations.

The probabilistic damage identification results are now compared to the actual damage that has been observed at damage state S4. Figures 6a and 6b show the damage at the bottom two stories of

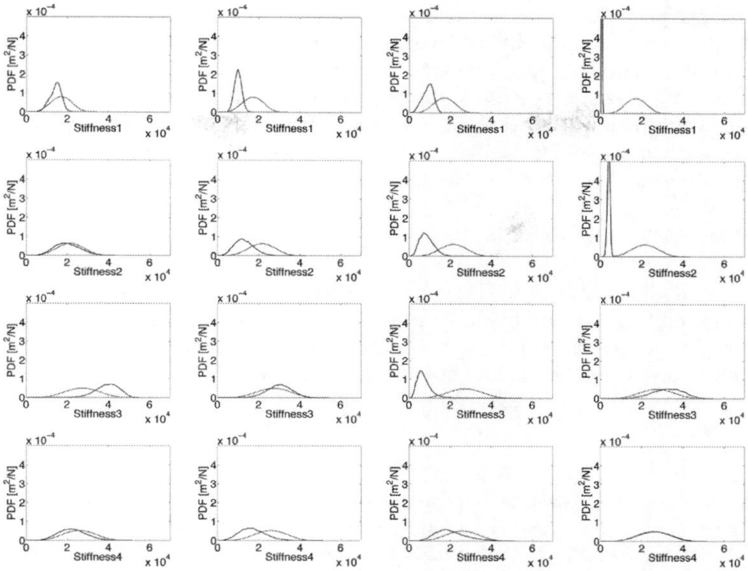

Figure 5. The posterior (blue) and prior (red) marginal probability density of the effective Young's modulus of the substructures 1 (top) up to 4 (bottom) in the damage states (a) S0, (b) S2, (c) S3, and (d) S4.

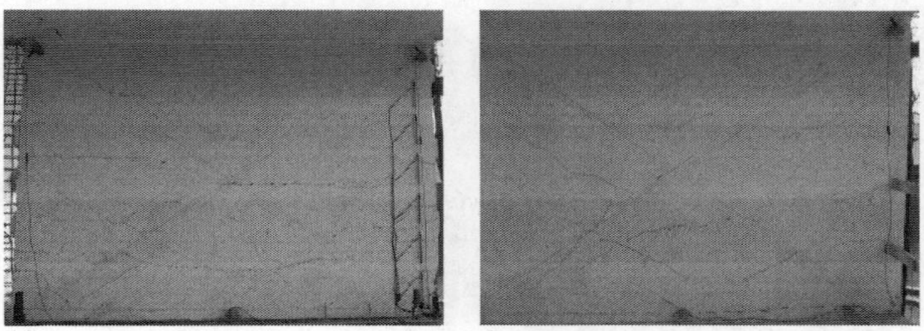

Figure 6. Observed cracks at (a) the first and (b) the second story of the web wall at damage state S4.

the web wall at state S4. From these figures, it can be observed that horizontal flexural cracks as well as inclined/diagonal cracks have developed. Furthermore, during the seismic test EQ4, a lap-splice failure (i.e., debonding between longitudinal steel reinforcement bars and the surrounding concrete) occurred in the web wall at the bottom of the second story. These results confirm the dramatic decrease of the wall stiffness in the lower and upper half of the first story (substructures 1 and 2), as identified from the Bayesian updating procedure at damage state S4. The cracks observed in the web wall at the second story in Figure 6 also explain the shift of the marginal PDF towards lower values of the stiffness in substructures 3 and 4 at damage state S3.

5 CONCLUSION

In this paper, Bayesian model updating is used to identify damage in a seven-story reinforced concrete test structure that has been tested on the UCSD-NEES shake table. This test structure has been submitted to a sequence of historical earthquake records that were designed so as to damage the building progressively. After each seismic excitation, vibration data from both white noise and ambient excitation were used to determine the modal characteristics of the structure. The modal characteristics of the first three longitudinal modes of the building are used for a probabilistic assessment of the damage by means of Bayesian FE model updating. The updating parameters are the scaling factors of the effective stiffness of 10 substructures in the FE model. The Bayesian inference scheme is used to compute the posterior joint probability density of the stiffness in the substructures. The maximum a posteriori (MAP) estimate of the stiffness has a low value for the substructures where the actual damage was observed in the test structure. The variance of the identified values tends to be lower for those substructures where the most severe decrease in stiffness occurs. A good agreement is obtained between the results of the Bayesian model updating and the actual damage observed in the test structure.

ACKNOWLEDGEMENTS

Partial supports of this research by Lawrence Livermore National Laboratory with Dr. David McCallen as Program Leader and by the Englekirk Center Industry Advisory Board are gratefully acknowledged. Any opinions, findings, and conclusions or recommendations expressed in this material are those of the authors and do not necessarily reflect those of the sponsors.

REFERENCES

Beck, J. and S.-K. Au (2002). Bayesian updating of structural models and reliability using Markov Chain Monte Carlo simulation. *ASCE Journal of Engineering Mechanics 128*(4), 380–391.

Beck, J. and L. Katafygiotis (1998). Updating models and their uncertainties. I: Bayesian statistical framework. *ASCE Journal of Engineering Mechanics 124*(4), 455–461.

Doebling, S., C. Farrar, M. Prime, and D. Shevitz (1996). Damage identification and health monitoring of structural and mechanical systems from changes in their vibration characteristics: a literature review. Report LA-13070-MS, Los Alamos National Laboratory, Los Alamos, NM.

Filippou, F. and M. Constantinides (2004). Fedeaslab getting started guide and simulation examples. Technical report NEESgrid-2004-22.

Friswell, M., J. Mottershead, and H. Ahmadian (2001). Finite-element model updating using experimental test data: parametrization and regularization. *Philosophical Transactions of the Royal Society 359*, 169–186.

Hastings, W. (1970). Monte Carlo sampling methods using markov chains and their applications. *Biometrika 57*(1), 97–109.

Jaynes, E. (2003). *Probability Theory. The Logic of Science*. Cambridge, UK: Cambridge University Press.

Moaveni, B. (2007). *System and Damage Identification of Civil Structures*. Ph. D. thesis, Department of Structural Engineering, University of California, San Diego.

Mosegaard, K. and A. Tarantola (1995). Monte Carlo sampling of solutions to inverse problems. *Journal of Geophysical Research 100*, 12431–12447.

Panagiotou, M., J. Restrepo, and J. Conte (2007). Shake table test of a 7-story full scale reinforced concrete structural wall building slice phase I: rectangular wall section. Report draft SSRP-07-07, Department of Structural Engineering, University of California, San Diego, La Jolla, CA.

Papadimitriou, C. (2004, July). Bayesian inference applied to structural model updating and damage detection. In *9th ASCE Specialty Conference on Probabilistic Mechanics and Structural Reliability*, Albuquerque, USA.

Robert, C. and G. Casella (2004). *Monte Carlo statistical methods* (2nd ed.). New York: Springer.

Sambridge, M. and K. Mosegaard (2002). Monte Carlo methods in geophysical inverse problems. *Reviews of Geophysics 40*(3), 1–29.

Schevenels, M., G. Lombaert, G. Degrande, and S. François (2008). A probabilistic assessment of resolution in the SASW test and its impact on the prediction of ground vibrations. *Geophysical Journal International 172*(1), 262–275.

Sohn, H., C. Farrar, F. Hemez, D. Shunk, D. Stinemates, and B. Nadler (2003). A review of structural health monitoring literature: 1996–2001. Report LA-13976-MS, Los Alamos National Laboratory, Los Alamos, NM.

Tarantola, A. (2005). *Inverse problem theory and methods for model parameter estimation*. Philadelphia, USA: SIAM.

Van Overschee, P. and B. De Moor (1996). *Subspace identification for linear systems*. Dordrecht, The Netherlands: Kluwer Academic Publishers.

Vanik, M., J. Beck, and S. Au (2000). Bayesian probabilistic approach to structural health monitoring. *ASCE Journal of Engineering Mechanics 126*(7), 738–745.

Reliability and Optimization of Structural Systems – Straub (ed)
© 2010 Taylor & Francis Group, London, ISBN 978-0-415-88179-1

Optimal structural control under stochastic uncertainty

Kurt Marti & Ina Stein
Aerospace Engineering and Technology, Federal Armed Forces University Munich, Neubiberg/Munich, Germany

ABSTRACT: In order to stabilize mechanical structures under dynamic applied loads, active control strategies are taken into account. The structures usually are stationary, safe and stable without external dynamic disturbances, such a strong earthquakes, wind turbulences, water waves, etc. Thus, in case of dynamic disturbances, additional control elements can be installed enabling active control actions. Active control strategies for mechanical structures are applied in order to counteract heavy applied dynamic loads, such as earthquakes, wind, water waves, etc., which would lead to large vibrations causing possible damages of the structure. Modeling the structural dynamics by means of a system of first order random differential equations for the state vector (displacement vector q and time derivative of a), stochastic optimal controls are determined in order to cope with the stochastic uncertainty involved in the dynamic parameters, the initial values and the applied loadings.

1 STOCHASTIC STRUCTURAL CONTROL SYSTEMS

1.1 *Stochastic structural control: Active control*

In active control of dynamic structures, cf. (Block 2008; Nagarajaiah and Narasimhan 2007; Soong, T.T. 1988; Soong, T.T. 1990; Soong, T.T. and Costantinou, M.C. 1994; Spencer and Nagarajaiah 2003; Yang, J.N. and Soong, T.T. 1988), the behavior of the m-vector $z=z(t)$ of displacements with respect to time t is described by a system of second order linear differential equations for $z(t)$ having a right hand side being the sum of the stochastic applied load process and the control force depending on a control n-vector function $u(t)$:

$$M\ddot{q} + D\dot{q} + Kq(t) = f\Big(t,\omega,u(t)\Big), t_0 \le t \le t_f. \tag{1a}$$

Hence, the force vector $f=f(t,\omega,u(t))$ on the right hand side of the dynamic equation (1a) is given by the sum

$$f(t,\omega,u) = f_0(t,\omega) + f_a(t,\omega,u) \tag{1b}$$

of the applied load $f_0=f_0(t,\omega)$ being a vector-valued stochastic process describing e.g. external loads or excitation of the structure caused by earthquakes, wind turbulences, water waves, etc., and the actuator or control force vector $f_a=f_a(t,\omega,u)$ depending on an input or control n-vector function $u=u(t), t_0 \le t \le t_f$. Here, ω denotes the random element, lying in a certain probability space (Ω, A, P), used to represent random variations. Furthermore, M, D, K, resp., denotes the $m\times m$ mass, damping and stiffness matrix. In many cases the actuator or control force f_a is linear, i.e.

$$f_a = \Gamma u \tag{1c}$$

with a certain $m\times n$ matrix Γ.

By introducing appropriate matrices, the linear system of second order differential equations (1a,b) can be represented by a system of first order differential equations as follows:

$$\dot{z} = Az(t,\omega) + Bu + b(t,\omega) \tag{2a}$$

with

$$A := \begin{pmatrix} 0 & I \\ -M^{-1}K & -M^{-1}D \end{pmatrix}, \qquad B := \begin{pmatrix} 0 \\ M^{-1}\Gamma \end{pmatrix}, \qquad b(t,\omega) := \begin{pmatrix} 0 \\ M^{-1}f_a(t,\omega) \end{pmatrix} \tag{2b}$$

Moreover, $z = z(t)$ is the $2m$-state vector defined by

$$z = \begin{pmatrix} q \\ q \end{pmatrix} = \begin{pmatrix} q \\ \dot{q} \end{pmatrix}. \tag{2c}$$

1.2 *Open-loop feedback control*

Assuming that the control force $f_a = \Gamma u$ is generated by means of a *PD*-controller, for the input n-vector function $u = u(t)$, we have

$$u(t) := \varphi\Big(t, q(t), \dot{q}(t)\Big) \tag{3a}$$

with a control law $\varphi = \varphi(t, q, \dot{q})$ which is determined in the following by means of the **open-loop feedback method** being the main tool in **model predictive control**, cf. (Marti 2008a):

In open-loop feedback control, from an arbitrary time point t_b, $t_0 \le t_b \le t_f$, to the final time point t_f, first an optimal open-loop control $u^* = u^*(t)$ is determined on the remaining time interval $[t_b; t_f]$:

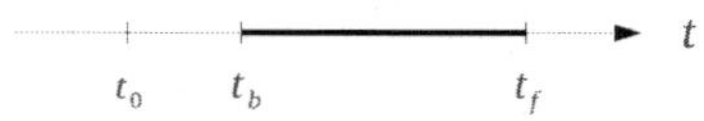

Moreover, it is supposed then that at each time point t_b the state $z_b = z(t_f)$ is available (*PD*-controller). Having then a stochastic optimal open-loop control $u^* = u^*(t), t_b \le t \le t_f$ on each remaining time interval $t_b \le t \le t_f$ with an arbitrary starting time t_b, $t_0 \le t_b \le t_f$, a stochastic optimal open-loop feedback control law is then defined by

$$\varphi(t_b, z(t_b)) := u^*(t_b), t_0 \le t_b \le t_f\,. \tag{3b}$$

2 EXPECTED TOTAL COST FUNCTION

The performance function F for active structural control systems is defined , cf. (Marti 2001; Marti 2004; Marti 2008b), by the conditional expectation of the total costs being the sum of costs L along the trajectory, arising from the displacements $z = z(t, \omega)$ and the control input $u = u(t, \omega)$, and possible terminal costs G arising at the final state z_f. Hence, on the remaining time interval $t_b \le t \le t_f$ we have the following conditional expectation of the total cost function:

$$F := \mathbb{E}\left(\int_{t_b}^{t_f} L\Big(t, \omega, z(t,\omega), u(t,\omega)\Big)dt + G(t_f, \omega, z(t_f, \omega)) \Big| \mathfrak{A}_{t_b} \right), \tag{4a}$$

Supposing quadratic costs along the trajectory, the function L is given by

$$L(t, \omega, z, u) := \frac{1}{2} z^T Q(t, \omega) z + \frac{1}{2} u^T R(t, \omega) u \tag{4b}$$

with positive (semi) definite $2m \times 2m, n \times n$, resp., matrix functions $Q = Q(t, \omega), R = R(t, \omega)$. Furthermore, $G = G(t_f, \omega, z(t_f, \omega))$ describes possible terminal costs. In case of endpoint control G is defined by

$$G(t_f, \omega, z(t_f, \omega)) := \frac{1}{2}(z(t_f, \omega) - z_f(\omega))^T S(z(t_f, \omega) - z_f(\omega)), \tag{4c}$$

where S is a positive definite (semi) weight matrix, and $z_f = z_f(\omega)$ denotes the (possible probabilistic) final state.

Remark 1 *Instead of* $\frac{1}{2}u^T Ru$*, in the following we also use a more general convex control cost function* $C = C(u)$.

3 OPEN-LOOP CONTROL PROBLEM ON THE REMAINING TIME INTERVAL $[t_b, t_f]$

Having the differential equation with random coefficients describing the behavior of the dynamic mechanical structure under uncertainty and the costs arising from displacements and at the terminal state, on a given remaining time interval $[t_b, t_f]$ an optimal open-loop control $u^* = u^*(t), t_b \le t \le t_f$, is a solution of the following stochastic optimal control problem:

$$\min \quad \mathbb{E}\left(\int_{t_b}^{t_f} \frac{1}{2}\left(z(t,\omega)^T Q z(t,\omega) + u(t)^T R u(t)\right)\, dt + G(t_f,\omega,z(t_f,\omega)) \Big| \mathfrak{A}_{t_b} \right) \tag{5a}$$

$$\text{s.t.} \quad \dot{z}(t,\omega) = Az(t,\omega) + Bu(t) + b(t,\omega), \text{ a.s.,} \qquad t \in [t_b; t_f] \tag{5b}$$

$$z(t_b,\omega) = z_b \text{ (given)} \tag{5c}$$

$$u(t) \in D_t, \qquad t \in [t_b; t_f]\,. \tag{5d}$$

An important property of (5a–d) is stated next:

Lemma 1 *If the terminal cost function $G = G(t_f, \omega, z)$ is convex in z, and the feasible domain D_t is convex for each time point t, $t_0 \le t \le t_f$, then the stochastic optimal control problem (5a–d) is a convex optimization problem.*

4 THE STOCHASTIC HAMILTONIAN OF (5a–d)

According to (Marti 2008a), the stochastic Hamiltonian H related to the stochastic optimal control problem (5a-d) reads:

$$H(t,\omega,z,y,u) := L(t,\omega,z,u) + y^T g(t,\omega,z,u). \tag{6a}$$

4.1 *Expected Hamiltonian (with respect to the time interval $[t_b, t_f]$ and information $\mathfrak{A}_{t_b}$)*

For the definition of a $H-$ minimum control the conditional expectation of the stochastic Hamiltonian is needed:

$$\overline{H}^{(b)} := \mathbb{E}\left(H(t,\omega,z,y,u)|\mathfrak{A}_{t_b}\right) = C(u) + q(t)^T u + \ldots \tag{6b}$$

$$q(t) := \mathbb{E}\left(B(\omega)^T y(t,\omega)|\mathfrak{A}_{t_b}\right) = q(t,t_b), \quad t \ge t_b\,. \tag{6c}$$

4.2 *H-minimum control on $[t_b, t_f]$*

In order to formulate the two-point boundary value problem for an optimal open-loop control $u^* = u^*(t), t \in [t_b; t_f]$, we need first a H-minimum control $\widetilde{u^*} = \widetilde{u^*}(t, z(t,\cdot), y(t,\cdot)), t \in [t_b; t_f]$, defined, cf. (Marti 2008a), for $t \in [t_b; t_f]$ as a solution of the following convex stochastic optimization problem , cf. (Marti 2008b):

$$\min \mathbb{E}\left(H(t,\omega,z(t,\omega),y(t,\omega),u)|\mathfrak{A}_{t_b}\right) \tag{7a}$$

s.t.

$$u \in D_t, \tag{7b}$$

where $z = z(t,\omega), y = y(t,\omega)$ are certain trajectories.

According to (7a, b) the H-minimum control

$$\widetilde{u^*} = \widetilde{u^*}\left(t, z(t,\cdot), y(t,\cdot)\right) = \widetilde{u^*}(t,q) \tag{8a}$$

is defined by

$$\widetilde{u^*}(t,q) := \underset{u \in D_t}{\operatorname{argmin}}\, C(u) + q(t)^T u \qquad \text{for } t \ge t_b\,. \tag{8b}$$

For strictly convex, differentiable cost functions $C = C(u)$, as e.g. $C(u) = \frac{1}{2}u^T Ru$ with positive definite matrix R, the necessary and sufficient condition for $\widetilde{u^*}$ reads in case $D_t = \mathbb{R}^n$

$$\nabla C(u) + q(t) = 0. \tag{9a}$$

If $u \mapsto \nabla C(u)$ is a 1–1-operator, then the solution of (9a) reads

$$u = v(q) := \nabla C^{-1}(-q). \tag{9b}$$

With (6c) and (8b) we then have

$$\widetilde{u^*}(t,q) = v(q(t)) = \nabla C^{-1}\left(-\mathbb{E}\left(B(\omega)^T y(t,\omega)|\mathfrak{A}_{t_b}\right)\right) = \widetilde{u^*}(q(t)). \tag{9c}$$

5 CANONICAL (HAMILTONIAN) SYSTEM

In the following we suppose that a H-minimum control $\widetilde{u^*} = \widetilde{u^*}(t, z(t,\cdot), y(t,\cdot))$, $t \in [t_b; t_f]$, i.e., a solution $\widetilde{u^*} = \widetilde{u^*}(t,q) = v(q(t)))$ of the stochastic optimization problem (7a,b) is available. According to (Marti 2008a), a stochastic optimal open-loop control $u^* = u^*(t), t \in [t_b; t_f]$,

$$u^*(t) = \widetilde{u^*}\Big(t, z^*(t,\cdot), y^*(t,\cdot)\Big), \qquad t \in [t_b; t_f] \tag{10}$$

can be obtained by solving the following stochastic two-point boundary value problem related to (5a–d):

Theorem 1 *If $z^* = z^*(t,\omega), y^* = y^*(t,\omega), t_0 \le t \le t_f$, is a solution of*

$$\dot{z}(t,\omega) = Az(t,\omega) + B\nabla C^{-1}\left(-\overline{B(\omega)^T y(t,\omega)}^{(b)}\right) + b(t,\omega), \quad t \in [t_b; t_f] \tag{11a}$$

$$z(t_b,\omega) = z_b \tag{11b}$$

$$\dot{y}(t,\omega) = -A^T y(t,\omega) - Qz(t,\omega) \tag{11c}$$

$$y(t_f,\omega) = \nabla G(t_f,\omega,z(t_f,\omega)) \tag{11d}$$

then the function $u^* = u^*(t), t \in [t_b; t_f]$, defined by (10) is a an optimal open-loop control for the remaining time interval.

6 MINIMUM ENERGY CONTROL

In this case we have $Q = 0$, i.e., there are no costs for the displacements $z = \begin{pmatrix} q \\ \dot{q} \end{pmatrix}$. Then the solution of (11c,d) reads

$$y(t,\omega) = e^{A^T(t_f - t)}\nabla_z G(t_f,\omega,z(t_f,\omega)), \qquad t \in [t_b; t_f]. \tag{12a}$$

This yields for fixed Matrix B

$$\widetilde{u^*}(t,q(t)) = v(q(t)) = \nabla C^{-1}\left(-B^T e^{A^T(t_f - t)}\overline{\nabla_z G(t_f,\omega,z(t_f,\omega))}^{(b)}\right), \qquad t \in [t_b; t_f] \tag{12b}$$

Having (12a, b), for the state trajectory $z = z(t,\omega)$ we get, see (11a, b), the following system of ordinary differential equations

$$\dot{z}(t,\omega) = Az(t,\omega) + B\nabla C^{-1}\left(-B^T e^{A^T(t_f - t)}\overline{\nabla_z G(t_f,\omega,z(t_f,\omega))}^{(b)}\right) + b(t,\omega), \quad t \in [t_b; t_f] \tag{13a}$$

$$z(t_b,\omega) = z_b \tag{13b}$$

The solution of system (13a,b) in $t \in [t_b; t_f]$ reads

$$z(t,\omega) = e^{A(t - t_b)}z_b + \int_{t_b}^{t} e^{A(t-s)}\left(b(s,\omega) + B\nabla C^{-1}\left(-B^T e^{A^T(t_f - s)}\overline{\nabla_z G(t_f,\omega,z(t_f,\omega))}^{(b)}\right)\right)ds. \tag{14}$$

For the final state $z = z(t_f, \omega)$ we get the relation:

$$z(t_f,\omega) = e^{A(t_f-t_b)} z_b + \int_{t_b}^{t_f} e^{A(t_f-s)} \left(b(s,\omega) + B\nabla C^{-1} \left(-B^T e^{A^T(t_f-s)} \overline{\nabla_z G(t_f,\omega,z(t_f,\omega))}^{(b)} \right) \right) ds\,. \quad (15)$$

6.1 *Endpoint control*

In the case of endpoint control, the terminal cost function is given by the following definition (16a), where $z_f = z_f(\omega)$ denotes the desired – possible random – final state:

$$G(t_f,\omega,z(t_f,\omega)) := \frac{1}{2}\|z(t_f,\omega) - z_f(\omega)\|^2\,. \quad (16a)$$

Hence,

$$\overline{\nabla G(t_f,\omega,z(t_f,\omega))}^{(b)} = \overline{z(t_f,\omega)}^{(b)} - \overline{z_f}^{(b)} = \mathbb{E}\left(z(t_f,\omega)\middle|\mathfrak{A}_{t_b}\right) - \mathbb{E}\left(z_f\middle|\mathfrak{A}_{t_b}\right)\,. \quad (16b)$$

Taking expectations $\mathbb{E}(\ldots|\mathfrak{A}_{t_b})$, from (15) we get the following condition for $\overline{z(t_f,\omega)}^{(b)}$:

$$\overline{z(t_f,\omega)}^{(b)} = e^{A(t_f-t_b)} z_b + \int_{t_b}^{t_f} e^{A(t_f-s)} \left(\overline{b(s,\omega)}^{(b)} + B\nabla C^{-1} \left(-B^T e^{A^T(t_f-s)} \left(\overline{z(t_f,\omega)}^{(b)} - \overline{z_f}^{(b)} \right) \right) \right) ds\,. \quad (17)$$

6.1.1 *Quadratic Control Costs*

Here, the control cost function $C = C(u)$ reads

$$C(u) = \frac{1}{2} u^T R u \quad (18a)$$

hence,

$$\nabla C^{-1}(w) = R^{-1} w\,. \quad (18b)$$

Consequently, (17) reads

$$\overline{z(t_f,\omega)}^{(b)} = e^{A(t_f-t_b)} z_b + \int_{t_b}^{t_f} e^{A(t_f-s)} \overline{b(s,\omega)}^{(b)}\, ds - \int_{t_b}^{t_f} e^{A(t_f-s)} B R^{-1} B^T e^{A^T(t_f-s)}\, ds \left(\overline{z(t_f,\omega)}^{(b)} - \overline{z_f(\omega)}^{(b)} \right)\,. \quad (19)$$

Define now

$$U := \int_{t_b}^{t_f} e^{A(t_f-s)} B R^{-1} B^T e^{A^T(t_f-s)}\, ds\,. \quad (20)$$

Lemma 2 *$I + U$ is regular.*

Therefore, from (19) we get

$$\overline{z(t_f,\omega)}^{(b)} = (I+U)^{-1} e^{A(t_f-t_b)} z_b + (I+U)^{-1} \int_{t_b}^{t_b} e^{A(t_f-s)} \overline{b(s,\omega)}^{(b)}\, ds + (I+U)^{-1} U \overline{z_f}^{(b)}\,. \quad (21)$$

Now, (21) and (16b) yield

$$\begin{aligned} \overline{\nabla_z G(t_f,\omega,z(t_f,\omega))} &= \overline{z(t_f,\omega) - z_f}^{(b)} = \overline{z(t_f,\omega)}^{(b)} - \overline{z_f}^{(b)} \\ &= (I+U)^{-1} \left(e^{A(t_f-t_b)} z_b + \int_{t_b}^{t_f} e^{A(t_f-s)} \overline{b(s,\omega)}^{(b)}\, ds \right) + \left((I+U)^{-1} U - I \right) \overline{z_f}^{(b)}\,. \end{aligned} \quad (22)$$

Thus, the stochastic optimal open loop control $u^*(t)$ on $[t_b, t_f]$ is given by, cf. (9b),

$$u^*(t) = -R^{-1}B^Te^{A^T(t_f-t)}\left((I+U)^{-1}\left(e^{A(t_f-t_b)}z_b + \int_{t_b}^{t_f} e^{A(t_f-s)}\overline{b(s,\omega)}^{(b)}\, ds\right) + \left((I+U)^{-1}U - I\right)\overline{z_f}^{(b)}\right) \tag{23}$$

Finally, for the stochastic optimal open loop feedback control law $\varphi = \varphi(t, z(t))$ we have the following result:

Theorem 2 *The stochastic optimal open loop feedback control law $\varphi = \varphi(t, z(t))$ is given by*

$$\begin{aligned}\varphi(t, z(t)) =& \underbrace{-R^{-1}B^Te^{A^T(t_f-t)}(I+U)^{-1}e^{A(t_f-t)}}_{G_0(t)}z\\ & \underbrace{-R^{-1}B^Te^{A^T(t_f-t)}(I+U)^{-1}\int_t^{t_f} e^{A(t_f-s)}\overline{b(s,\omega)}^{(t)}\, ds}_{G_1(t,\overline{b(\cdot,\omega)}^{(t)})}\\ & \underbrace{-R^{-1}B^Te^{A^T(t_f-t)}\left((I+U)^{-1}U - I\right)}_{G_2(t)}\overline{z_f}^{(t)},\end{aligned} \tag{24a}$$

hence,

$$\varphi(t, z) = G_0(t)z + G_1(t, \overline{b(\cdot,\omega)}^{(t)}) + G_2(t)\overline{z_f}^{(t)}. \tag{24b}$$

7 NONZERO WEIGHT MATRIX Q

According to (11a–d), for the adjoint trajectory $y = y(t, \omega)$ we have the system of differential equations

$$\begin{aligned}\dot{y}(t,\omega) =& -A^Ty(t,\omega) - Qz(t,\omega)\\ y(t_f,\omega) =& \nabla G(t_f,\omega,z(t_f,\omega)),\end{aligned}$$

which has the following solution for given $z(t, \omega)$ and $\nabla G(t_f, \omega, z(t_f, \omega))$:

$$y(t,\omega) = \int_t^{t_f} e^{A^T(s-t)}Qz(s,\omega)\, ds + e^{A^T(t_f-t)}\nabla G(t_f,\omega,z(t_f,\omega)). \tag{25}$$

From (25) we then get

$$\overline{y}^{(b)}(t) = \mathbb{E}^{(b)}(y(t,\omega)) = \mathbb{E}\left(y(t,\omega)|\mathfrak{A}_{t_b}\right) = \int_t^{t_f} e^{A^T(s-t)}Q\overline{z}^{(b)}(s)\, ds + e^{A^T(t_f-t)}\overline{\nabla G(t_f,\omega,z(t_f,\omega))}^{(b)} \tag{26}$$

The unknown function $\overline{z(t)}^{(b)}$, and the vector $z(t_f, \omega)$ in this equation are both given, based on $\overline{y(t)}^{(b)}$, by the initial value problem, see (11a,b),

$$\dot{z}(t,\omega) = Az(t,\omega) + B\nabla C^{-1}\left(-B^T\overline{y}^{(b)}(t)\right) + b(t,\omega) \tag{27a}$$

$$z(t_b,\omega) = z_b. \tag{27b}$$

Taking expectations, considering the state vector at the final time point t_f, resp., yields the expressions:

$$\overline{z}^{(b)}(t) = e^{A(t-t_b)}z_b + \int_{t_b}^{t} e^{A(t-s)}\left(\overline{b(s)}^{(b)} + B\nabla C^{-1}\left(-B^T\overline{y}^{(b)}(s)\right)\right) ds \tag{28a}$$

$$z(t_f,\omega) = e^{A(t_f-t_b)}z_b + \int_{t_b}^{t_f} e^{A(t_f-s)}\left(b(s,\omega) + B\nabla C^{-1}\left(-B^T\overline{y}^{(b)}(s)\right)\right) ds \tag{28b}$$

7.1 *Quadratic control and terminal costs*

Corresponding to (16a,b) and (18a,b), suppose

$$\nabla G(t_f,\omega,z(t_f,\omega)) = z(t_f,\omega) - z_f(\omega),$$

$$\nabla C^{-1}(w) = R^{-1}w.$$

According to (10) and (9c), in the present case the optimal open-loop control is given by

$$u^*(t) = \widetilde{u^*}(t,q(t)) = R^{-1}\left(-\mathbb{E}\left(B^T y(t,\omega)|\mathfrak{A}_{t_b}\right)\right) = -R^{-1}B^T\overline{y}^{(b)}(t). \tag{29a}$$

Hence, we need the function $\overline{y}^{(b)} = \overline{y}^{(b)}(t)$. From (26) and (16a,b) we have

$$\overline{y}^{(b)}(t) = e^{A^T(t_f-t)}\left(\overline{z(t_f)}^{(b)} - \overline{z_f}^{b}\right) + \int_t^{t_f} e^{A^T(s-t)}Q\overline{z}^{(b)}(s)\, ds. \tag{29b}$$

Inserting (28a,b) into (29b), we have

$$\begin{aligned}\overline{y}^{(b)}(t) =& e^{A^T(t_f-t)}\left(e^{A(t_f-t_b)}z_b - \overline{z_f}^{(b)} + \int_{t_b}^{t_f} e^{A(t_f-s)}\left(\overline{b(s)}^{(b)} - BR^{-1}B^T\overline{y}^{(b)}(s)\right) ds\right)\\ &+ \int_t^{t_f} e^{A^T(s-t)}Q\left(e^{A(s-t_b)}z_b + \int_{t_b}^{s} e^{A(s-\tau)}\left(\overline{b(\tau)}^{(b)} - BR^{-1}B^T\overline{y}^{(b)}(\tau)\right) d\tau\right) ds.\end{aligned} \tag{29c}$$

In the following we develop a condition that guarantees the existence and uniqueness of a solution $\overline{y}^{b} = \overline{y}^{(b)}(t)$ of equation (29c):

Theorem 3 *In the space of continuous functions, the above equation (29c) has a unique solution if*

$$c_B < \frac{1}{c_A\sqrt{c_{R^{-1}}(t_f-t_0)\left(1+\frac{(t_f-t_0)c_Q}{2}\right)}}. \tag{30}$$

Here, $c_A := \sup_{t_b\le t\le s\le t_f} \|e^{A(t-s)}\|_F$, $c_B := \|B\|_F$, $c_{R^{-1}} := \|R^{-1}\|_F$, $c_Q := \|Q\|_F$, *and the index F denotes the Frobenius-Norm.*

Remark 2 *Condition (30) holds if the matrix* Γ *in (1c) has a sufficiently small Frobenius norm.*

Theorem 4 *Having* $\overline{y}^{(b)}(t)$*, according to (29a) the stochastic optimal open-loop control* $u^*(t)$*,* $t_b \le t \le t_f$*, reads:*

$$u^*(t) = -R^{-1}B^T\overline{y}^{(b)}(t). \tag{31a}$$

Moreover,

$$\varphi(t_b,z_b) := u^*(t_b), \qquad t_0 \le t_b \le t_f, \tag{31b}$$

is then the stochastic optimal open-loop feedback control law.

Remark 3 *Putting* $Q=0$ *in (25), we again obtain the open-loop feedback control law (24b) in Section 6.1.1.*

8 STOCHASTIC WEIGHT MATRIX $Q = Q(t, \omega)$

In the following we consider the case that, cf.(4a–d), the weight matrix for the evaluation of the displacements $z = z(t, \omega)$ is stochastic and may depend also on time t, $t_0 \le t \le t_f$. In order to take into account especially the size of the applied load $f_0 = f_0(t, \omega)$, cf.(11a,b), in the following we consider the stochastic weight matrix $Q(t, \omega) : = \|f_0(t, \omega)\|^2 Q$, where Q is again a positive (semi)definite $2m \times 2m$ matrix, and $\|\cdot\|$ denotes the Euclidian norm. According to (11a–d), for the adjoint variable $y = y(t, \omega)$ we then have the system of differential equations

$$\dot{y}(t,\omega) = -A^T y(t,\omega) - \eta(t,\omega) Q z(t,\omega)$$

$$y(t_f,\omega) = \nabla G(t_f, \omega, z(t_f,\omega)),$$

where $\eta(t, \omega) := \|f_0(t, \omega\|^2$. Assuming that we have the weighted terminal costs

$$G(t_f, \omega, z(t_f,\omega)) := \frac{1}{2}\eta(t_f,\omega)\|z(t_f,\omega) - z_f(\omega)\|^2,$$

for the adjoint variable $y = y(t, \omega)$, we have the boundary value problem

$$\dot{y}(t,\omega) = -A^T y(t,\omega) - \eta(t,\omega) Q z(t,\omega) \tag{32a}$$

$$y(t_f,\omega) = \eta(t_f,\omega)\left(z(t_f,\omega) - z_f(\omega)\right) \tag{32b}$$

Corresponding to (25), from (32a,b) we then get the solution

$$y(t,\omega) = \int_t^{t_f} e^{A^T(s-t)} Q \eta(s,\omega) z(s,\omega)\, ds + e^{A^T(t_f-t)}\left(\eta(t_f,\omega) z(t_f,\omega) - \eta(t_f,\omega) z_f(\omega)\right), \quad t \in [t_b; t_f] \tag{33a}$$

Taking conditional expectations of (33a) with respect to $\mathfrak{A}_{t_b}$, corresponding to (29b) we obtain

$$\overline{y}^{(b)}(t) = e^{A^T(t_f-t)}\left(\overline{\eta(t_f,\cdot)z(t_f,\cdot)}^{(b)} - \overline{\eta(t_f,\cdot)z_f(\cdot)}^{(b)}\right) + \int_t^{t_f} e^{A^T(s-t)} Q \overline{\eta(s,\cdot)z(s,\cdot)}^{(b)}\, ds, \quad t \in [t_b; t_f] \tag{33b}$$

Assuming again that the matrix B, see (2b), is fixed, corresponding to (29a), the stochastic optimal open-loop control $u^* = u^*(t)$ on $[t_b; t_f]$ is then given, cf. (6c) and (33b), by

$$u^*(t) = R^{-1}(-q(t))$$

$$= -R^{-1}B^T\left(e^{A^T(t_f-t)}\left(\overline{\eta(t_f,\cdot)z(t_f,\cdot)}^{(b)} - \overline{\eta(t_f,\cdot)z_f(\cdot)}^{(b)}\right) + \int_t^{t_f} e^{A^T(s-t)} Q \overline{\eta(s,\cdot)z(s,\cdot)}^{(b)}\, ds\right), \tag{34}$$

where R is the positive definite matrix arising in the definition of the control costs, cf. (4b). Moreover, the weighted conditional mean trajectory

$$t \mapsto \overline{\eta(t,\cdot)z(t,\cdot)}^{(b)} = \mathbb{E}\left(\eta(t,\omega)z(t,\omega)|\mathfrak{A}_{t_b}\right), \qquad t \in [t_b; t_f], \tag{35a}$$

is determined in the present open-loop feedback approach as follows. We first remember that the optimal trajectory is defined by the initial value problem (11a,b) or (27a,b). Approximating the weighted conditional mean trajectory (35a) by

$$t \mapsto \mathbb{E}\left(\eta(t_b,\omega)z(t,\omega)|\mathfrak{A}_{t_b}\right), \qquad t \in [t_b; t_f], \tag{35b}$$

we multiply (27a,b) by $\eta(t_b, \omega)$. Thus, the trajectory $t \mapsto \eta(t_b, \omega)z(t, \omega)$, $t \ge t_b$, is the solution of the initial value problem

$$\frac{d}{dt}\eta(t_b,\omega)z(t,\omega) = A\eta(t_b,\omega)z(t,\omega) - BR^{-1}B\eta(t_b,\omega)\overline{y}^{(b)}(t) + \eta(t_b,\omega)b(t,\omega) \tag{36a}$$

$$\eta(t_b,\omega)z(t_b,\omega) = \eta(t_b,\omega)z_b\,. \tag{36b}$$

Taking conditional expectations of (3a,b) with respect to $\mathfrak{U}_{t_b}$, for the approximate weighted conditional mean trajectory (35b) we obtain the initial value problem

$$\frac{d}{dt}\overline{\eta(t_b,\cdot)z(t,\cdot)}^{(b)} = A\overline{\eta(t_b,\cdot)z(t,\cdot)}^{(b)} - BR^{-1}B\overline{\eta}^{(b)}(t_b)\overline{y}^{(b)}(t) + \overline{\eta(t_b,\cdot)b(t,\cdot)}^{(b)} \tag{37a}$$

$$\overline{\eta(t_b,\cdot)z(t_b,\cdot)}^{(b)} = \overline{\eta}^{(b)}(t_b)z_b\,, \tag{37b}$$

where $\overline{\eta}^{(b)}(t) := \mathbb{E}(\eta(t,\omega)|\,\mathfrak{U}_{t_b}),\ t \geq t_b$. Consequently, the approximate weighted conditional mean trajectory (35b) can be represented, cf. (28a,b), by

$$\overline{\eta(t_b,\cdot)z(t,\cdot)}^{(b)} = e^{A(t-t_b)}\overline{\eta}^{(b)}(t_b)z_b + \int_{t_b}^{t} e^{A(t-s)}\left(\overline{\eta(t_b,\cdot)b(s,\cdot)}^{(b)} - BR^{-1}B^T\overline{\eta}^{(b)}(t_b)\overline{y}^{(b)}(s)\right)ds, \quad t \in [t_b;t_f]\,. \tag{38}$$

Obviously, a corresponding approximative representation for

$$t \mapsto \overline{\eta(t_f,\cdot)z(t_f,\cdot)}^{(b)}$$

can be obtained, cf. (28b), by using (38) for $t = t_f$.
Inserting now (38) into (33b), corresponding to (29c) we find the following approximate fixed point condition for the conditional mean adjoint trajectory $t \mapsto \overline{y}^{(b)}(t),\ t_b \leq t \leq t_f$, needed in the representation (??) of the optimal open-loop control $u^* = u^*(t),\ t_b \leq t \leq t_f$:

$$\begin{aligned}
\overline{y}^{(b)}(t) =& e^{A^T(t_f-t)}\left(\overline{\eta(t_f,\cdot)z(t_f,\cdot)}^{(b)} - \overline{\eta(t_f,\cdot)z_f(\cdot)}^{(b)}\right) + \int_{t_b}^{t_f} e^{A^T(s-t)}Q\overline{\eta(s,\cdot)z(s,\cdot)}^{(b)}\,ds \\
=& e^{A^T(t_f-t)}\Bigg(e^{A(t_f-t_b)}\overline{\eta}^{(b)}(t_b)z_b - \overline{\eta(t_f,\cdot)z_f(\cdot)}^{(b)} \\
&+ \int_{t_b}^{t_f} e^{A(t_f-s)}\left(\overline{\eta(t_b,\cdot)b(s,\cdot)}^{(b)} - BR^{-1}B^T\overline{\eta}^{(b)}(t_b)\overline{y}^{(b)}(s)\right)ds\Bigg) \\
&+ \int_{t}^{t_f} e^{A^T(s-t)}Q\Bigg(e^{A(s-t_b)}\overline{\eta}^{(b)}(t_b)z_b \\
&+ \int_{t_b}^{s} e^{A(s-\tau)}\left(\overline{\eta(t_b,\cdot)b(\tau,\cdot)}^{(b)} - BR^{-1}B^T\overline{\eta}^{(b)}(t_b)\overline{y}^{(b)}(\tau)\right)d\tau\Bigg)\,ds
\end{aligned} \tag{39}$$

Corresponding to Theorem 3 we can also develop a condition that guarantees the existence and uniqueness of a solution $\overline{y}^b = \overline{y}^{(b)}(t)$ of equation (39):

Theorem 5 *In the space of continuous functions, equation (39) has a unique solution if*

$$c_B < \frac{1}{c_A\sqrt{c_{R^{-1}}\overline{\eta}^{(b)}(t_b)(t_f-t_0)\left(1+\frac{(t_f-t_0)c_Q}{2}\right)}}\,. \tag{40}$$

Here again, $c_A := \sup_{t_b \leq t \leq s \leq t_f} \|e^{A(t-s)}\|_F,\ c_B := \|B\|_F,\ c_{R^{-1}} := \|R^{-1}\|_F,\ c_Q := \|Q\|_F$, *and the index F denotes the Frobenius-Norm.*

Theorem 6 *Having $\overline{y}^{(b)}(t)$ fulfilling the fixed point condition (39), according to (29a) the stochastic optimal open-loop control $u^*(t)$, $t_b \leq t \leq t_f$, reads:*

$$u^*(t) = -R^{-1}B^T\overline{y}^{(b)}(t). \tag{41a}$$

Moreover,

$$\varphi(t_b, z_b) := u^*(t_b), t_0 \leq t_b \leq t_f, \tag{41b}$$

is then the stochastic optimal open-loop feedback control law.

9 CONCLUSION

Due to the great advantages of open-loop feedback controls, *stochastic optimal open-loop feedback controls* have been constructed by taking into account the random parameter variations in the stochastic structural control problem. For finding stochastic optimal open-loop controls, on the remaining time intervals $t_b \leq t \leq t_f$ with $t_0 \leq t_b \leq t_f$, the stochastic Hamilton function of the control problem has been introduced. Then, the class of $H-$ minimum controls can be determined by solving a finite-dimensional stochastic optimization problem for minimizing the conditional expectation of the stochastic Hamiltonian subject to the remaining deterministic control constraints at each time point t. Having a $H-$ minimum control, the related two-point boundary value problem with random parameters can be formulated for the computation of the stochastic optimal optimal state- and costate-trajectory. Due to the linear-quadratic structure of the underlying control problem, the state and costate trajectory can be determined analytically to a large extent. Inserting then these trajectories into the H-minimum control, stochastic optimal open-loop controls are found on an arbitrary remaining time interval. These controls yield then immediately a stochastic optimal open-loop feedback control law.

REFERENCES

Block, C. (2008). *Aktive Minderung personeninduzierter Schwingungen an weit gespannten Strukturen im Bauwesen*. Number 336 in Fortschrittberichte VDI, Reihe 11, Schwingungstechnik. Düsseldorf: VDI-Verlag GmbH.

Marti, K. (2001). Stochastic optimization methods in robuts adaptive control of robots. In M. Groetschel et al. (Eds.), *Online Optimization of Large Scale Systems*, pp. 545–577. (Berlin-Heidelberg-New York: Springer-Verlag).

Marti, K. (2004). Adaptive optimal stochastic trajectory planning and control (AOSTPC) for robots. In K. Marti et al. (Eds.), *Dynamic Stochastic Optimization*, pp. 155–206. (Berlin-Heidelberg: Springer-Verlag).

Marti, K. (2008a). Approximate solutions of stochastic control problems by means of convex approximations. In Topping, B.H.V et al. (Eds.), *Proceedings of the 9th Int. Conference on Computational Structures Technology (CST08)*. (Stirlingshire, UK: Civil-Comp Press). paper No. 52.

Marti, K. (2008b). *Stochastic Optimization Problems, 2nd edition.* (Berlin-Heidelberg: Springer–Verlag).

Nagarajaiah, S. and S. Narasimhan (2007). Optimal control of structures. In Arora, J.S. (Ed.), *Optimization of Structural and Mechanical Systems*, pp. 221 – 244. (New Jersey [etc.]: World Scientific).

Soong, T.T. (1988). Active structural control in civil engineering. *Engineering Structures 10*, pp. 74–84.

Soong, T.T. (1990). *Active Structural Control: Theory and Practice.* (New York: Longman Scientific and Technical, J. Wiley).

Soong, T.T. and Costantinou, M.C. (1994). *Passive and Active Structural Vibration Control in Civil Engineering.* CISM Courses and Lectures No. 345. (Wien-New York: Springer–Verlag).

Spencer, B. and S. Nagarajaiah (2003). State of the art of structural control. *Journal of Structural Engineering, ASCE 129 (7)*, pp. 845–856.

Yang, J.N. and Soong, T.T. (1988). Recent advance in active control of civil engineering structures. *Probabilistic engineering mechanics 3 (4)*, pp. 179–188.

Reliability and Optimization of Structural Systems – Straub (ed)
© 2010 Taylor & Francis Group, London, ISBN 978-0-415-88179-1

Identification of the parameters of complex constitutive models: Least squares minimization vs. Bayesian updating

Thomas Most
Research Training Group 1462, Bauhaus-Universität Weimar, Germany

ABSTRACT: In this study the common least-squares minimization approach is compared to the Bayesian updating procedure. In the content of material parameter identification the posterior parameter density function is obtained from its prior and the likelihood function of the measurements. By using Markov Chain Monte Carlo methods, such as the Metropolis-Hastings algorithm (Hastings 1970), the global density function including local peaks can be computed. Thus this procedure enables an accurate evaluation of the global parameter quality. However, the computational effort is remarkable larger compared to the minimization approach. Thus several methodologies for an efficient approximation of the likelihood function are discussed in the present study.

1 INTRODUCTION

In all fields of computational mechanics numerical models are used to analyze engineering problems. In material modeling many types of constitutive formulations exist for a wide range of materials. Generally the parameters of these models are identified from experimental data. Beside manual identification by try and error or based on experiences fully automatic identification procedures using optimization strategies have become popular in the last years. Nevertheless, not only the parameter values itself are of interest for a later numerical analysis, but also their accuracy which is mainly influenced by the measurement errors and the type of experiments. For very complex constitutive models it is very difficult to judge in advance which experiments are required to enable a sufficient identifiability of all parameters. Therefor an estimation of the parameter uncertainties is very useful. Based on the determined optimal parameter set as the result of a least squares minimization, where the gap between experimental and numerical response data is minimized, the parameter covariances can be estimated based on the measurement errors and a linearization of the model (Ledesma et al. 1996). This approach was derived for a convex optimization problem. Applications for non-convex problems using complex material models in geomechanical applications can be found in (Hofmann et al. 2009), (Knabe et al. 2009).

Alternatively to the ordinary least squares formulation Kalman filter identification has been introduced (Cividini et al. 1983), (Bittani et al. 1984). This technique is a Bayesian estimator using prior information to identify the parameters. Similar to the covariance analysis in least squares minimization the model itself is linearized and the parameter distributions are taken as Gaussian which implies a convex optimization problem.

Another method uses Bayesian inference to estimate the parameters including their statistical properties (Beck and Arnold 1977). This approach can be applied for non-convex problems and arbitrary parameter and measurement error distributions and works without any linearization of the numerical model.

Recently inverse approximation schemes using neural networks have been proposed to identify the constitutive parameters (Novák and Lehký 2006), (Most et al. 2007). In (Unger and Könke 2010) this approach has been extended using Bayesian neural networks in order to estimate the parameter accuracy. In opposite to the other procedures there measurement errors are neglected and the parameter uncertainty arises only from the identification procedure. This results in a vanishing uncertainty if the number of training samples is increased dramatically. Therefor this procedure is not discussed further in this study.

2 LEAST SQUARES MINIMIZATION

2.1 *Maximum likelihood formulation*

A deterministic models which relates a set of responses $\mathbf{x} = \mathbf{x}(\mathbf{p})$ with the model parameters $\mathbf{p}$ is assumed to be given. In the parameter identification procedure the measurements $\mathbf{x^*} = [x_1^* \dots x_m^*]^T$ are obtained by experiments.

The likelihood of the parameters is proportional to the conditional probability of measurements $\mathbf{x}^*$ from a given parameter set $\mathbf{p}$ (Ledesma et al. 1996):

$$L = k \cdot f(\mathbf{x}^* | \mathbf{p}). \tag{1}$$

If the chosen model is correct the gap between numerical responses and measurements $(\mathbf{x}^* - \mathbf{x})$ is caused only by the measurement errors. Thus $P(\mathbf{x}^* | \mathbf{p})$ is equivalent to the probability of reproducing the measurement errors. Assuming a multivariate normal distribution we obtain

$$P(\mathbf{x}^* - \mathbf{x}) = \frac{1}{\sqrt{(2\pi)^m |\mathbf{C_{xx}}|}} \exp\left[-\frac{1}{2} (\mathbf{x}^* - \mathbf{x})^T \mathbf{C_{xx}^{-1}} (\mathbf{x}^* - \mathbf{x}) \right]. \tag{2}$$

Maximizing the likelihood L is equivalent to minimize $S = -2 \ln L$. This yields to the well known least squares objective function:

$$J = (\mathbf{x}^* - \mathbf{x})^T \mathbf{C_{xx}^{-1}} (\mathbf{x}^* - \mathbf{x}), \tag{3}$$

where $\mathbf{C_{xx}}$ is the covariance matrix of the measurements. In (Ledesma et al. 1996) it is shown, that the optimal weights used in the objective function are

$$\mathbf{W} = \mathbf{C_{xx}^{-1}}. \tag{4}$$

If the objective function is linearized, the following update scheme is obtained

$$\Delta \mathbf{p} = \left(\mathbf{A}^T \mathbf{C_{xx}^{-1}} \mathbf{A} \right)^{-1} \mathbf{A}^T \mathbf{C_{xx}^{-1}} \Delta \mathbf{x}, \tag{5}$$

where $\mathbf{A}$ is the sensitivity matrix of the parameters with respect to the model responses

$$\mathbf{A} = \frac{\partial \mathbf{x}}{\partial \mathbf{p}}. \tag{6}$$

The update scheme can be used to obtain the optimal parameter set with the minimum objective function value. This approach is a gradient based method which leads to the global minimum without fail only for a convex optimization problem. If this method is applied for a non-convex problem, it may stuck in a local minimum. This is the case for more complex constitutive models. For this reason often global optimization schemes are used to identify the optimal parameter set. One of these methods is presented in the next section.

If the optimal parameter set is finally identified, the covariances of the parameters $\mathbf{C_{pp}}$ can be estimated by using the linearized relation from Eq.(5) which leads to the so-called Markov estimator

$$\mathbf{C_{pp}} = \left(\mathbf{A}_{opt}^T \mathbf{C_{xx}^{-1}} \mathbf{A}_{opt} \right)^{-1}. \tag{7}$$

In (Hofmann et al. 2009) and (Knabe et al. 2009) this estimator is applied also for non-convex optimization problem, which means that around the global optimum a local convex problem is assumed and the model is linearized at this optimum. Both, parameter and measurement distributions are assumed to be Gaussian within this approach.

2.2 *Particle swarm optimization*

In our study a population based global optimization algorithm, the particle swarm optimization (PSO), is applied which enables detecting the global optimum even if several local minima exist. Each population consists of a given number of particles where each particle position is equivalent to a set of parameters $\mathbf{p}_i^k$ and is updated in each iteration step k by the simple scheme (Kennedy and Eberhart 1995)

$$\begin{aligned} \mathbf{p}_i^{k+1} &= \mathbf{p}_i^k + \mathbf{v}_i^{k+1} \\ \mathbf{v}_i^{k+1} &= \omega \mathbf{v}_i^k + c_1 \mathbf{r}_1 (\mathbf{P}_i^k - \mathbf{p}_i^k) + c_2 \mathbf{r}_2 (\mathbf{P}_g^k - \mathbf{p}_i^k) \end{aligned} \tag{8}$$

where c_1 and c_2 are constants and $\mathbf{r}_1$ and $\mathbf{r}_2$ are vectors of uniformly distributed random variables between zero and one. $\mathbf{P}_i^k$ is the best position of a single particle and $\mathbf{P}_g^k$ is the global best position. We extend the original concept with passive congregation (He et al. 2004)

$$\mathbf{v}_i^{k+1} = \omega \mathbf{v}_i^k + c_1 \mathbf{r}_1 (\mathbf{P}_i^k - \mathbf{p}_i^k) + c_2 \mathbf{r}_2 (\mathbf{P}_g^k - \mathbf{p}_i^k) + c_3 \mathbf{r}_3 (\mathbf{R}_i^k - \mathbf{p}_i^k) \tag{9}$$

where an additional term is added to the original scheme to decrease the risk of running into a local minimum. This term $\mathbf{R}_i^k$ is defined as the best position of a randomly chosen particle. The required constants c_1, c_2 and c_3 are given in (He et al. 2004). If new particle positions are outside the parameter boundaries they are corrected by a harmony search scheme according to (Li et al. 2007).

3 BAYESIAN ESTIMATORS

3.1 *Kalman filter*

Kalman filter technology was firstly applied for parameter identification in (Cividini et al. 1983), (Bittani et al. 1984). More recent applications can be found in (Bolzon et al. 2002), (Fedele et al. 2006), (Furukawa and Pan 2009). In this approach the measurement noise is assumed to be a time-dependent white Gaussian random process characterized by zero mean, zero cross-correlation, and a time-dependent covariance matrix

$$\begin{aligned} E(\boldsymbol{\epsilon}^{(t)}) &= \mathbf{0}, \\ E(\boldsymbol{\epsilon}^{(t)} \boldsymbol{\epsilon}^{(s)^T}) &= \mathbf{0}, s \neq t \\ E(\boldsymbol{\epsilon}^{(t)} \boldsymbol{\epsilon}^{(t)^T}) &= \mathbf{C}_{\mathbf{xx}}^{(t)}. \end{aligned} \tag{10}$$

Using a linearization of the model the following update scheme can be derived

$$\begin{aligned} \hat{\mathbf{p}}^{(t)} &= \hat{\mathbf{p}}^{(t-1)} + \mathbf{K}^{(t)} (\mathbf{x}^* - \mathbf{x}(\hat{\mathbf{p}}^{(t-1)})) \\ \hat{\mathbf{C}}_{\mathbf{pp}}^{(t)} &= \hat{\mathbf{C}}_{\mathbf{pp}}^{(t-1)} - \mathbf{K}^{(t)} \mathbf{A}^{(t)} \hat{\mathbf{C}}_{\mathbf{pp}}^{(t-1)} \end{aligned} \tag{11}$$

where $\mathbf{K}$ is the Kalman gain matrix

$$\mathbf{K}^{(t)} = \hat{\mathbf{C}}_{\mathbf{pp}}^{(t-1)} \mathbf{A}^{(t)^T} \left[\mathbf{A}^{(t)} \hat{\mathbf{C}}_{\mathbf{pp}}^{(t-1)} \mathbf{A}^{(t)^T} + \mathbf{C}_{\mathbf{xx}}^{(t)} \right]^{-1} \tag{12}$$

and $\mathbf{A}$ is again the sensitivity matrix. As initial setting generally the prior information is taken

$$\hat{\mathbf{p}}^{(0)} = \mathbf{p}_0, \quad \hat{\mathbf{C}}_{\mathbf{pp}}^{(0)} = \mathbf{C}_{\mathbf{pp}_0}. \tag{13}$$

Due to the linearization of the model and the assumed Gaussian measurement noise, the parameter distributions are implicitly assumed to be Gaussian. In a static approach the time-step t is equivalent to the iteration step and the measurement covariance matrix remains constant.

(Cividini et al. 1983) has shown, that for vanishing prior information, where $(\mathbf{C}_{\mathbf{pp}_0})^{-1}$ tends to zero, the estimated parameter covariance is equivalent to the Markov estimator.

3.2 *Bayesian updating*

Bayesian updating is based on the well-known Bayes' theorem which shows the relation between one conditional probability and its inverse

$$P(A|B) = \frac{P(B|A)\,P(A)}{P(B)}. \tag{14}$$

Assuming a prior multivariate distribution of the parameters $\pi_{prior}(\mathbf{p})$ and a likelihood-function $\pi(\mathbf{x}^*|\mathbf{p})$ for the distribution of the measurements the conditional probability of the parameters can be obtained as (Beck and Arnold 1977)

$$\pi_{posterior}(\mathbf{p}) = \pi(\mathbf{p}|\mathbf{x}^*) = \frac{\pi_{prior}(\mathbf{p}) \cdot \pi(\mathbf{x}^*|\mathbf{p})}{\pi(\mathbf{x}^*)}. \tag{15}$$

Since the normalization constant $\pi(\mathbf{x}^*)$ is difficult to determine due to the only implicitly given distribution function, realizations of $\pi_{posterior}(\mathbf{p})$ are obtained by Markov-Chain Monte Carlo Simulation (Hastings 1970). For jointly Gaussian measurement errors the likelihood function is equivalent to Eq.(2).

3.3 Metropolis-Hastings algorithm

The Metropolis-Hastings algorithm (Hastings 1970) enables the sampling of an implicitly given not necessarily scaled density function. Therefor it is very important in interferential problems. This method is a Monte Carlo Simulation using a first order Markov chain, where the next sample depends only on the current state.

The algorithm generates a sequence of samples $\{\mathbf{p}^{(1)}, \mathbf{p}^{(2)}, \ldots, \mathbf{p}^{(n)}\}$ which statistically converge to the given distribution, which is in our case the Bayesian posterior multivariate parameter distribution. The algorithm works as follows:

1. Sample candidate $\mathbf{p}^*$ from jumping distribution $q(\mathbf{p}^*, \mathbf{p}^{(t-1)})$
2. Calculate

$$\alpha = \min\left[1, \frac{\pi(\mathbf{p}^*|\mathbf{x}^*)}{\pi(\mathbf{p}^{(t-1)}|\mathbf{x}^*)}\frac{q(\mathbf{p}^{(t-1)}, \mathbf{p}^*)}{q(\mathbf{p}^*, \mathbf{p}^{(t-1)})}\right] \tag{16}$$

3. Sample uniformly distributed $U \in (0.0; 1.0)$
4. If $U \leq \alpha$ accept $\mathbf{p}^{(t)} = \mathbf{p}^*$, otherwise $\mathbf{p}^{(t)} = \mathbf{p}^{(t-1)}$
5. Return to step 1

The jumping distribution is taken generally to be a Gaussian or uniform distribution. If it is symmetric $q(\mathbf{p}^{(t-1)}, \mathbf{p}^*)/q(\mathbf{p}^*, \mathbf{p}^{(t-1)})$ is equal to one. The initial parameter set is chosen randomly in the given parameter ranges. In this study the prior parameter distribution is taken as a uniform distribution in these parameter ranges. Thus the Metropolis-Hastings algorithm will only accept samples inside this boundaries. At the beginning the generated sequence is biased depending on the starting point and the jumping distribution. But after a certain burn-in phase it turns into an ergodic process. Thus the statistical evaluation of the parameter distributions is done be neglecting a given number of the first samples.

The variances of the jumping distribution strongly influences acceptance rate (AR) and convergence speed. If the distribution variances are taken too small, the convergence is very low and almost all samples will be accepted. An optimal statistical evaluation is enabled if the acceptance rate is between 10 and 30% as shown in (Unger 2009).

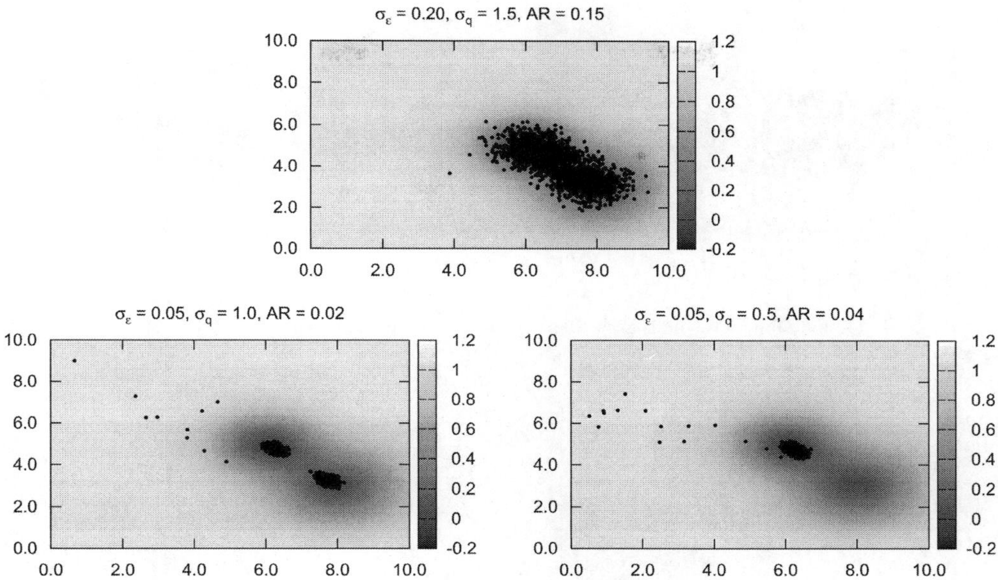

Figure 1. Samples from Metropolis-Hastings algorithm for a non-convex problem with different measurement errors.

Table 1. Estimated parameter variation for a non-convex problem with different measurement errors and variances of the jumping distribution.

Bayes' updating					
σ_ϵ	σ_q	AR	No. samples	$\tilde{\sigma}_{X_1}$	$\tilde{\sigma}_{X_2}$
0.05	1.0	0.02	100000	0.81	0.80
0.10	1.5	0.05	44000	0.77	0.76
0.20	1.5	0.15	13000	0.79	0.81
0.05	0.5	0.04	44000	0.13	0.12
0.10	0.5	0.20	10000	0.78	0.78
0.20	0.5	0.50	4000	0.79	0.79

For non-convex problems the variances of the jumping distribution have to be chosen in that way, that the peaks of the likelihood function are sufficiently covered. This is shown in Figure 1 and Table 1 depending on the measurement errors. If the measurement errors are small and the variances of the jumping distribution are to small, only one local optimum is covered by the sampling procedure. But if the variances are chosen large enough the acceptance rate is very small and a large number of samples is necessary to obtain statistically accurate results.

3.4 *Scaling of measurement errors*

In the previous section it was mentioned, that for small measurement errors a large number of samples may be required if the optimization problem is non-convex. For this reason two possibilities to reduce this numerical effort are discussed.

The first idea is to scale the covariances of the measurement errors by a constant factor a

$$\mathbf{C}_{\mathbf{xx}}^{scaled} = a \cdot \mathbf{C}_{\mathbf{xx}}^{orig}. \tag{17}$$

Based on the scaled likelihood function a sequence of samples is computed with the Metropolis-Hastings algorithm. The estimates of the parameter mean values and covariances of the original

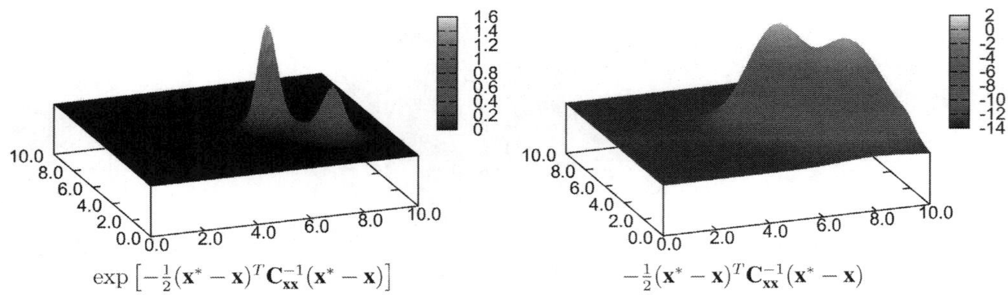

Figure 2. Likelihood function with sharp peaks and corresponding exponential term for small measurement errors.

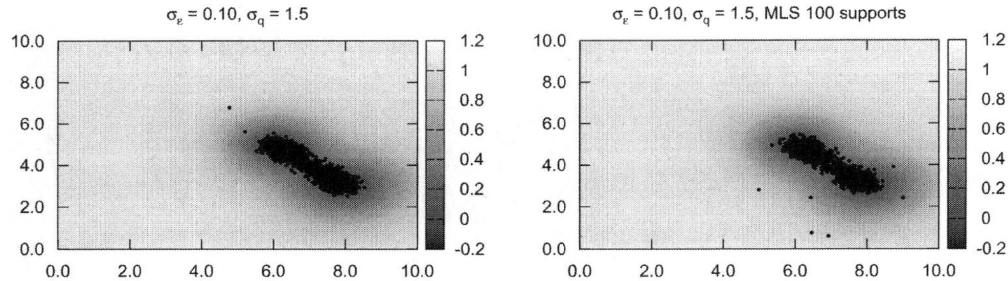

Figure 3. Computed parameter samples by using the original and approximated likelihood function.

distribution can be computed as follows

$$\bar{p}_k = \frac{1}{n}\sum_{i=1}^{n} p_k^{(i)} \frac{f_{orig}(\mathbf{p}^{(i)})}{f_{scaled}(\mathbf{p}^{(i)})}$$

$$C_{kl} = \frac{1}{n-1}\sum_{i=1}^{n}(p_k^{(i)} - \bar{p}_k)(p_l^{(i)} - \bar{p}_l)\frac{f_{orig}(\mathbf{p}^{(i)})}{f_{scaled}(\mathbf{p}^{(i)})} \tag{18}$$

where the joint densities functions f_{orig} and f_{scaled} have to be known. Both are equivalent to the corresponding likelihood function scaled by the normalization constant. Since this normalization constant is not known it has to be estimated from the samples. In our opinion this is only possible by assuming a certain distribution type. This would suspend the advances of the Bayesian updating approach and would limit the method again to convex optimization problems. Therefor this scaling procedure is not investigated further in this study.

3.5 *Approximation of the likelihood function*

Another approach to reduce the numerical effort is to approximate the likelihood function by a suitable meta-model and using this approximation function instead of real model solutions. In (Orlande et al. 2008) this was shown by using radial basis function approximation. Dependent on the variance of the measurement noise the likelihood function contains sharp peaks around the local optima as shown in Figure 2. This requires an increased number of support points in the meta-model approach. For this reason in this study only the exponential term is approximated, which is equivalent to the objective function in Eq.(3). Then the measurement noise variance is only a scaling factor with no influence on the approximation quality. As meta-model approach interpolating Moving Least Squares (Most and Bucher 2008) is applied, since it can handle clustered support points and high gradients in the approximation function.

In Figure 3 and Table 2 the results with original and approximated likelihood function are compared for the non-convex problem. The results indicate a very good agreement in the parameter

Table 2. Estimated parameters by using the original and approximated likelihood function.

MLS approximation $\sigma_\epsilon = 0.10$, $\sigma_q = 1.5$				
	$\tilde{\mu}_{X_1}$	$\tilde{\mu}_{X_2}$	$\tilde{\sigma}_{X_1}$	$\tilde{\sigma}_{X_2}$
Original likelihood	7.22	3.83	0.77	0.76
MLS 100 supports	7.23	3.98	0.71	0.70
MLS 200 supports	6.91	4.08	0.78	0.73
MLS 500 supports	7.05	3.95	0.73	0.73

Table 3. Estimated parameter variations for the 1D elasto-plastic model.

			$\sigma_\epsilon[10^8 \text{ N/m}^2]$		
			0.05	0.10	0.20
Sample variation	σ_E	$[10^{11}\text{N/m}^2]$	0.025	0.052	0.157
	σ_{σ_Y}	$[10^8\text{N/m}^2]$	0.019	0.036	0.117
Markov estimator	$\tilde{\sigma}_E$	$[10^{11}\text{N/m}^2]$	0.025	0.050	0.151
	$\tilde{\sigma}_{\sigma_Y}$	$[10^8\text{N/m}^2]$	0.019	0.038	0.113
Kalman filter	$\tilde{\sigma}_E$	$[10^{11}\text{N/m}^2]$	0.025	0.050	0.151
	$\tilde{\sigma}_{\sigma_Y}$	$[10^8\text{N/m}^2]$	0.019	0.038	0.113
Bayes' updating	$\tilde{\sigma}_E$	$[10^{11}\text{N/m}^2]$	0.025	0.050	0.145
	$\tilde{\sigma}_{\sigma_Y}$	$[10^8\text{N/m}^2]$	0.019	0.038	0.115

estimates. The numerical effort is reduced dramatically, since only 100 evaluations of the real model are required instead of 44000 by using the original likelihood function.

4 NUMERICAL EXAMPLES

4.1 *1D elasto-plastic model without hardening*

In this analytical example an one-dimensional elasto-plastic model with two parameters, the Young's modulus E and the yield stress σ_Y is investigated. As reference values $E_{ref} = 2.1 \cdot 10^{11}$ N/m^2 and $\sigma_{Y,ref} = 5.0 \cdot 10^8$ N/m^2 are chosen. By applying constant Gaussian measurement noise σ_ϵ for all observations 1000 samples are generated and for each sample the optimal parameter set is obtained by a gradient-based optimization method. The resulting variations of the identified parameters are shown in Table 3.

Additionally the Markov estimator is used at the parameter reference values and the Kalman filter approach is applied by using a large prior covariance $\mathbf{C}^0_{\mathbf{pp}} = diag[100, 100] \cdot 10^8$ N/m^2 in order to reduce the effect of the prior information. The results of both methods coincide excellent with the sample values due to the convexity of the problem. In Figure 5 the resulting parameter samples are shown.

Finally the Bayesian updating procedure is used whereby the prior distributions are chosen to be uniform between the optimization bounds $E \in (1.0; 3.0)$, $\sigma_Y \in (2.0; 10.0)$. The resulting estimates are given in Table 3. In Figure 5 the samples from the Metropolis-Hastings algorithm are shown, where the burn-in phase can be seen. The first 1000 samples are not considered in the statistical evaluation in order to get unbiased estimates. The results agree very well with the other approaches. Only for a larger measurement noise minor deviation can been observed which might be caused by the linearization in the Markov estimator and Kalman filter approaches.

4.2 *Bilinear interface model for concrete cracking*

In this example the parameters of a cohesive crack model are identified. The investigations are based on the experimental program of (Trunk 1999). In Figure 6 the concrete specimen is shown

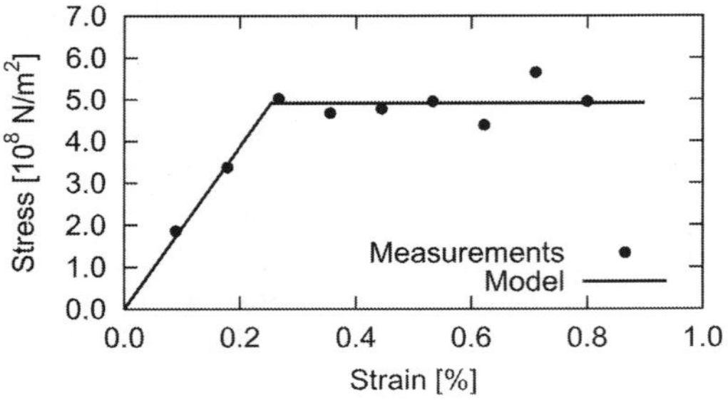

Figure 4. Stress-strain dependence of the 1D elasto-plastic model.

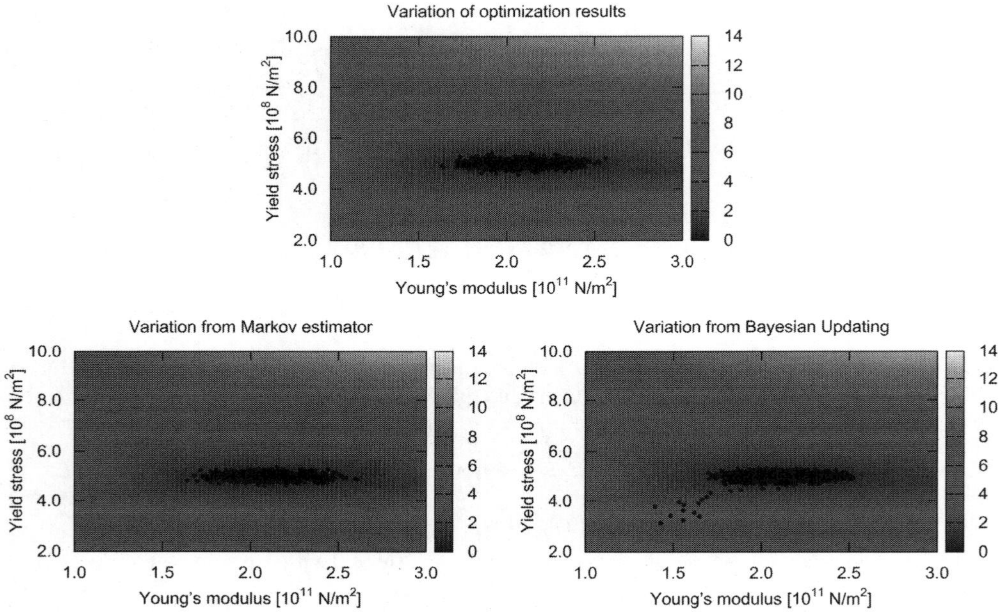

Figure 5. Deterministic objective function of the 1D elasto-plastic model including the parameter samples from the different approaches for $\sigma_\epsilon = 0.2 \cdot 10^8$ N/m^2.

with the corresponding dimensions. The numerical analysis is carried out using a finite element model with a predefined crack discretized by finite interface elements. Further details about the numerical suimulation can be found in (Most 2005). For the interface elements a bilinear softening law is used shown in principle in Figure 6. The remaining base material is assumed to be linear elastic. The following parameters have been indentified within the experimental program

- Young's modulus $E = 2.83 \cdot 10^{10}$ N/m^2
- Tensile strength $f_t = 2.27 \cdot 10^6$ N/m^2
- Specific fracture energy $G_f = 285$ Nm/m^2
- Softening shape parameters $\alpha_\sigma = \sigma_1 / f_t = 0.163$, $\alpha_u = \Delta u_{N1} / \Delta u_{Nc} = 0.242$

The corresponding measurements and the numerical load displacement curve is shown in Figure 7.

In a first step the measurements are taken as deterministic and 100 optimization runs with random start points using a gradient based and the particle swarm approach are carried out. The results are given in Table 4. Due to the existence of several local minima the gradient based methods does not converge to the global optimum in almost the half of the runs, whereby the PSO method

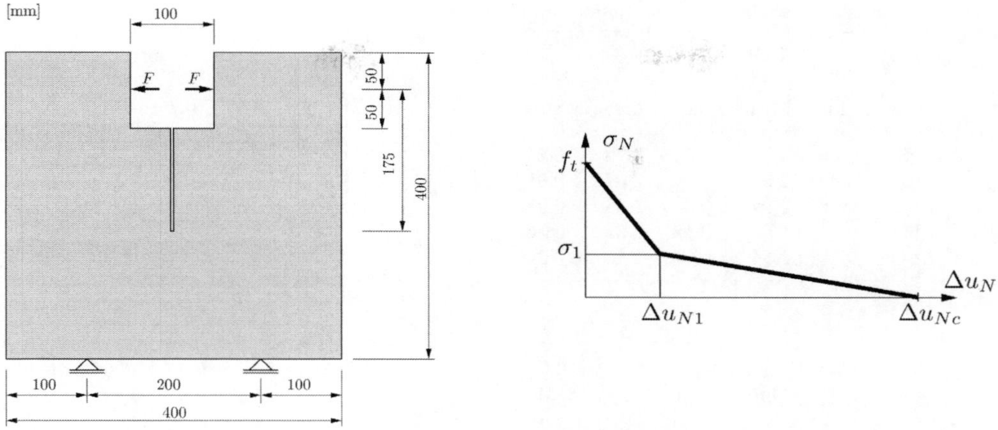

Figure 6. Wedge splitting test with system geometry and bilinear interface model.

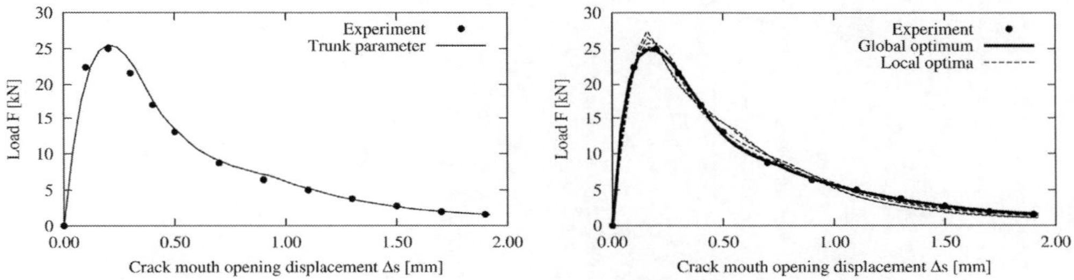

Figure 7. Load displacement curves for the wedge splitting test using the identified parameters.

Table 4. Optimization results for the wedge splitting test.

				Global		Total	
			Best	Mean	Std.	Mean	Std.
Gradient (success rate 51%)	E	$[10^{10}\ \mathrm{N/m^2}]$	3.85	3.89	0.11	3.72	0.39
	f_t	$[10^6\ \mathrm{N/m^2}]$	2.04	2.03	0.02	2.18	0.38
	G_f	$[10^2\ \mathrm{Nm/m^2}]$	2.84	2.84	0.01	2.81	0.07
	α_σ	$[10^{-1}]$	1.57	1.58	0.05	1.86	0.51
	α_u	$[10^{-1}]$	2.54	2.55	0.07	2.46	0.41
PSO (success rate 89%)	E	$[10^{10}\ \mathrm{N/m^2}]$	3.85	3.82	0.05	3.74	0.28
	f_t	$[10^6\ \mathrm{N/m^2}]$	2.04	2.05	0.01	2.13	0.31
	G_f	$[10^2\ \mathrm{Nm/m^2}]$	2.84	2.84	0.00	2.82	0.05
	α_σ	$[10^{-1}]$	1.57	1.57	0.02	1.70	0.42
	α_u	$[10^{-1}]$	2.54	2.53	0.03	2.44	0.32

is successfull in almost all cases. In Table 4 the statistics of the runs ending close to the global optimum and of all runs are compared. The results indicate a very good accuracy of the PSO method which is even better than the gradient based approach.

In the next step the parameter uncertainties are estimated with the different approaches by assuming a constant measurement noise of $\sigma_\epsilon = 0.5 \cdot 10^6\ \mathrm{N/m^2}$. In Table 5 and Figure 8 the results of these analyses are shown. They indicate are very good agreement of the Bayes' estimates with

Table 5. Parameter estimates and correlations for the wedge splitting test.

	Samples		Markov		Bayes	
	Mean	Std.	Mean	Std.	Mean	Std.
E	3.78	0.30	3.85	0.24	3.79	0.24
f_t	2.07	0.07	2.04	0.06	2.07	0.05
G_f	2.84	0.08	2.84	0.07	2.83	0.08
α_σ	1.62	0.32	1.57	0.18	1.62	0.26
α_u	2.51	0.26	2.54	0.17	2.52	0.24

$$Corr_{\mathbf{pp}}^{Samples} = \begin{bmatrix} 1.00 & -0.65 & 0.29 & -0.26 & 0.16 \\ & 1.00 & -0.13 & 0.40 & -0.34 \\ & & 1.00 & -0.31 & -0.65 \\ & & & 1.00 & 0.41 \\ & & & & 1.00 \end{bmatrix}$$

$$Corr_{\mathbf{pp}}^{Markov} = \begin{bmatrix} 1.00 & -0.59 & 0.29 & -0.83 & -0.40 \\ & 1.00 & -0.05 & 0.57 & -0.09 \\ & & 1.00 & 0.02 & -0.79 \\ & & & 1.00 & 0.00 \\ & & & & 1.00 \end{bmatrix}$$

$$Corr_{\mathbf{pp}}^{Bayes} = \begin{bmatrix} 1.00 & -0.56 & 0.17 & -0.09 & 0.12 \\ & 1.00 & -0.02 & 0.21 & -0.24 \\ & & 1.00 & -0.21 & -0.69 \\ & & & 1.00 & 0.49 \\ & & & & 1.00 \end{bmatrix}$$

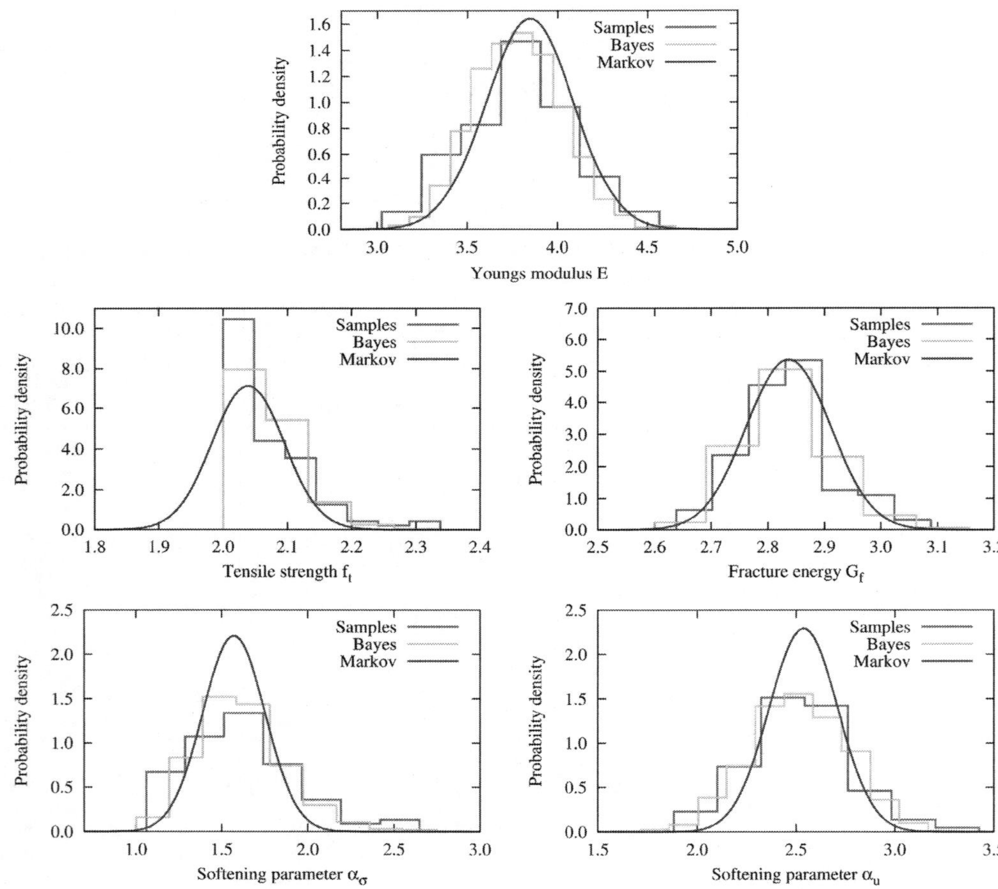

Figure 8. Estimated parameter distributions for the wedge splitting test.

the variation from the sample analysis. The Markov estimator gives a good approximation of the variation, but some parameter dependencies are estimated not correct.

5 CONCLUSIONS

The approximation of parameter uncertainties and dependencies by Bayes' updating is more accurate then using the Markov estimator. However, the Markov estimator is suitable as a rough estimate even for non-convex problems. A remarkable reduction of the numerical effort is possible by using meta-model approach to represent the likelihood function.

ACKNOWLEDGMENT

This research has been supported by the German Research Council (DFG) through Research Training Group 1462, which is gratefully acknowledged by the author.

REFERENCES

Beck, J. V. and K. Arnold (1977). *Parameter estimation in engineering and science*. New York: Wiley Interscience.

Bittani, S., G. Maier, and A. Nappi (1984). Inverse problems in structural elasto-plasticity: a Kalman filter approach. In A. Sawczuk and G. Bianchi (Eds.), *Plasticity Today*. London: Elsevier Applied Science.

Bolzon, G., R. Fedele, and G. Maier (2002). Parameter identification of a cohesive crack model by Kalman filter. *Computer Methods in Applied Mechanics and Engineering 191*, 2847–2871.

Cividini, A., G. Maier, and A. Nappi (1983). Parameter estimation of a static geotechnical model using a Bayes' approach. *International Journal of Rock Mechanics and Mining Sciences & Geomechanics Abstracts 20*, 215–226.

Fedele, R., G. Maier, and M. Whelan (2006). Stochastic calibration of local constitutive models through measurements at the macroscale in heterogeneous media. *Computer Methods in Applied Mechanics and Engineering 195*, 4971–4990.

Furukawa, T. and J. W. Pan (2009). Stochastic identification of elastic constants for anisotropic materials. *International Journal for Numerical Methods in Engineering 81*, 429–452.

Hastings, W. (1970). Monte Carlo sampling methods using Markov chains and their applications. *Biometrika 57*, 97–109.

He, S., Q. Wu, J. Wen, J. Saunders, and R. Paton (2004). A particle swarm optimizer with passive congregation. *BioSystems 78*, 135–147.

Hofmann, M., T. Most, and G. Hofstetter (2009). Parameter identification for partially saturated soil models. In *Proc. 2nd Intern. Conf. Computational Methods in Tunnelling, Ruhr University Bochum, Germany, September 9–11, 2009*.

Kennedy, J. and R. Eberhart (1995). Particle swarm optimization. In *Proceedings of the IEEE International Conference on Neural Networks*, Volume 4.

Knabe, T., M. Zimmerer, T. Most, and T. Schanz (2009). Identification of constitutive parameters for geomaterials modeling using an optimization strategy. In *Proc. 2nd Intern. Conf. Computational Methods in Tunnelling, Ruhr University Bochum, Germany, September 9–11, 2009*.

Ledesma, A., A. Gens, and E. E. Alonso (1996). Estimation of parameters in geotechnical backanalysis – I. Maximum likelihood approach. *Computers and Geotechnics 18*, 1–27.

Li, L., Z. Huang, F. Liu, and Q. Wu (2007). A heuristic particle swarm optimizer for optimization of pin connected structures. *Computers and Structures 85*, 340–349.

Most, T. (2005). *Stochastic crack growth simulation in reinforced concrete structures by means of coupled finite element and meshless methods*. Ph. D. thesis, Bauhaus-Universität Weimar, Germany.

Most, T. and C. Bucher (2008). New concepts for moving least squares: An interpolating non-singular weighting function and weighted nodal least squares. *Engineering Analysis with Boundary Elements 32*, 461–470.

Most, T., G. Hofstetter, M. Hofmann, Novàk, and D. Lehky (2007). Approximation of constitutive parameters of material models by artificial neural networks. In B. H. V. Topping (Ed.), *Proc. 9th Intern. Conf. on the Application of Artificial Intelligence to Civil, Structural and Environmental Engineering, St. Julians, Malta, September 18–21, 2007*. Civil-Comp Press.

Novák, D. and D. Lehký (2006). ANN inverse analysis based on stochastic small-sample training set simulation. *Engineering Applications of Artificial Intelligence 19*, 731–740.

Orlande, H. R. B., M. J. Colaco, and G. S. Dulikravich (2008). Approximation of the likelihood function in the Bayesian technique for the solution of inverse problems. *Inverse Problems in Science and Engineering 16*, 677–692.

Trunk, B. (1999). *Einfluss der Bauteilgrösse auf die Bruchenergie von Beton*. Ph. D. thesis, Eidgenössische Technische Hochschule, Zürich, Switzerland.

Unger, J. (2009). *Neural networks in a multiscale approach for concrete*. Ph. D. thesis, Bauhaus-Universität Weimar.

Unger, J. F. and C. Könke (2010). An inverse parameter identification procedure assessing the quality of the estimates using Bayesian neural networks. *Applied Soft Computing Journal, submitted.*

Reliability and Optimization of Structural Systems – Straub (ed)
© 2010 Taylor & Francis Group, London, ISBN 978-0-415-88179-1

Risk-based operation and maintenance planning for offshore wind turbines

Jannie Jessen Nielsen & John Dalsgaard Sørensen
Department of Civil Engineering, Aalborg University, Denmark

ABSTRACT: Operation and Maintenance (O&M) are large contributors to the cost of energy for offshore wind turbines. Optimal planning of O&M should include use of inspections and monitoring results to make decisions that minimize the expected costs through the lifetime of the structures. For offshore structures it is especially important because of the dependence on weather windows for inspections and repairs to be possible. A model has been developed to estimate the expected costs to corrective and condition based maintenance for a wind turbine with a single component. The deterioration of the component is simulated, and the expected costs are found for different strategies. An application example shows that condition based maintenance has the potential of reducing the costs, and a risk based approach can be used to find the optimal strategy for O&M. Further the influence of failure rate and damage parameters are evaluated.

1 INTRODUCTION

Operation and Maintenance (O&M) make up around 25–30% of the cost of energy for offshore wind turbines. A major contributor to this amount is the large number of failures that forces the wind turbine to stop producing power until the damage has been repaired (Wiggelinkhuizen, Verbruggen, Braam, Rademakers, Xiang, and Watson 2008). Some failures are severe failures of major components, where the direct costs to spare parts and vessels are large, but also the indirect costs caused by lost production are large because of a large number of days to mobilize a crew with the required spare parts and vessel and additional days to finish the repair. These failures are not expected as the design life of these components often equals the life of the wind turbines, and these failures are rare. Another critical source is a large number of failures of minor components that have significantly smaller design life than the entire wind turbine. The repair itself is often relatively cheap, but associated downtime can make up large costs, especially if repair is delayed by bad weather. Also the failures of minor components can be critical if they result in serial damage to larger components.

1.1 *Condition based maintenance*

In order to reduce the costs to O&M the number of failures can be reduced by using preventive maintenance strategies, where repairs are performed before an actual failure see e.g. (Walford 2006). For condition based maintenance repairs are decided based on the actual condition of the components. This requires understanding of the origin and development of the different failure types, and research is going on it this area see e.g. (Faulstich, Hahn, Jung, Rafik, and Ringhandt 2008) and (Giebhardt 2004). Many components are exposed to deterioration precesses e.g. fatigue, wear, corrosion, and erosion, and eventually this leads to failure. An understanding of the source of damage and development of failure can be combined to an appropriate damage model that can be used to plan repairs in advance.

The damage models will always be encumbered with uncertainties, and condition monitoring measurements can be used to update the models and reduce the uncertainty. Condition monitoring covers both offline inspections and online monitoring, where a large number of techniques are available see e.g (Nilsson and Bertling 2007) and (Wiggelinkhuizen, Verbruggen, Braam, Rademakers, Xiang, and Watson 2008). These methods are also subject to uncertainty, and this should be taken

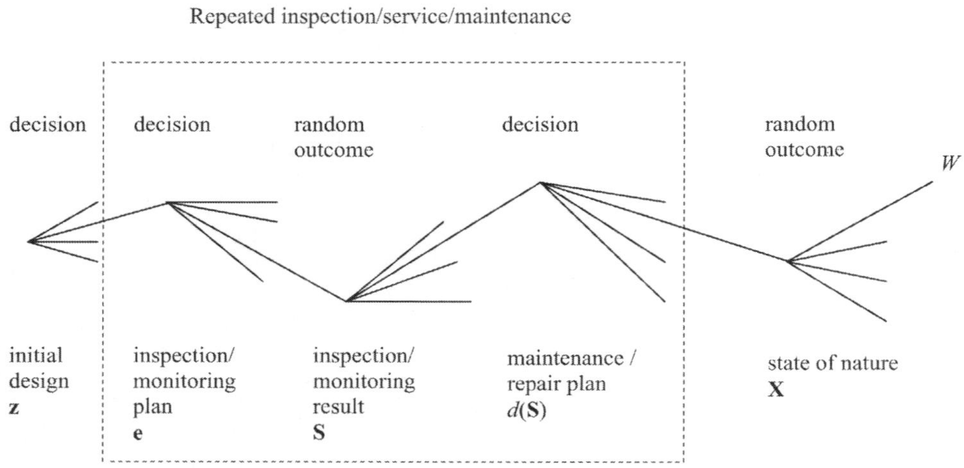

Figure 1. Decision tree for optimal planning of operation and maintenance. (Sørensen 2009).

into account. Monitoring methods might also help to determine the events leading up to failure, e.g. acoustic emission sensors can detect a crack before it is visible.

1.2 *Optimal planning of O&M*

Through the life time of a wind turbine there are decisions to make that affects the total costs to O&M. These decisions should be made using risk-based methods such that the expected total costs through the life time are minimized, see the wind turbine framework in (Sørensen 2009) and the theoretical basis in (Raiffa and Schlaifer 1961). Figure 1 shows a decision tree for the life cycle approach to planning of O&M.

There are three types of decisions which considers in turn the initial design, inspections/ monitoring, and repairs. This correspond to a pre-posterior decision problem, where the state of nature at one point is affected by previous decisions. Some decisions on the initial design, **z**, has influence on the O&M costs, e.g. the reliability level of components and decisions that influence the maintainability of the wind turbine.

The decision that is most important in order to avoid that damages lead to failures, is the decision on how and when to make repairs, given by the decision rule $d(\mathbf{S})$. The premise for this decision is the inspection/monitoring result, **S**, that is dependent on the decisions made regarding inspection/monitoring, **e**. An increased monitoring effort will have a higher cost, but the decision on repairs can be made with less uncertainty about the damage state.

The realizations of the stochastic variables, **X**, including damage development and possible failure, will depend on the decisions made. The optimal strategy for a fixed initial design, z, can be found as the one with minimum expected costs, W^*:

$$W^* = \min_e \min_d E_X E_{S|X} W(\mathbf{e}, \mathbf{S}, d(\mathbf{S}), \mathbf{X}) \tag{1}$$

E denotes expectation with respect to the subscribed variables.

In this paper simulation has been used to evaluate the expected costs for different decisions on maintenance strategies, both corrective maintenance and condition based maintenance with different inspection/repair decisions. The optimal solution has been found as the one with smallest expected costs, and the influence of the damage parameters on the optimal strategy and costs has been investigated.

2 MODEL

In this study the expected costs for different O&M strategies are found for a design lifetime of 20 years. For simplicity only one wind turbine with a single component is considered. This

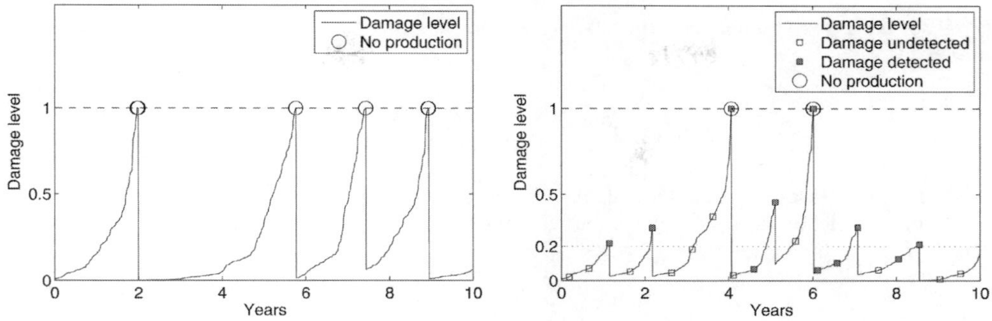

Figure 2. Damage development for corrective (left) and condition based maintenance (right).

method allows an easy understandable model to be used and clearly illustrates the main decision mechanisms.

The damage development is governed by the weather, and repairs require appropriate weather windows. Further the lost power production depends on the wind speed. Met Ocean weather data from Horns Reef 2 in Denmark has been provided by Danish Hydraulic Institute, and has been bootstrapped on a yearly basis, which gives pseudo random weather. The used weather data consist of hourly mean values of the significant wave height and wind speed.

The damage size is measured on a relative scale where failure occurs at the value 1. If this value is reached, the component needs a corrective repair in order to restart production of power, as illustrated to the left in figure 2. Condition based maintenance is implemented based on imperfect offline inspections at a fixed interval. The decision rule for repairs is a limit value of the damage size. If a damage larger than this is found at an inspection, a condition based repair is performed, as illustrated to the right in figure 2, where the damage limit is 0.2.

2.1 *Damage model*

The component considered in this analysis is assumed to be exposed to deterioration, where an exponential damage model is appropriate. The increase in damage is found using Paris' law for crack growth, see e.g. (Straub 2004). The model is general and can be calibrated to be used in many different applications. The increase in damage is considered to be dependent on both the wind and wave load and the influence of the control system. For simplicity only the mean significant wave height H_s in included directly in the model, as the wind speed and wave height are approximately proportional.

The increase in damage dD/dt is calculated for each 3 hour interval, and is obtained from:

$$\frac{dD}{dt} = C \cdot \Delta K^m \tag{2}$$

C is the damage coefficient, and m is the damage exponent, that gives the shape of the damage curve. ΔK is the damage intensity factor range, and is given by:

$$\Delta K = \Delta s \cdot \sqrt{\pi D} \tag{3}$$

Δs is the cyclic damage range that is considered to the proportional to the mean significant wave height H_s in the three hour period with the proportionality factor x_s. The uncertainty in the damage model is modeled by considering x_s and C as stochastic variables, which each takes a new value after each repair. The initial damage size D_0 is also considered being a stochastic variable.

2.2 *Inspections*

For condition based maintenance inspections are performed at a fixed interval. The inspections are imperfect, so there is a probability that the damage is not detected. The probability of detection

(*PoD*) of damage is higher for larger damages, and is in this analysis modeled by an exponential model:

$$PoD(D) = 1 - \exp\left(\frac{-D}{\lambda}\right) \tag{4}$$

λ is the expected value of detectable damages, and depends on the inspection method. In general a more reliable method has a smaller λ, but is more expensive. Figure 3 shows a *PoD*-curve, where λ is 0.4.

2.3 *Repairs*

Both corrective and condition based repairs are included in the model. A corrective repair is necessary if failure has occurred, and the duration of a repair is modeled stochastic. Repairs can be performed by either boat or helicopter, where a boat is assumed to require a significant wave height smaller than 1.5 m and wind speed smaller than 10 m/s in the repair period (one hour mean values). A helicopter only requires a wind speed smaller than 20 m/s, but is more expensive. Three different transport strategies have been implemented in the model:

- Boat: Always use boat for repairs.
- ASAP: Repair as soon as possible. Boat if possible right away, else helicopter.
- "Risk": The cheapest alternative is always chosen, when lost production is taken into account, calculated based on a perfect weather forecast.

Condition based repairs are assumed to be performed at the same time as inspections. They are performed if a damage size larger than a repair limit is found at the inspection.

2.4 *Costs*

The costs included in the model are costs to inspections, repairs, and lost production. For condition based maintenance there are costs to inspections and to repairs. Both are assumed to be performed by boat and have fixed prices. Corrective repairs imply costs to lost production and repairs. Costs to lost production are calculated using a power curve for a 5 MW wind turbine. The cost of corrective repairs depends of the duration of the repair (repair days) and the transport type, as shown in table 1. All costs, C, are discounted into present values, C_0, using $C_0 = C/(1+r)^T$ where T is the time difference in years and r is the real rate of interest.

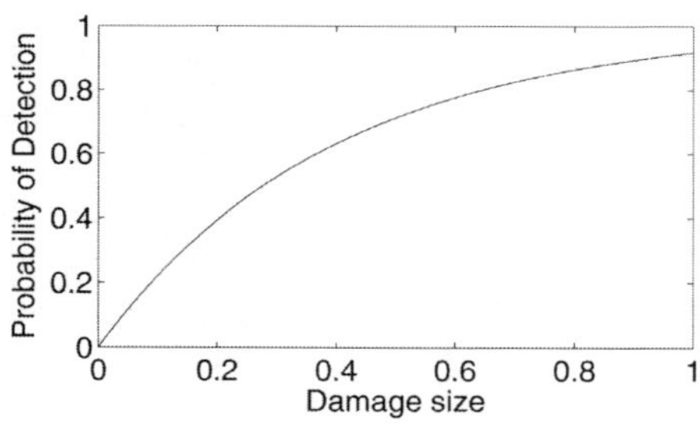

Figure 3. Probability of detection as function of damage size.

Table 1. Cost of repairs.

Repair type	Transport	Costs
Condition based	Boat	C_R
Corrective	Boat	$C_R(C_{B1} + C_{B2} \cdot \text{repair days})$
Corrective	Heli	$C_R(C_{H1} + C_{H2} \cdot \text{repair days})$

3 APPLICATION EXAMPLE

The described model is generic, and other damage, inspection, and repair models can easily be implemented to represent a specific case. In this section the model has been used to find the optimal inspection plan and repair limit for a case where the costs are given table 2 and the damage parameters are given in table 3. The damage parameters are balanced for H_s inserted in meters, and are calibrated to a mean time between failures (MTBF) of 2 years, which is typical for some components (Faulstich, Hahn, Jung, Rafik, and Ringhandt 2008).

3.1 *Corrective maintenance*

For corrective maintenance the only decision concerns the transport strategy. Figure 4 shows the expected number of repairs and expected costs for the three different transport strategies. When only a boat is used there are large costs to lost production. When ASAP is used these costs are reduced

Table 2. Specific costs.

Description	Symbol	Value
Power price	b	0.04 euro/kWh
Inspection cost	C_{INSP}	2500 euro
Repair cost	C_R	10000 euro
Boat cost factors	C_{B1}/C_{B2}	1.0/0.5
Helicopter cost factors	C_{H1}/C_{H2}	1.5/0.75
Real rate of interest	r	5%

Table 3. Means, coefficients of variation, and distribution types for damage parameters.

Symbol	Mean	COV	Distribution
m	2	–	Deterministic
C	10^{-6}	0.2	Lognormal
x_s	11.5	0.1	Lognormal
D_0	0.02	0.02	Exponential
Repair days	3 days	0.5	Lognormal

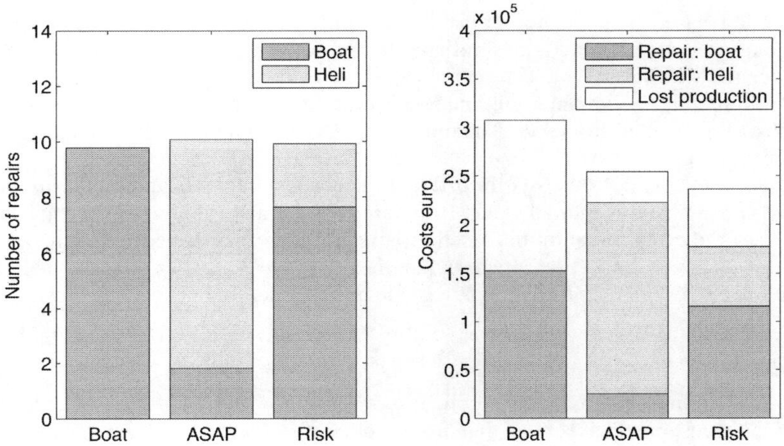

Figure 4. Influence of transport strategy.

Table 4. Total expected costs in 1000 euro.

		Repair limit		
		0.0	0.1	0.2
Inspection interval months	3	251	235	230
	6	199	192	195
	9	196	195	199

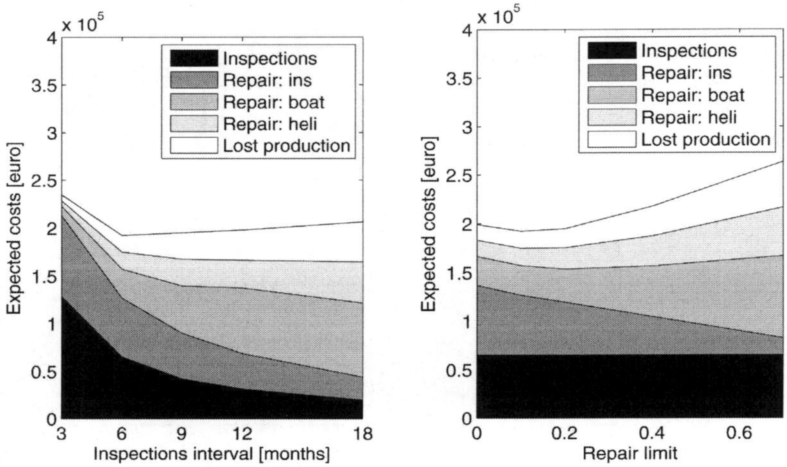

Figure 5. Expected costs for different inspection intervals and repair limits.

significantly, but the costs to expensive repairs by helicopter are increased. The risk-based option with perfect weather forecast gives the perfect balance between lost production and fast expensive repairs. In reality this option is not realistic as the forecasts will always be uncertain, and the costs to corrective maintenance with best realistic transport strategy will be between the options ASAP and risk-based. In the following the risk-based transport strategy is used for all corrective repairs.

3.2 *Condition based maintenance*

When condition based maintenance is used there will still be a need for corrective repairs, if the condition based maintenance does not succeed in avoiding all failures, but the total costs to corrective repairs will be smaller. The optimal strategy is the one with smallest total costs. In this case study different inspection intervals and repair limits are considered, and table 4 shows the total expected costs for selected strategies. The minimum is found at an inspection interval of 6 months and a repair limit at 0.1.

Figure 5 shows the expected costs for different inspection intervals and repair limits, when the other is hold constant at the optimal value. In general a larger inspection interval and a larger repair limit decreases the costs to condition based repairs and increases the costs to corrective repairs. The curve for different inspection intervals is rather flat, and in reality other considerations will affect the choice between 6 and 12 month, e.g. the need for service visits.

3.3 *Damage exponent*

In the previous example the damage exponent was set to $m = 2.0$. A higher value of the damage exponent will correspond to a damage that more suddenly increases from a low to a high value, and is therefore harder to detect using condition based maintenance. For a damage exponent $m = 2.6$ and the mean value of x_s calibrated to make the MTBF remain two years, the optimal strategy has

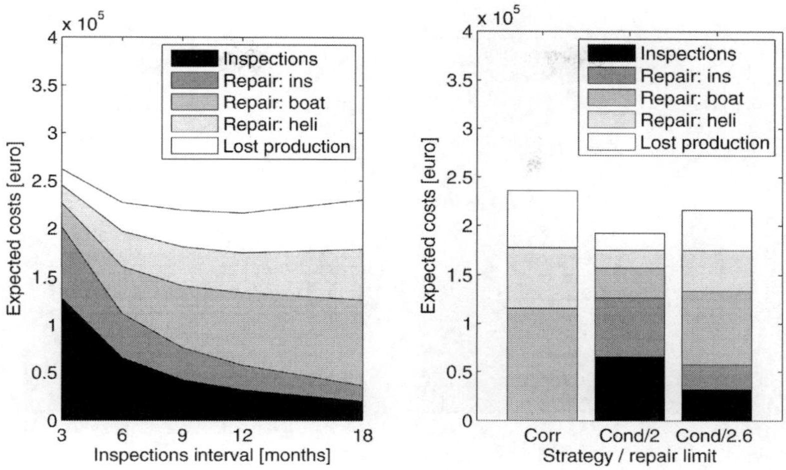

Figure 6. Expected costs for a damage exponent $m = 2.6$ to the left, and for different strategies to the right.

been found to be an inspection interval of 12 months and repair limit at 0. This repair limit implies that a detected damage should always be repaired.

The diagram to the left in figure 6 shows the expected costs for different inspection intervals. Comparison with figure 5 shows that a smaller inspection interval does not reduce the costs to of corrective repairs as much for a higher damage exponent, and thus the optimal inspection interval is larger. The diagram to the right in figure 6 shows the expected costs when only corrective repairs are used, and when also condition based repairs are used for two different damage exponents. The costs to corrective maintenance are not affected by the damage exponent when MTBF is hold constant, but condition based maintenance is less effective in reducing total costs for a larger damage exponent.

3.4 *Influence of failure rate*

If the the mean value of x_s is calibrated to a MTBF of 4 years instead of 2, and the damage exponent is $m = 2.0$, the optimal strategy has been found to be an inspection interval of 12 month and a repair limit at 0.1. The diagram to the left in figure 7 shows the expected costs for different inspection intervals. The optimal inspection interval is larger compared to the case with MTBF of 2 years because a smaller interval will give too large costs to inspections compared to the decrease in corrective repairs.

The diagram to the right in figure 7 shows the expected costs normalized with respect to the total costs for corrective maintenance for a MTBF at 2 and 4 years respectively. Condition based maintenance is equally good at reducing the total costs to O&M for both MTBF, if the optimal strategy is chosen in each case.

4 DISCUSSION

The generic model presented in this paper can in principle be modified to reflect any situation, if knowledge is available about an appropriate damage and inspection model for the specific case. The method allows the evaluation of expected costs for a discrete set of different strategies. In the shown example the strategies only considered inspection interval and repair limit, but other parameters could be included as well, e.g. different inspection procedures with different reliability levels and costs, and other types of decision rules for repairs.

The present model only considers a single wind turbine with one component, but in reality a wind turbine has many components with different MTBF. Here it could be relevant to find a joint optimal strategy, where the same inspection interval is used, but the repair limit could differ for different components. For two components with MTBF of 2 and 4 years respectively the joint

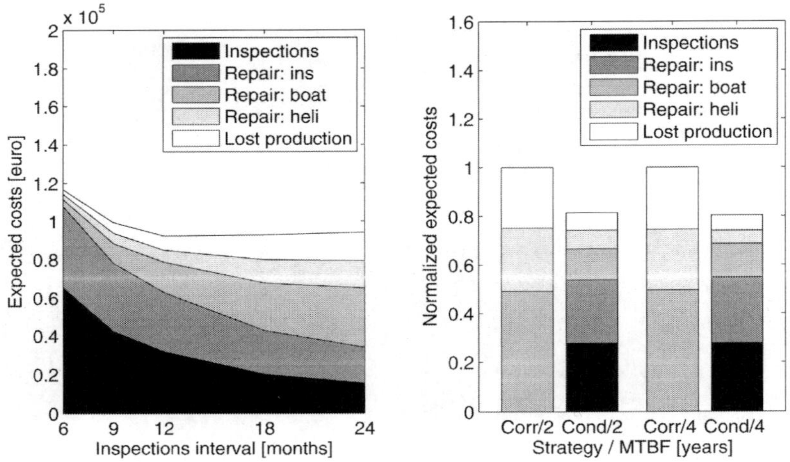

Figure 7. Expected costs for MTBF = 4 years to the left, and expected costs normalized to the costs for corrective maintenance for each MTBF to the right.

optimal strategy seems to have an inspection interval of around 12 months, because the cost curve is most steep on the lower side of the optimal point, see figure 5 and 7.

A high value of the damage exponent significantly reduces the gain from condition based maintenance, which means that the method is most effective for curves with more constant slope.

ACKNOWLEDGEMENTS

The work presented in this paper is part of the project "Reliability-based analysis applied for reduction of cost of energy for offshore wind turbines" supported by the Danish Council for Strategic Research, grant no. 2104-08-0014. The financial support is greatly appreciated.

REFERENCES

Faulstich, S., B. Hahn, H. Jung, K. Rafik, and A. Ringhandt (2008). Appropriate failure statistics and reliability characteristics. This paper had been presented at the DEWEK 2008.

Giebhardt, J. (2004). Predictive condition monitoring for offshore wind energy converters with respect to the IEC61400-25 standard. This paper had been presented at the DEWEK 2004.

Nilsson, J. and L. Bertling (2007). Maintenance management of wind power systems using condition monitoring systems. *IEEE Transactions on Energy Conversion* 22(1).

Raiffa, H. and R. Schlaifer (1961). *Applied statistical decision theory*. Cambridge: Harvard University Press/Cambridge University Press.

Sørensen, J. D. (2009). Framework for risk-based planning of operation and maintenance for off-shore wind turbines. *Wind Energy 12(5)*, 493–506.

Straub, D. (2004). *Generic approaches to risk based inspection planning for steel structures*. Ph.D. thesis, Swiss Federal Institute of Technology, ETH Zurich.

Walford, C. (2006). Wind turbine reliability: Understanding and minimizing wind turbine operation and maintenance costs. Technical report, SANDIA Report. SAND2006-1100.

Wiggelinkhuizen, E., T. Verbruggen, H. Braam, L. Rademakers, J. Xiang, and S. Watson (2008). Assessment of condition monitoring techniques for offshore wind farms. *Journal of Solar Energy Engineering, Transactions of the ASME 130*(3), 0310041–0310049.

Reliability and Optimization of Structural Systems – Straub (ed)
© 2010 Taylor & Francis Group, London, ISBN 978-0-415-88179-1

A scalable parametric approximation to multi-normal probability integrals

K. Nishijima, J. Qin & M.H. Faber
Institute of Structural Engineering, ETH Zurich, Switzerland

ABSTRACT: The evaluation of probability integrals remains a central technical issue in reliability assessment of engineered systems. Recently, a new class of approaches has been proposed by Bucher (2009) and Naess et al. (2009). The underlying idea is to formulate a scalable parameterization of the integrals in such a way that the scale factor is uniquely related to the probability. By calculating the integrals for a range of scale factors chosen such that e.g. Monte Carlo sampling is efficient for their solution (for large probabilities), the parameters in the parameterization may be readily estimated. Thereafter, the parameterization may be applied for the scale factors corresponding to the integral of interest which typically corresponds to a small probability. The performance of such approaches thus depends on the choice of parameterization of the probability integral and the functional form utilized for the extrapolation. The present paper proposes a scheme for the parameterization and the corresponding functional forms for the extrapolation, considering multi-normal probability integrals. The proposed scheme is illustrated on four principal examples including series and parallel systems and a case with multiple design points. Through the examples the performance of the proposed scheme is investigated, and the advantages are emphasized.

1 INTRODUCTION

1.1 *Background*

In the assessment of the reliability of engineered systems, the efficient evaluation of probability integrals remains a technical challenging issue. Modern structural reliability theory has provided a number of efficient approaches to approach this challenge including First and Second Order Reliability Method (FORM/SORM), probability bounds, and various Monte Carlo based sampling techniques.

In practical applications the best choice of approach depends on the specifics of the problem, and to identify this in general it is necessary to understand the strengths and weaknesses of the different approaches in some detail. As a consequence there is a tendency, when analyzing the reliability of rather complex engineered facilities such as offshore platforms, infrastructure networks and industrial facilities, to use the easily applicable but not always efficient Monte Carlo based probability integrations techniques. Even though various variance reduction schemes exist, the use of these in general adds to the need of expertise of the user. The user, on the other hand, also needs to master the engineering aspects of the modeling of the considered facility and cannot be expected also to be an expert on probability integration techniques. This leaves the generally very inefficient so-called crude Monte Carlo scheme as a preferred approach for probability evaluations in practical applications.

1.2 *Probability integral solution proposed by Bucher (2009) and Naess et al. (2009)*

Bucher (2009) and Naess et al. (2009) propose a novel class of approaches for evaluating probability integrals. The fundamental idea behind this class of approaches is to formulate a scalable parameterization of the probability integral. The parameters in the parameterization of the probability integral is then estimated by calculating the integrals for choices of the scale factor for which the integral may be readily estimated using e.g. crude Monte Carlo simulation. The integral of interest is then approximated using the parameterization and by extrapolation in terms of the scale factor: Bucher (2009) proposes to parameterize probability integrals by scaling of the standard deviation of the basic (normal) random variables; Naess et al. (2009) propose to parameterize probability integrals by scaling of the mean values of the basic random variables. Commonly for the two approaches; by choosing the values of the scale factor in such a way that the corresponding probabilities become large, the probability integrals can be accurately estimated with small computational efforts. Finally, based on the evaluation of these probability integrals together with the assumed functional form in the parameterization representing the parameterized probability integrals, the probability integral of interest can be obtained by extrapolation.

Bucher (2009) proposes such a scheme in the domain of normal variables for the case of component reliability assessment. Naess et al. (2009) propose another but related scheme in the original domain of the basic random variables for the case of reliability assessment of components and systems. In these papers, the performance and advantages of each scheme are illustrated with several examples.

1.3 *Motivation and focus of the present paper*

The scheme proposed by Naess et al. (2009) which addresses component and systems reliability analysis takes basis in a postulated functional form for the scalable parameterization of the probability integral, however without any specific theoretical justification. The scheme proposed by Bucher (2009), on the other hand, specifically addresses components reliability analysis. Since the derivations of the functional form for the scalable parameterization of the probability integrals in both mentioned approaches take basis in the asymptotic characteristics of the probability integrals, the applications of these schemes are generally restricted to situations where the corresponding underlying assumption is valid. In general, however, a-priori there is no way to check whether this assumption is valid for a given problem.

The main focus of the present paper is, thus, to investigate and derive scalable parameterizations of the multi-normal probability integral which are applicable for both component and systems reliability analysis problems and which on the one hand satisfies the known asymptotic solutions but moreover also apply for situations where the asymptotic characteristics of the probability integral solutions cannot be anticipated.

2 PROBABILITY INTEGRAL SOLUTION BY EXTRAPOLATION

2.1 *Problem setting*

The objective of the present paper concerns the evaluation of the probability integral I, represented in the following form:

$$I = \int_D f_{\mathbf{X}_n}(\mathbf{x}_n) d\mathbf{x}_n \tag{1}$$

where $f_{\mathbf{X}_n}(\mathbf{x}_n)$ is the joint probability density function of random variables $\mathbf{X}_n = (X_1, X_2, \ldots, X_n)^T$ and the generally complexly bounded integral domain D is given as:

$$D = \left\{ \bigcup_{m=1}^{M} \bigcap_{j=1}^{J_m} g_{mj}(\mathbf{x}_n) \leq 0 \right\} \tag{2}$$

$g_{mj}(\cdot)$ $(j = 1, \ldots, J_m,\ m = 1, \ldots, M)$ are functions that jointly characterize the integral domain. I, $g_{mj}(\cdot)$ and D respectively correspond to the probability of failure of a system, the performance of individual components of the system (limit state functions) and the domain representing the system failure.

Since it is in principle possible to transform the non-standard normal random variables into independent standard normal random variables by an appropriate transformation, hereafter the random variables are assumed to be independent standard normal random variables. General cases where the random variables are not standard normal are investigated with practical examples of complex engineered systems in Nishijima et al. (2010).

2.2 *Concept of probability integral solution by extrapolation*

The approach proposed by Bucher (2009) and Naess et al. (2009) may be stated in a slightly generalized form as follows: *Consider* a scalable parametric representation of the probability integral $I(\lambda)$ where the scalar λ is the scale factor so that for a certain value λ_0 of λ the probability integral $I(\lambda_0)$ corresponds to the probability integral of interest. *Assume* a functional form $f(\lambda|\mathbf{q})$ characterized by parameters $\mathbf{q}$, which approximates the probability integrals $I(\lambda)$, i.e. $I(\lambda) \approx f(\lambda|\mathbf{q})$. Then, *estimate* the parameters $\mathbf{q}$ by using the estimated values $I(\lambda)$ of the probability integrals by (crude) Monte Carlo sampling for several values of λ that correspond to relatively high probabilities. Substituting the estimated parameters $\hat{\mathbf{q}}$ into the functional form $f(\lambda|\mathbf{q})$, the probability integral $I(\lambda_0)$ of interest is obtained by *extrapolation* as $I(\lambda_0) = f(\lambda_0|\hat{\mathbf{q}})$.

The main contribution of the present paper is to propose such a scalable parametric representation of multi-normal probability integral $I(\lambda)$ with functional forms $f(\lambda|\mathbf{q})$ for the extrapolation. The proposed parametric representation and corresponding functional forms are presented in the subsequent section.

3 PROPOSED SCHEME

3.1 *Parametric representation of integral*

Breitung (1984), Hohenbichler et al. (1987) and Breitung and Hohenbichler (1989) consider, in the context of structural reliability analysis, a multi-normal probability integral $I(\lambda)$ whose integral domain is characterized by a real-valued positive factor λ as:

$$I(\lambda) = \int_{\lambda D} \phi_n(\mathbf{u}_n) d\mathbf{u}_n \tag{3}$$

where $\mathbf{u}_n = (u_1, u_2, \ldots, u_n)^T$, $\varphi_n(\mathbf{u}_n)$ is the joint probability density function of the independent standard multi-normal variables $\mathbf{U}_n = (U_1, U_2, \ldots, U_n)^T$, and the domain λD is the scaled domain of D using the scale factor λ, see Figure 1 (left). The original integral domain D is given in the form of Equation (2) and the scaled domain λD is written as:

$$\lambda D = \left\{ \bigcup_{m=1}^{M} \bigcap_{j=1}^{J_m} g_{mj}\left(\mathbf{u}_n / \lambda\right) \le 0 \right\} \tag{4}$$

The case of $\lambda = 1$ corresponds to the original integral domain D. In the papers (Breitung (1984), Hohenbichler et al. (1987) and Breitung and Hohenbichler (1989)), it is pointed out that the integral above may be reduced to the following form:

$$I(\lambda) = \int_{D} \phi_n(\lambda \mathbf{u}_n) d\mathbf{u}_n = (2\pi)^{-n/2} \lambda^n \int_{D} \exp\left(-\frac{\lambda^2}{2} |\mathbf{u}_n|^2 \right) d\mathbf{u}_n \tag{5}$$

and the asymptotic solution for the integral $I(\lambda)$ (as $\lambda \to \infty$) is derived. It should be mentioned here that the integral form in the right hand side is a special case of the so-called Laplace-type

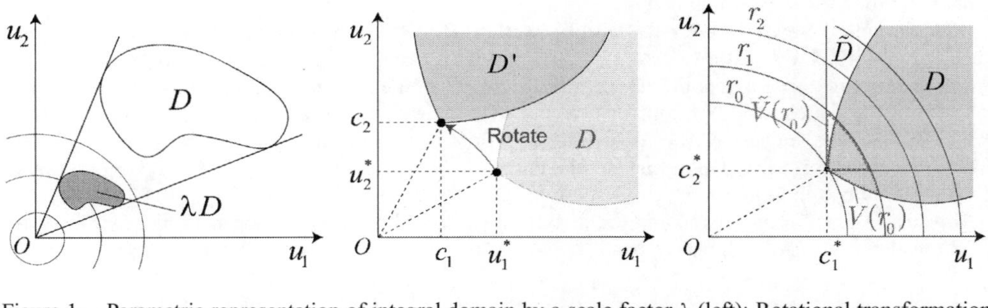

Figure 1. Parametric representation of integral domain by a scale factor λ (left); Rotational transformation of integral domain (middle); Discretization of integrals (right).

integral, for which asymptotic approximations have been extensively investigated, see e.g. Bleistein and Handelsman (1975).

The present paper employs this parametric representation of integral domain as the basis for the extrapolation, whereby the domain D should be chosen so that for a certain value λ_0 of λ the scaled domain $\lambda_0 D$ corresponds to the integral domain of interest. Note that the scale factor λ plays the equivalent role as the scale factor f introduced in Bucher (2009) in regard to the parameterization of the probability integral.

3.2 *Asymptotic form*

Without going into details (see Breitung (1994) for the basis of the theoretical justification), it is found that the integral $I(\lambda)$ can be approximated asymptotically by:

$$I(\lambda) \sim \kappa_\infty \prod_{i=1}^{k} c_i \Phi(-c_i \lambda) \tag{6}$$

with some constants $\kappa_\infty(>0)$ and $c_i(>0)$ $(i = 1, \ldots, k)$. Here, "$\sim$" signifies that for two functions f_1 and f_2 $\lim_{\lambda\to\infty} f_1(\lambda)/f_2(\lambda) = 1$ holds, k represents the number of the "active" limit state functions, i.e. the number of the functions that satisfy $g_{mj}(\mathbf{u}_n^*/\lambda) = 0$ at the design point $\mathbf{u}_n^*$, where $|\mathbf{u}_n^*| = \min_{\mathbf{u}_n \in D} |\mathbf{u}_n|$. Here, it is assumed that there exists only one design point. The geometrical interpretations of the constants c_i $(i = 1, \ldots, k)$ and κ_∞ are given in the following.

The values of the constants c_i are related to the distance from the origin to the design point $\mathbf{u}_n^*$ as:

$$\sum_{i=1}^{k} c_i^2 = |\mathbf{u}_n^*|^2 \tag{7}$$

However, the set of the values of the constants c_i $(i = 1, \ldots, k)$ do not necessarily uniquely identify the design point. On the other hand, given the design point $\mathbf{u}_n^* = (u_1^*, \ldots, u_n^*)$ and the number k of the active limit state functions, it is possible to choose a set of the values of the constants $c_i(>0)$ $(i = 1, \ldots, k)$ such that the relation in Equation (7) holds. Then, the point defined by the set $(c_1, \ldots, c_k, 0, \ldots, 0)$ in the n dimensional space can be regarded as the design point with a rotational transformation of the original integral domain, see Figure 1 (middle) for illustration. Note for any rotational transformation in the standard normal space, the value of the integral $I(\lambda)$ is invariant. It should be mentioned here that the asymptotic form of the integral $I(\lambda)$ is essentially characterized only by $|\mathbf{u}_n^*|$, k and κ_∞ as:

$$\kappa_\infty \prod_{i=1}^{k} c_i \Phi(-c_i \lambda) \sim \kappa_\infty \prod_{i=1}^{k} c_i (2\pi)^{-1/2} (c_i \lambda)^{-1} \exp\left(-\frac{c_i^2 \lambda^2}{2}\right) = (2\pi)^{-k/2} \kappa_\infty \lambda^{-k} \exp\left(-\frac{\lambda^2}{2} |\mathbf{u}_n^*|^2\right) \tag{8}$$

Here, the relation $\Phi(-x) \sim (2\pi)^{-1/2} x^{-1} \exp(-x^2/2)$ as $x \to \infty$ is utilized. Thus, individual values of the constants c_i are asymptotically not relevant for the integral $I(\lambda)$.

The constant κ_∞ is characterized by the first and second partial derivatives of the limit state functions $g_{mj}(\cdot)$ at the design point $\mathbf{u}_n^*$, which, however, is invariant in regard to rotational transformations, see Breitung (1994).

The so-called Second Order Reliability Method (SORM) provides algorithms to identify the design point $\mathbf{u}_n^*$ as well as the constant κ_∞ for complexly bounded integral domains e.g. corresponding to the failure domain of engineered systems; however, the implementation of the algorithms into programming codes is not straightforward in the cases where the integral domains are complexly bounded, and the success of the identification of the design point often requires trial-and-errors. Therefore, in the present paper, instead, the constants κ_∞ and c_i $(i = 1, \ldots, k)$ are estimated using the concept described in the previous section.

Taking basis in Equation (6) and the discussion above, the functional form for the extrapolation in the following form is thus proposed:

$$f(\lambda \mid \mathbf{q}) = a_\infty \prod_{i=1}^{K} \Phi(-c_i \lambda) \tag{9}$$

where $\mathbf{q} = (a_\infty, c_1, c_2, \ldots, c_K)$. For convenience, $\kappa_\infty \prod_{i=1}^{K} c_i$ in Equation (6) is replaced by a_∞, and without loss of generality it is assumed that $c_1 \geq c_2 \geq \ldots \geq c_K$. Since the number k of the active limit state functions is in general unknown a-priori, a sufficiently large number K is assumed here, which, however, does not exceed the number n of the dimension of the integral; by doing so, it is anticipated that the values of c_i for $i > K$ are estimated either close to zero or for negative values so that the term $\prod_{i=K+1}^{m} \Phi(-c_i\lambda)$ does not significantly contribute to the change of the value of $f(\lambda|\mathbf{q})$ as a function of λ.

Finally, it should be noted that the individual estimated values of the constants c_i $(i = 1, \ldots, k)$ are not relevant, but only the squared sum $\sum_{i=1}^{k} c_i^2$ is relevant (refer to the sentence below Equation (8)); another set of differently estimated values of the individual constants c_i which yet has the same value of the squared sum could correspond to an equivalent integral domain by a rotational transformation, see Figure 1 (middle).

3.3 *Non-asymptotic form*

The functional form presented in the previous section takes basis in the asymptotic characteristics of the integral $I(\lambda)$. It is, however, in general not known if and to which degree the assumption of the asymptotic characteristics is valid for the integral domains of interest; a critical example is shown in Section 4. Thus, a functional form of the extrapolation is required which can be applied for a broader range of the factor λ, beyond the range where the asymptotic characteristics is valid. For this purpose, the functional form presented in the previous section is extended.

Consider the integral $I(\lambda)$ defined by Equation (3). With a rotational transformation of the integral domain, it is possible that the integral $I(\lambda)$ is expressed as:

$$I(\lambda) = \int_{\lambda D^*} \phi_n(\mathbf{u}_n) d\mathbf{u}_n \tag{10}$$

where the design point of the transformed integral domain λD^* is $(\lambda c_1^*, \ldots, \lambda c_k^*, 0, \ldots, 0)$ satisfying $\sum_{i=1}^{k} c_i^2 = |\mathbf{u}_n^*|^2$ and $c_i > 0$ $(i = 1, \ldots, k)$. Define an integral $J(\lambda)$ as:

$$J(\lambda) = \int_{\lambda \tilde{D}^*} \phi_n(\mathbf{u}_n) d\mathbf{u}_n \tag{11}$$

where the integral domain $\lambda \tilde{D}^*$ is given as:

$$\lambda \tilde{D}^* = \left\{ (u_1, \ldots, u_n) \,|\, \bigcap_{i=1}^{k} \lambda c_i^* < u_i \right\} \tag{12}$$

That is, the two integral domains share the same design point $(\lambda c_1^*, \ldots, \lambda c_k^*, 0, \ldots, 0)$. Further, it should be realized that the integral $J(\lambda)$ is written as:

$$J(\lambda) = \int_{-\infty}^{\infty} \cdots \int_{-\infty}^{\infty} \int_{\lambda c_k^*}^{\infty} \int_{\lambda c_{k-1}^*}^{\infty} \cdots \int_{\lambda c_1^*}^{\infty} \phi_n(\mathbf{u}_n) d\mathbf{u}_n = \prod_{i=1}^{k} \Phi\left(-c_i^* \lambda\right) \tag{13}$$

Denote by $a(\lambda)$ the ratio between the two integrals $I(\lambda)$ and $J(\lambda)$ as: $a(\lambda) = I(\lambda)/J(\lambda)$. From the definition of the ratio $a(\lambda)$, it follows:

$$I(\lambda) = a(\lambda) J(\lambda) = a(\lambda) \prod_{i=1}^{k} \Phi\left(-c_i^* \lambda\right) \tag{14}$$

Thus, the introduction of the ratio $a(\lambda)$ can be seen as an extension of Equation (6). Note that Equation (14) is not an approximation but an identity. In what follows, the approximation of the functional form of the ratio $a(\lambda)$ is investigated, which does not depend on the asymptotic characteristics of the integral $I(\lambda)$.

The ratio $a(\lambda)$ can be reformulated and approximated as follows:

$$
\begin{aligned}
a(\lambda) &= \frac{\int_{\lambda D^*} \phi_n(\mathbf{u}_n) d\mathbf{u}_n}{\int_{\lambda \tilde{D}^*} \phi_n(\mathbf{u}_n) d\mathbf{u}_n} = \frac{\lambda^n \int_{D^*} \phi_n(\lambda \mathbf{u}_n) d\mathbf{u}_n}{\lambda^n \int_{\tilde{D}^*} \phi_n(\lambda \mathbf{u}_n) d\mathbf{u}_n} = \frac{\int_{D^*} \exp\left(-\lambda^2 |\mathbf{u}_n|^2/2\right) d\mathbf{u}_n}{\int_{\tilde{D}^*} \exp\left(-\lambda^2 |\mathbf{u}_n|^2/2\right) d\mathbf{u}_n} \\
&\approx \frac{\sum_{l=0}^{\infty} \exp\left(-\lambda^2 r_l^2/2\right) V(r_l)}{\sum_{l=0}^{\infty} \exp\left(-\lambda^2 r_l^2/2\right) \tilde{V}(r_l)} = \frac{\exp\left(-\lambda^2 r_0^2/2\right)\left[V(r_0) + \sum_{l=1}^{\infty} \exp\left(-\lambda^2 (r_l^2 - r_0^2)/2\right) V(r_l)\right]}{\exp\left(-\lambda^2 r_0^2/2\right)\left[\tilde{V}(r_0) + \sum_{l=1}^{\infty} \exp\left(-\lambda^2 (r_l^2 - r_0^2)/2\right) \tilde{V}(r_l)\right]} \\
&= \frac{V(r_0) + \sum_{l=1}^{\infty} \exp\left(-\lambda^2 (r_l^2 - r_0^2)/2\right) V(r_l)}{\tilde{V}(r_0) + \sum_{l=1}^{\infty} \exp\left(-\lambda^2 (r_l^2 - r_0^2)/2\right) \tilde{V}(r_l)} = \frac{V(r_0) + \sum_{l=1}^{\infty} \exp\left(-\lambda^2 R_l\right) V(r_l)}{\tilde{V}(r_0) + \sum_{l=1}^{\infty} \exp\left(-\lambda^2 R_l\right) \tilde{V}(r_l)}
\end{aligned}
\tag{15}
$$

Here, for the discretization in the second line, it is utilized that the integrands in the last term of the first line are only functions of the distance $|\mathbf{u}_n|$ from the origin. $\{r_l\}_{l=0}^{\infty}$ is the sequence of the distances from the origin at which the integrals are discretized, and $r_0 = |\mathbf{u}_n^*| > 0, r_l = r_0 + l\Delta r$, $\Delta r > 0$.$\{V(r_l)\}_{l=0}^{\infty}$ and $\{\tilde{V}(r_l)\}_{l=0}^{\infty}$ are the sequences of the corresponding volumes at the discretized distances, see Figure 1 (right). $R_l = (r_l^2 - r_0^2)/2$ and it can be approximated as: $R_l = ((r_0 + l\Delta r)^2 - r_0^2)/2 \approx l\Delta r$. In the following, the approximation of the summations in the last term of Equation (15) is considered.

The values of the terms $V(r_l)$ and $\tilde{V}(r_l)$ grow at most with the polynomial order of $n-1$ with respect to l, whereas the values of the exponents exponentially decrease as a function of l. Thus, the summations in the last term can be approximated by a finite number of terms in the summations. Consider two extreme situations where the values of $V(r_l)$ are constant and change quasi-exponentially. These two situations can be represented as: $V(r_l) \approx zw^l$ $(z > 0, w > 0)$. Assuming the same situations for $\tilde{V}(r_l)$ and substituting these approximations into the last term of Equation (15), it is found that the ratio $a(\lambda)$ may be approximated by:

$$
a(\lambda) \approx a_\infty \frac{1 - b_1 \exp(-b_2 \lambda^2)}{1 - b_3 \exp(-b_4 \lambda^2)} \tag{16}
$$

where the parameters $a_\infty, b_1, b_2, b_3, b_4$ are positive. Note that $a(\lambda) \to a_\infty$ (as $\lambda \to \infty$), which is consistent with the discussion on the asymptotic characteristics of the probability integral in the previous section. Finally, the functional form for the extrapolation is proposed for the non-asymptotic (general) case as:

$$
f(\lambda \mid \mathbf{q}) = a_\infty \frac{1 - b_1 \exp(-b_2 \lambda^2)}{1 - b_3 \exp(-b_4 \lambda^2)} \prod_{i=1}^{K} \Phi\left(-c_i \lambda\right) \tag{17}
$$

where $\mathbf{q} = (a_\infty, b_1, b_2, b_3, b_4, c_1, \ldots, c_K)$.

4 EXAMPLES

In order to illustrate the performance of the proposed scheme, four principal examples are considered in the following. The integral domains of the four examples are defined as (see also Figure 2):

$$
\begin{aligned}
&\text{Example 1}: D(\lambda) = \left\{(u_1, u_2) \mid u_1 > \lambda/\sqrt{2} \cap u_2 > \lambda/\sqrt{2}\right\} \\
&\text{Example 2}: D(\lambda) = \left\{(u_1, u_2) \mid \lambda/\sqrt{2} < u_1 < \sqrt{2}\lambda \cap \lambda/\sqrt{2} < u_2 < \sqrt{2}\lambda\right\} \\
&\text{Example 3}: D(\lambda) = \left\{(u_1, u_2) \mid u_1 > \lambda \cup u_2 > 1.1\lambda\right\} \\
&\text{Example 4}: D(\lambda) = \left\{(u_1, u_2) \mid u_1^2 + u_2^2 > \lambda^2\right\}
\end{aligned}
\tag{18}
$$

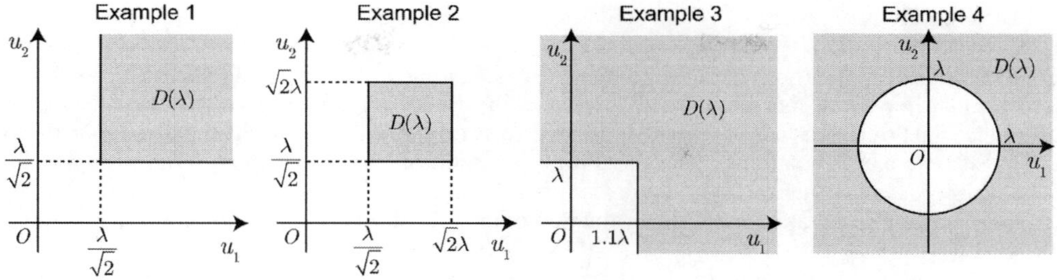

Figure 2. Integral domains in the examples.

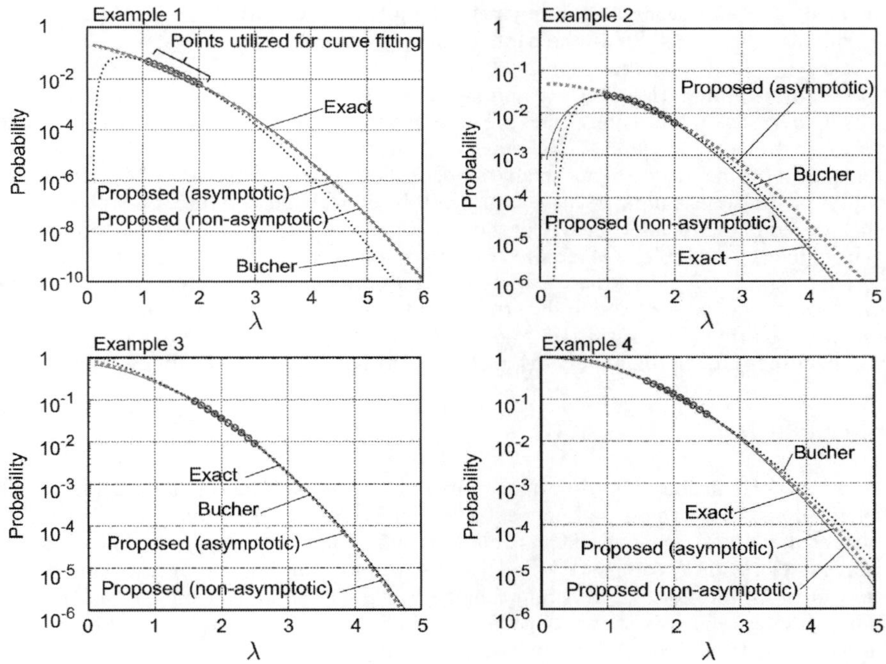

Figure 3. Comparisons of the probability integrals obtained by different extrapolation schemes.

In each example, the values of the probability integral $I(\lambda)$ for the range $\lambda > \lambda_U$ are evaluated by the extrapolation with the proposed functional forms (asymptotic and non-asymptotic cases) using the values of the probability integral $I(\lambda)$ for certain range $[\lambda_L, \lambda_U]$ of the factor λ. These ranges are illustrated in Figure 3 by "points utilized for curve fitting". Here, 10 points are utilized for the curve-fitting. In these examples, the values for the probability integral $I(\lambda)$ for the ranges $[\lambda_L, \lambda_U]$ are evaluated numerically, not by means of sampling. By doing so, the performance of the proposed scheme is examined without possible effects of the estimation errors of the probability integrals by sampling; in practical applications, however, these should be evaluated by sampling. For the purpose of comparison, the probability integral $I(\lambda)$ for the same range are also evaluated with the functional form proposed by Bucher (2009); however, it should be emphasized that Bucher (2009) proposes the use of the functional form only for the case of component reliability assessment.

The results are shown in Figure 3. It is seen that in all the example the proposed functional form (Equation (17)) for the non-asymptotic case succeeds in evaluating the probability integrals $I(\lambda)$ for the range $\lambda > \lambda_U$ with high precision. The proposed functional form (Equation (9)) for the asymptotic case fails to evaluate the probability integral in Example 2. The functional form proposed by Bucher (2009) performs well except for the case of the parallel system (see Figure 3 (top, left)).

5 DISCUSSION

For the practical applications of the approaches for evaluating probability integrals by extrapolation, it is utmost important that the extrapolation scheme works for all cases of practical relevance; otherwise the limitations of the scheme must be clearly provided in such a way that the applicability of the scheme can be examined a-priori. From this point of view, even though the extrapolation scheme proposed by Bucher (2009) works for many cases including some cases of parallel systems, it cannot be practically applied for the evaluation of the probability integrals for the case of parallel systems.

Whereas the approach for estimating the parameters in the functional form for the extrapolation is not discussed in the examples, it is important to choose suitable methods for the estimation and set appropriate conditions required for the methods. In the example above, the parameters are estimated by a non-linear curve fitting method implemented in MATLAB® (`lsqcurvefit`) with the initial values of 0.8 for all the parameters to be estimated.

In practical applications of the proposed scheme, the probability integrals for certain ranges must be estimated by sampling. However, the optimal setting is not yet clear in regard to (1) the number of the discrete values of the factor λ and the corresponding probability integrals $I(\lambda)$ utilized for the estimation of the parameters, (2) the range of the factor to be utilized for the estimation of the parameters, and (3) the degree of acceptable statistical errors associated with the estimation of the probability integrals by sampling. These issues will be considered in future research.

Finally, the computational efficiency of the presented approach in regard to sampling should be emphasized; in each example, it is required only to perform 10 evaluations of probability integrals of the order of $[10^{-1}, 10^{-2}]$ by Monte Carlo simulation, which approximately reduces the required number of samples in comparison to the crude Monte Carlo simulation by a factor of $10^3 \sim 10^4$ if the probability of original interest is 10^{-6}; assuming the number of samples required for evaluation of the probabilities are approximately inversely proportional to the probabilities.

6 CONCLUSION

The present paper considers multi-normal probability integral solutions by the use of scalable parameterizations, the fundamental concept of which are originally proposed by Bucher (2009) and Naess et al. (2009). Taking basis in the concept underlying in these papers, a formulation for a scalable parameterization is proposed based on the asymptotic theory as well as the non-asymptotic characteristics of multi-normal probability integrals. The proposed scheme is illustrated on four principal examples, and it is shown that the proposed scheme work well for series and parallel systems as well as for the case of multiple design points.

REFERENCES

Bleistein, N., and Handelsman, R. A. (1975). Asymptotic expansions of integrals, Holt, Rinehart and Winston.

Breitung, K. (1984). Asymptotic approximations for multinormal integrals. Journal of Engineering Mechanics Division, 110(3), 357–366.

Breitung, K., and Hohenbichler, M. (1989). Asymptotic Approximations for Multivariate Integrals with an Application to Multinormal Probabilities. Journal of Multivariate Analysis, 30(1), 80–97.

Breitung, K. W. (1994). Asymptotic Approximations for Probability Integrals, Springer-Verlag.

Bucher, C. (2009). Asymptotic sampling for high-dimensional reliability analysis. Probabilistic Engineering Mechanics, 24, 504–510.

Hohenbichler, M., Gollwitzer, S., Kruse, W., and Rackwitz, R. (1987). New light on first- and second-order reliability methods. Structural Safety, 4, 267–284.

Naess, A., Leira, B. J., and Batsevych, O. (2009). System reliability analysis by enhanced Monte Carlo simulation. Structural Safety, 31(5), 349–355.

Nishijima, K., Qin, J., and Faber, M. H. (2010). Probability integral solution by extrapolation for system reliability assessment. International Symposium on Reliability Engineering and Risk Management (ISRERM2010), Shanghai, China, (to be presented).

Reliability and Optimization of Structural Systems – Straub (ed)
© 2010 Taylor & Francis Group, London, ISBN 978-0-415-88179-1

Stochastic life-cycle cost analysis of structures in a finite time planning horizon

Mahesh D. Pandey & Tianjin Cheng
University of Waterloo, Waterloo, Ontario, Canada

ABSTRACT: The paper presents a stochastic model of structural degradation and maintenance cost optimization over the life cycle of a structure. The maintenance optimization models reported in the literature mostly minimize the asymptotic expected cost rate computed using the classical renewal theorem. Since practical engineering projects have relatively short and finite operating life and financing time horizon, the application of asymptotic formulation in such cases becomes questionable. This paper presents a new finite time model for computing the expected life cycle cost with and without discounting, and highlights its advantages through a practical case study.

1 INTRODUCTION

The safety of critical civil infrastructure systems, such as bridges, dams, dikes, is adversely affected by aging related degradation of structural components. In this context, inspection and maintenance programs are implemented for timely detection and mitigation of aging effects in order to comply with the regulatory limits of safety and reliability. The theory of stochastic processes serves as a fundamental basis for modeling uncertain and dynamic process of structural degradation. Maintenance is also a stochastic process, because the number and types of future maintenance actions are uncertain variables.

Probabilistic modeling of structural degradation has been well developed in traditional civil engineering literature (Mori & Ellingwood 1994). However, the Life Cycle Cost (LCC) analysis in most studies has been implicitly limited to the first failure of structure. It is evident from the lack of applications of stochastic renewal theory to LCC analysis problems. In conventional reliability engineering, the "first-failure" problem is referred to as "non-repairable" system analysis, which is conceptually straightforward. Van Noortwijk and his co-workers (van Noortwijk et al. 1995, van Noortwijk & van Gelder 1996, van Noortwijk & Klatter 1999) applied the renewal theory for LCC analysis in a formal sense, and used stochastic gamma process for modeling degradation. More comprehensive examples of stochastic maintenance cost analysis are presented in a series of recent papers (Streicher & Rackwitz 2004, Rackwitz et al. 2005, Streicher et al, 2008, Rackwitz & Joanni 2009). Various types of maintenance policies are studied in industrial engineering and operations research literature, depending on whether or not the inspection schedule is periodic, inspection tools are perfect, failure detection is immediate, or repair duration is finite. Grall et al. (2002) studied condition-based maintenance with non-periodic inspection. Castanier et al. (2003) analyzed a policy in which both the future maintenance (replacement or imperfect repair) and the inspection schedule depend on the current degradation.

The most commonly used criterion for optimizing LCC is based on minimizing the expected cost per unit time, i.e. the *cost rate*. The computation of the cost rate is difficult as it involves computation of convolutions of different probability distributions. The renewal theorem provides a simple alternative to compute long term or asymptotic value of the expected cost rate (Cox 1962). The asymptotic rate is the expected cost in one renewal cycle divided by the expected duration of the renewal cycle. In a technical sense, the asymptotic approach reduces a stochastic renewal problem to a much simpler "non-repairable" or first-failure problem. Therefore, most of the literature for maintenance optimization relies on asymptotic cost rate minimization.

Many practical engineering systems have relatively finite operating life and financing horizon in which the applicability of asymptotic formula can become questionable. This paper presents a

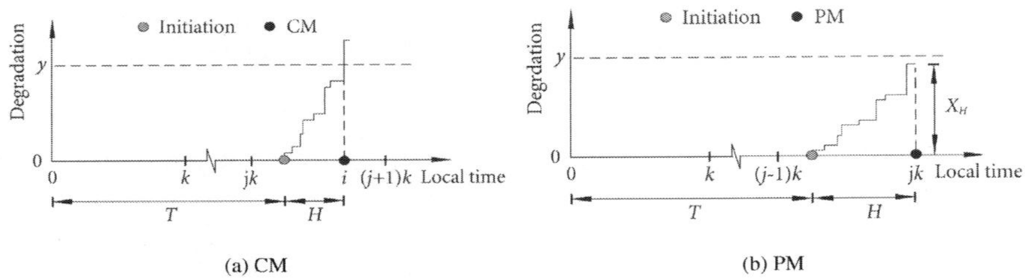

Figure 1. Illustration of CM and PM.

more rigorous finite time model for evaluating the expected LC cost of a maintenance program, as a refinement of asymptotic approach. The proposed approach is based on formulating the expected cost as a generalized renewal equation, which can be easily solved in a discrete time setting. The model is extended to include discounting and compute the net present value of LCC.

The paper considers a structural degradation process that initiates randomly and then it grows in a stochastic manner over time. The model uses two stochastic processes to describe: (1) random initiation of degradation as a geometric process, and (2) cumulative growth as a stochastic gamma process. the structure fails when the damage exceeds a threshold y, and corrective maintenance (CM, Figure 1(a)) is required to renew the component. To reduce the risk of failure, periodic inspections are carried out at a regular interval k. At the time of inspection, preventive maintenance (PM, Figure 1(b)) is done to repair any detected damage and restore the component to as good as new condition. The LCC of a maintenance program includes costs of inspection, PM and CM. In this paper, PM cost is considered as a function of the amount of degradation experienced by the component at the time of inspection. This model is of generic nature, and it can be used to analyze degradation of various nuclear reactor components, which operate in aggressive environment of high pressure, temperature and radiation field.

The paper presents a case study considering the maintenance of berm breakwater, as given in (van Noortwijk & van Gelder 1996). It is illustrated that the asymptotic formula over predicts the life cycle cost as compared to that obtained from the proposed finite time model.

2 STOCHASTIC LCA: BASIC CONCEPTS

In this section, basic concepts underlying the evaluation of the asymptotic and the finite time cost rate are presented. For the sake of simplicity, discounting of cost is not considered in this section.

2.1 *Proposed finite time cost analysis*

In case of a repairable system, multiple renewals can take place in a time interval $(0, t]$, as shown in Figure 2. Let $C(t_1, t_2)$ be the total Life Cycle (LC) cost of maintenance in the time interval $(t_1, t_2]$, $t_1 \leq t_2$. The cost $C(0, t)$, compactly denoted as $C(t)$, is a random variable, since the number of renewals in $(0, t]$ is a random variable and also the type of renewals, corrective or preventive, is not known a priori. Similarly, the duration of a renewal cycle, denoted as L, is also a random variable, because the time of renewal is not known due to stochastic nature of degradation initiation and growth. Let the probability mass function(PMF) of L be $Q(i) = P\{L = i\}$. The expected life cycle cost, $E\{C(t)\}$, is derived by conditioning on the time of first renewal, $L = i$, and then using the law of total expectation. Write $E\{C(t)\}$ in a more compact notation $U(t)$, then

$$U(t) = \sum_{i=1}^{t} E\{C(t)|L = i\}\, Q(i) + C_I(t)P\{L > t\}, \tag{1}$$

where $C_I(t)$ is the cost up to t when $L > t$. Obviously $C_I(t) = [t/k]c_I$ since only inspection cost is incurred in $(0, t]$ then.

On the other hand, if $L = i < t$, the total cost can be split into two parts: $C_1(i)$ and $C(i, t)$ (see Figure 2). $C_1(i)$ is the cost in a renewal cycle of length $L = i$, which can be $C_F(i)$ if it is a CM or $C_P(j), j = i/k$, if it is a PM.

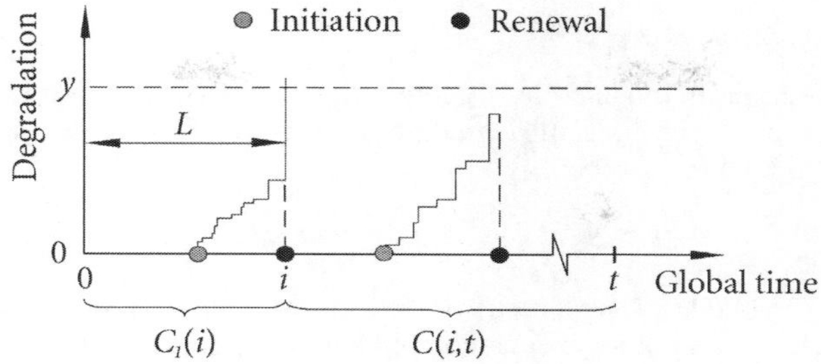

Figure 2. Sample path of renewal process.

An important point is that the cost in the remaining time horizon, $C(i, t)$, is stochastically similar to the cost $C(0, t-i)$ or $C(t-i)$. The reason is that the degradation process is independent of the first renewal, and the same maintenance policy is adopted in the remaining interval (i, t). Therefore, we can consider time i as a new origin of the renewal process and analyze the cost in the interval $(0, t-i)$. Thus,

$$E\{C(t)|L=i\} = C_1(i) + E\{C(i,t)\} = C_1(i) + E\{C(t-i)\} = C_1(i) + U(t-i). \tag{2}$$

Substituting this result into Eq. (1) gives

$$U(t) = \sum_{i=1}^{t} U(t-i)Q(i) + G(t), \tag{3}$$

where $G(t) = \sum_{i=1}^{t} C_1(i)Q(i) + C_I(t)P\{L > t\}$. Equation (3) is a generalized renewal equation, which can be recursively solved in a discrete time setting. The solution of (3) requires probabilistic information about $Q(i)$ and $C_1(i)$ with reference to the first renewal cycle only. The expected cost rate in a finite time horizon is defined as

$$u(t) = U(t)/t. \tag{4}$$

The above analysis can be extended to the case of discounted cost, as shown later in the paper.

2.2 *Asymptotic cost analysis*

In a time interval $(0, t)$, the stochastic life cycle model consists of a sequence of renewal cycles, $L_1, L_2, \ldots, L_{N(t)}$, and associated costs incurred are $C_1, C_2, \ldots, C_{N(t)}$. The total cost corresponding to $N(t)$ number of renewals (a random variable) is given as $C(t) = \sum_{i=1}^{N(t)} C_i$. Using the renewal theorem, it can be proved that as $t \to \infty$ the cost rate $C(t)/t$ asymptotically converges to a value

$$u_\infty = \mu_C/\mu_L, \tag{5}$$

where μ_C and μ_L are the expected cost in a single renewal cycle and the expected length of a single renewal cycle, respectively, given as $\mu_C = \sum_{i=1}^{\infty} C_1(i)Q(i)$ and $\mu_L = \sum_{i=1}^{\infty} iQ(i)$.

Since the evaluation of u_∞ is quite simple, most papers in the literature use asymptotic cost rate as an objective function for life cycle cost optimization. The asymptotic formulation basically reduces the stochastic renewal process model to a single cycle or first failure problem. This approach is simplistic, because most structural components have finite service life. As the financing of an infrastructure is done in a finite planning horizon, it is important to derive accurate LCC in a finite time setting without relying on asymptotic solutions.

3 ANALYSIS OF A RENEWAL CYCLE

In this section, the probability distribution of the length of a single renewal cycle is analyzed. Unless explicitly stated, the time here refers to the local time, starting from the origin of a renewal cycle.

3.1 *Degradation initiation process*

The degradation initiation is modeled as a geometric process with a parameter ϕ, the initiation probability per unit time. The time of degradation initiation, T, is a geometrically distributed random variable with PMF

$$q_i^T = P\{T = i\} = \phi(1-\phi)^i. \tag{6}$$

where $i = 0, 1, 2 \cdots$. Although the geometric process is the simplest model of initiation, proposed formulation can include non-homogeneous Poisson and renewal processes as well.

3.2 *Degradation growth process*

The degradation growth is modeled as a cumulative gamma process. Let $X(\tau)$ denote the magnitude of degradation at time τ, after its initiation. The PDF of $X(\tau)$ is given as

$$f_\tau(x) = \frac{x^{\alpha\tau-1}}{\Gamma(\alpha\tau)}\left(\frac{\alpha}{\theta}\right)^{\alpha\tau}\exp\left(-\frac{\alpha x}{\theta}\right), \tag{7}$$

where α is the shape parameter and θ is the average growth rate. Note that the scale parameter of gamma process is given as $\beta = \theta/\alpha$. The continuous time gamma process can be transformed into a discrete time process by defining a unit time interval of $\Delta = 1/\alpha$. Thus the time axis is discretized as $h\Delta$ for $h = 1, 2, \cdots$. The degradation at any discrete time, $X(h\Delta)$, is compactly denoted as X_h and its PDF is obtained from Eq. (7) as

$$f_h(x) = \frac{x^{h-1}}{(h-1)!(\theta\Delta)^h}\exp\left(-\frac{x}{\theta\Delta}\right). \tag{8}$$

The failure is assumed to take place when degradation exceeds a critical threshold, y, which is followed by the corrective maintenance (CM) of the structure. The probability of CM at time $H = h$, starting from initiation, is given as

$$\begin{aligned} q_h^H = P\{X_{h-1} \le y, X_h > y\} &= \int_0^y f_{h-1}(x_1)\left[\int_{y-x_1}^{\infty} f_1(x_2)dx_2\right]dx_1 \\ &= \frac{1}{(h-1)!}\left(\frac{y}{\theta\Delta}\right)^{h-1}\exp\left(-\frac{y}{\theta\Delta}\right). \end{aligned} \tag{9}$$

3.3 *Distribution of renewal cycle length*

The structure is renewed by CM if failure takes place at any time due to degradation exceeding the limit y or by PM at the time of periodic inspection that takes place at an interval of k. A renewal cycle is denoted as $\{L, M\}$, where L is (random) cycle length and M is either CM or PM.

3.3.1 *Renewal by CM*

In this section, probability of CM at time i ($Q_F(i)$) is derived, which is the same as the probability of failure. Suppose the failure time i lies between an arbitrary inspection interval jk to $(j+1)k$. The local time elapsed from jk to i is given as k', where $0 < k' \le k$, so that $i = jk + k'$. Conditioning on initiation time T and using the law of the total probability lead to

$$Q_F(i) = \sum_{h=1}^{k'} P\{T = i-h, X_{h-1} \le y, X_h > y\} = \sum_{h=1}^{k'} q_{i-h}^T q_h^H. \tag{10}$$

Substituting Eqn. (6) and (9) into the above equation results in

$$Q_F(i) = \left[\sum_{h=1}^{k'} \frac{1}{(h-1)!}\left(\frac{y}{\theta\Delta}\right)^{h-1}(1-\phi)^{i-h}\right]\phi\exp\left(-\frac{y}{\theta\Delta}\right). \tag{11}$$

3.3.2 *Renewal by PM*

The PM takes place at discrete times jk $j = 1, 2, \cdots$, only. Using the conditioning argument, the probability of PM at j^{th} inspection, $Q_P(j)$, is given as

$$Q_P(j) = \sum_{h=1}^{k} P\{T = jk - h, X_h \leq y\} = \sum_{h=1}^{k} q_{jk-h}^{T} \int_0^y f_h(x)dx. \tag{12}$$

Substituting Eqn. (6) and (8) into the above equation gives

$$Q_P(j) = \left[\sum_{h=1}^{k} \frac{\gamma\left(h, \frac{y}{\theta\Delta}\right)}{(h-1)!(1-\Phi)^h}\right] \phi(1-\phi)^{jk}, \tag{13}$$

where $\gamma(v, u) = \int_0^u x^{v-1} e^{-x} dx$ is the lower incomplete gamma function.

3.3.3 *PMF of renewal cycle*

The PMF of the renewal cycle is finally derived as

$$Q(i) = Q_F(i) + \begin{cases} 0, & \text{if } i \neq k, 2k, \cdots, \\ Q_P(j), & \text{if } i = jk,\ j = 1, 2, \cdots \end{cases} \tag{14}$$

The survival or reliability function of L is

$$A(i) = 1 - \sum_{i'=1}^{i} Q(i') = 1 - \sum_{i'=1}^{i} Q_F(i') - \sum_{j=1}^{\lfloor i/k \rfloor} Q_P(j). \tag{15}$$

4 LIFE CYCLE COST ANALYSIS

The unit costs of CM and inspection are denoted as c_F and c_I, respectively. The cost of PM consists of a fixed cost c_P and a variable cost, c_V per unit of the degradation. This section presents expected cost rate evaluation with and without discounting, which serves as a basis for optimizing the inspection interval.

4.1 *Expected cost analysis*

In case of CM at time i, the cost in such a renewal cycle is given as

$$C_F(i) = n_I c_I + c_F, \tag{16}$$

where $n_I = \lfloor i/k \rfloor$ is the number of inspection until time i. PM takes place only at periodic inspection times $jk, j = 1, 2, \cdots$. The PM renewal cost includes a variable cost component, $c_V X_H$, where X_H is the degradation detected at time jk and $X_H \leq y$. Since X_H is a random variable, the expected cost of PM renewal cycle is calculated by substituting expected value of X_H, which is denoted as $B(j)$. Then the expected cost in a PM renewal cycle with length $L = jk$ is

$$C_P(j) = jc_I + c_P + B(j)c_V. \tag{17}$$

The derivation of $B(j)$ is given in Appendix. Since CM and PM are mutually exclusive ways of renewal, the expected cost in a renewal cycle and the expected length of a renewal cycle can be in general written as, respectively,

$$\mu_C = \sum_{i=1}^{\infty} C_F(i)Q_F(i) + \sum_{j=1}^{\infty} C_P(j)Q_P(j), \quad \mu_L = \sum_{i=1}^{\infty} iQ_F(i) + k\sum_{j=1}^{\infty} jQ_P(j). \tag{18}$$

Then the asymptotic cost rate is obtained as $u_\infty = \mu_C/\mu_L$ (van Noortwijk & van Gelder 1996).

The finite time formulation for LCC is given in the form of a generalized renewal equation

$$U(t) = \sum_{i=1}^{t} U(t-i)Q(i) + G(t), \tag{19}$$

where

$$G(t) = \sum_{i=1}^{t} C_F(i)Q_F(i) + \sum_{j=1}^{\lfloor t/k \rfloor} C_P(j)Q_P(j) + C_I(t)A(t). \tag{20}$$

4.2 *Discounted LCC*

The net present value of expected LC cost is computed by discounting using a discount factor, $\rho = 1/(1+r)$, where r is interest rate. Costs of CM and PM renewal cycles are written as

$$C_F(i) = \sum_{j'=1}^{j} \rho^{j'k} c_I + \rho^i c_F = \frac{\rho^k(1-\rho^{jk})}{1-\rho^k} c_I + \rho^i c_F, \tag{21}$$

$$C_P(j) = \sum_{j'=1}^{j} \rho^{j'k} c_I + \rho^{jk}\left[c_F + Bc_V\right] = \frac{\rho^k(1-\rho^{jk})}{1-\rho^k} c_I + \rho^{jk}\left(c_F + Bc_V\right). \tag{22}$$

The expected discounted cost over an infinite time horizon can be obtained as (van Noortwijk & van Gelder 1996)

$$U_\infty = \frac{\sum_{i=0}^{\infty} C_F(i)Q_F(i) + \sum_{j=1}^{\infty} C_P(j)Q_P(j)}{1 - \sum_{i=1}^{\infty} \rho^i Q(i)}. \tag{23}$$

The finite time formulation for discounted cost is given as

$$U(t) = \sum_{i=1}^{t} \rho^i Q(i)U(t-i) + G(t), \tag{24}$$

which is similar to Eqn. (19). $G(t)$ is given as in Eq. (20), except that C_F, C_P, and C_I should be taken in the discounted forms (21) and (22). The discounted inspection cost is given as

$$C_I(t) = \sum_{j=1}^{\lfloor t/k \rfloor} \rho^{jk} c_I = \frac{\rho^k(1-\rho^{\lfloor t/k \rfloor k})}{1-\rho^k} c_I, \tag{25}$$

The equivalent average cost $u^e(t)$ is often used as an optimization criterion, which is determined by averaging the discounted cost over a discounted time horizon, i.e.

$$U(t) = u^e(t)\sum_{i=1}^{t} \rho^i = \frac{\rho(1-\rho^t)}{1-\rho} u_e(t) \quad \Longrightarrow \quad u^e(t) = \frac{1-\rho}{\rho(1-\rho^t)} U(t). \tag{26}$$

Letting $t \to \infty$, the asymptotic equivalent average cost can be obtained as

$$u^e_\infty = \left(\rho^{-1} - 1\right) U_\infty. \tag{27}$$

5 EXAMPLE

This section considers an example of maintenance of berm breakwater, in which long shore transport of rocks after the breach of protection leads to a failure of the breakwater. Therefore, this structure has to be maintained at a regular interval. This problem was analyzed in (van Noortwijk & van Gelder 1996) based on asymptotic cost formulation only. Here the same data is used, as shown in

Table 1. Data for numerical example.

Parameter	Description	Value	Dimension
Δ	discrete time interval	1	year
θ	average rate of longshore rock transport	70	stone/year
ϕ	frequency of armour breaches	0.9	breach/year
r	discount rate per year	5	%
ρ	discount factor per year	0.9524	
c_I	cost of inspection	1000	euro
c_P	fixed PM cost	10,000	euro
c_V	variable PM cost	100	euro/stone
c_F	CM cost	2,500,000	euro
y	failure threshold	2500	stone
t	planning horizon	40	year

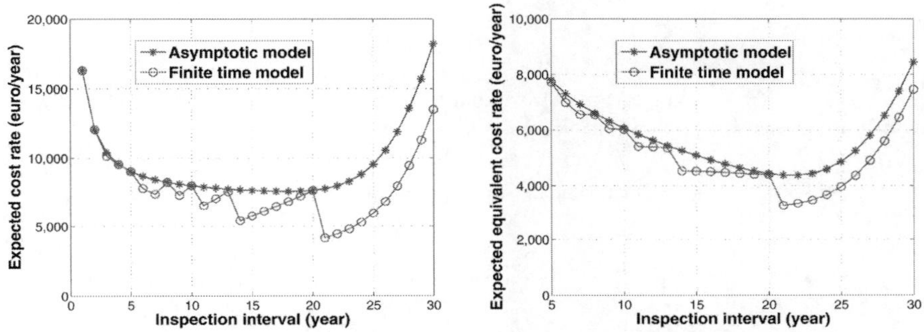

(a) Without discounting (b) With discounting

Figure 3. Expected cost rate vs. inspection interval.

Table 1, and discounting as well as finite time formulations are included in analysis. The comparison of results illustrates the impact of refinements in the proposed analysis.

The variations of cost rate and the expected equivalent cost rate with inspection interval are shown in Figure 3(a) and 3(b), respectively. Based on asymptotic analysis without discounting, an optimal inspection interval is found as 18 years with corresponding cost rate of 7610 euros/year, while the finite time model leads to a longer optimal interval of 21 years at a lower cost rate of 4140 euros/year. The asymptotic model over-predicts the optimal cost by 80%. In the case of discounting, the asymptotic model leads to an optimal interval of 22 years at an equivalent cost rate of 4590 euros/year, whereas optimal solution obtained from finite time model is 21 year inspection interval and equivalent cost rate of 3450 euros/year. Here, the asymptotic model over-predicts the optimal cost by 30%.

This section clearly shows that the use of a refined approach based on finite time formulation is necessary to obtain realistic estimates of life cycle cost associated with a maintenance program.

6 CONCLUSIONS

The optimization of life cycle cost of maintenance programs for an engineering system or structure has been an active area of research. This paper presents the derivation of the expected life-cycle cost, with and without discounting, associated with a maintenance program in a finite time horizon. The degradation initiation and growth in a structure are modeled as stochastic processes. The derivation is based on formulating a generalized renewal equation for the expected cost and numerically solving it.

The paper highlights the fact that the asymptotic expected cost can be a rather crude approximation of the real cost in a finite time horizon. The paper presents a case study involving periodic inspection and maintenance of a berm breakwater, which was analyzed in (van Noortwijk & van

Gelder 1996). This example illustrates that the asymptotic formula significantly over predicts the life cycle cost as compared to that obtained from the proposed finite time model. In case of a large fleet of components, the over prediction by asymptotic formula can be substantial, and it paints a pessimistic picture of the life cycle cost. It is concluded that the finite time model should be used for a realistic evaluation and optimization of the maintenance policy for critical infrastructure systems. The formulation presented in the paper can be extended to other types of maintenance policies.

APPENDIX: EXPECTED DEGRADATION IN A RENEWAL CYCLE

Note that

$$B(j) = \int_0^\infty x f(x|\text{PM at } jk)dx = \frac{1}{Q_P(j)} \int_0^\infty x f(x; \text{PM at } jk)dx, \tag{28}$$

where $f(x|\text{PM at } jk)$ is the probability density of X_H given the condition of PM at jk and $f(x; \text{PM at } jk)$ the joint probability density of X_H in the case of PM at jk, i.e.

$$f(x; \text{PM at } jk) = \frac{1}{dx} P\{X_H \in (x, x+dx], \text{PM at } jk\}. \tag{29}$$

Obviously, $P\{X_H \in (x, x+dx), \text{PM at } jk\}$ is equal to 0 when $x < 0$ or $x > y$. For $0 \leq x \leq y$, using the law of total probability by conditioning on T, we have

$$P\{X_H \in (x, x+dx), \text{PM at } jk\} = \sum_{h=1}^{k} P\{X_h \in (x, x+dx), T = jk - h\}$$

$$= \sum_{h=1}^{k} P\{T = jk - h\} P\{X_h \in (x, x+dx)\} = \sum_{h=1}^{k} q_{jk-h}^T f_h(x)dx. \tag{30}$$

Then $f(x; \text{PM at } jk) = \sum_{h=1}^{k} q_{jk-h}^T f_h(x)$, for $0 \leq x \leq y$. Hence

$$B(j) = \frac{1}{Q_P(j)} \sum_{h=1}^{k} q_{jk-h}^T \int_0^y x f_h(x)dx. \tag{31}$$

REFERENCES

Castanier, B. et al. 2003. A sequential condition-based repair/replacement policy with non-periodic inspections for a system subjected to continuous wear. *Applied Stochastic Models in Business and Industry 19*(4), 327–347.

Cox, D. 1962. *Renewal Theory*. London: Methuen.

Grall, A. et al. 2002. A condition-based maintenance policy for stochastically deteriorating systems. *Reliability Engineering and System Safety 76*, 167–180.

Mori, Y. & Ellingwood, B. 1994. Maintaining reliability of concrete structures. i: Role of inspection/repair. *Journal of Structural Engineering 120*(3), 834–845.

Rackwitz, R. et al. 2005. Socio-economically sustainable civil engineering infrastructures by optimization. *Structural Safety 27*(3), 187–229.

Rackwitz, R. & Joanni, A. 2009. Risk acceptance and maintenance optimization of aging civil engineering infrastructures. *Structural Safety 31*(3), 251–259.

Streicher, H. & Rackwitz, R. 2004. Time-variant reliability-oriented structural optimization and a renewal model for life-cycle costing. *Probabilistic Engineering Mechanics 19*(1–2), 171–183.

Streicher, H. et al 2008. Cost-benefit optimization and risk acceptability for existing, aging but maintained structures. *Structural Safety 30*(5), 375–393.

van Noortwijk, J. et al. 1995. A bayesian failure model based on isotropic deterioration. *European Journal of Operational Research* (2), 270–282.

van Noortwijk, J. & van Gelder, P. 1996. Optimal maintenance decisions for berm breakwaters. *Structural safety* (4), 293–309.

van Noortwijk, J. & Klatter, H. 1999. Optimal inspection decisions for the block mats of the eastern-scheldt barrier. *Reliability Engineering and System Safety*, 203–211.

Reliability and Optimization of Structural Systems – Straub (ed)
© 2010 Taylor & Francis Group, London, ISBN 978-0-415-88179-1

Geotechnical reliability updating using stochastic FEM

I. Papaioannou
Computation in Engineering, TU München, Germany

D. Straub
Engineering Risk Analysis Group, TU München, Germany

ABSTRACT: In geotechnical sites, deformation measurements are commonly made during the construction process. In this paper, it is shown how information obtained during such measurements can be utilized to update the estimate of the reliability at future construction stages. A recently proposed method for Bayesian updating of the reliability with information of equality type is successfully applied in conjunction with a stochastic nonlinear geotechnical finite element model. Therein uncertainty in the soil material properties is modelled by non-Gaussian random fields. The conditional reliability is evaluated by means of an efficient adaptive Monte Carlo method, known as subset simulation. The approach is demonstrated through an application to a sheet pile wall at a deformation-sensitive geotechnical construction site.

1 INTRODUCTION

Geotechnical engineers often use the observation method to deal with uncertainties in site conditions or performance behavior. The engineer collects a few hypotheses about site conditions and then gathers field observations to identify the correct hypothesis. In probability theory, Bayesian updating formalizes this procedure, by quantifying the effect of an observation on the probability of a prior hypothesis. Bayesian updating is applied here within the structural reliability concept. To this end, a recently proposed methodology (Straub 2010) that allows updating of the reliability given measurement information is applied. Hence, the effect of measurements at a geotechnical construction site on the failure probability of the system is quantified.

In geotechnical reliability analysis, material nonlinearities are often taken into account, leading to failure events defined by nonlinear performance functions. The largest uncertainties in the analysis stem from the mechanical properties of the soil materials. The inherent spatial variability of the soil necessitates the consideration of random fields for the accurate modeling of these uncertainties. Moreover, the consideration of the spatial variability seems to have an important influence on the computed reliability (Rackwitz 2000).

In the context of finite element (FE) reliability analysis, the stochastic FE discretization of the random fields may lead to a large number of random variables. The reliability of such high-dimensional nonlinear problems can be solved efficiently by either first or second order reliability methods combined with the direct differentiation method for the evaluation of the derivatives (Zhang & Der Kiureghian 1993, Der Kiureghian & Zhang 1999) or by certain simulation approaches, such as the subset simulation, the spherical subset simulation and recently the asymptotic sampling method (Au & Beck 2001, Katafygiotis & Cheung 2007, Bucher 2009). The former case requires alterations at the FE code level, while the latter may be easily coupled with a "black-box" FE code.

In this work, the subset simulation is applied for the reliability updating of a geotechnical FE model, involving an excavation in sand with a sheet pile retaining wall, whereby the underlying non-Gaussian random fields are discretized using the stochastic FE method.

2 STOCHASTIC FEM

In computational mechanics, the continuous nature is approximated by a discrete representation through the discretization procedure. Among the different discretization methods, the FE method is considered the most suitable for application in geotechnical engineering (Zienkiewicz & Taylor 2000). The FE method requires the splitting of the continuous medium into an assembly of individual elements, i.e. the FE mesh. Moreover, nonlinear FE analysis computes the deformations and stresses through the solution of a nonlinear system of equations, wherein the nonlinear effects of the soil material properties as well as the interaction between the soil and adjacent retaining structures are taken into account. For the simulation of construction sites, usually a sequence of nonlinear FE analyses is performed to model the different stages of the construction process.

The consideration of the stochastic nature of the mechanical properties of soil requires an additional discretization of the relevant random fields. A random field $X(\mathbf{t})$ may be loosely defined as a random function, which describes a random variable at each point $\mathbf{t}$ of a continuous space, where $\mathbf{t}$ is a location vector. To completely define a random field, the joint distribution of the random variables $\{X(\mathbf{t}_1), X(\mathbf{t}_2) \ldots, X(\mathbf{t}_n)\}$ for any $\{n, \mathbf{t}_1, \mathbf{t}_2, \ldots, \mathbf{t}_n\}$ must be specified. In most engineering applications, however, the joint distribution is modeled by a Gaussian copula (the Nataf distribution, Der Kiureghian & Liu 1986), in which case the distribution is completely described by second moment information and the marginal distribution. Denoting the expectation operator by E[.], we define the mean and auto-covariance function of the random field respectively by:

$$\mu_X(\mathbf{t}) = \mathrm{E}[X(\mathbf{t})] \tag{1}$$

$$\gamma_{XX}(\mathbf{t},\mathbf{s}) = \mathrm{E}[\{X(\mathbf{t}) - \mu(\mathbf{t})\}\{X(\mathbf{s}) - \mu(\mathbf{s})\}] \tag{2}$$

The variance function is expressed as:

$$\sigma_X^2(\mathbf{t}) = \gamma_{XX}(\mathbf{t},\mathbf{t}) \tag{3}$$

The auto-correlation coefficient function may then be defined by:

$$\rho_{XX}(\mathbf{t},\mathbf{s}) = \frac{\gamma_{XX}(\mathbf{t},\mathbf{s})}{\sigma_X(\mathbf{t})\sigma_X(\mathbf{s})} \tag{4}$$

A random field is said to be second-order homogeneous if its second-order properties are invariant to a shift of the origin. A direct consequence of this is that the mean and variance are constant over space and the auto-covariance and auto-correlation coefficient functions depend only on the difference in location $\boldsymbol{\tau}$ between two points:

$$\gamma_{XX}(\boldsymbol{\tau}) = \mathrm{E}[\{X(\mathbf{t}+\boldsymbol{\tau}) - \mu_X\}\{X(\mathbf{t}) - \mu_X\}] \tag{5}$$

$$\rho_{XX}(\boldsymbol{\tau}) = \frac{\gamma_{XX}(\boldsymbol{\tau})}{\sigma_X^2} \tag{6}$$

The continuous random field $X(\mathbf{t})$ may be approximated by a discrete $\hat{X}(\mathbf{t})$, defined by means of a finite set of random variables X_i. Several methods have been proposed for the discretization of random fields – a comprehensive review is given by Sudret & Der Kiureghian (2000). In this study, the midpoint method (Der Kiureghian & Ke 1988) is adopted. This method requires a stochastic finite element (SFE) mesh, which can be independent of the FE mesh used in the deterministic part of the analysis. According to this method, the value of the random field $X(\mathbf{t})$ over each SFE i is represented by its value at the midpoint $\mathbf{t}_i$ of the element:

$$X_i = X(\mathbf{t}_i) \tag{7}$$

If the random field is second-order homogeneous, the mean and the variance of the random variables X_i are not affected by the discretization and therefore are equal to their constant values over the entire field. The correlation matrix of the random variables is generated by computing the distances τ_{ij} between the midpoints of each element and inserting into Equation 6. The random field is thus reduced to a vector $\mathbf{X}$ of correlated random variables with known marginal distributions and correlation coefficients.

3 RELIABILITY ANALYSIS

Let $\mathbf{X}$ be a vector of random variables with joint probability density function (PDF) $f_{\mathbf{X}}(\mathbf{x})$. A failure event F is defined by a domain $\Omega_F = \{g(\mathbf{x}) \leq 0\}$ in the outcome space of $\mathbf{X}$, with $g(\mathbf{x})$ being a performance function with the following general form:

$$g(\mathbf{x}) = \min_{1 \leq k \leq K}\left[\max_{i \in C_1} g_i(\mathbf{x}) \ldots, \max_{i \in C_K} g_i(\mathbf{x})\right] \tag{8}$$

where $\{g_i(\mathbf{x}),\ i = 1, \ldots, m\}$ is a set of component performance functions and C_k is an index set denoting the k-th cut set. The component-reliability problem may be derived by setting $i = 1$, the parallel-system reliability problem by setting $K = 1$, and the series-system reliability problem by defining each cut set by a single index. The probability of failure for the general reliability problem is evaluated by:

$$\Pr(F) = \Pr\{g(\mathbf{X}) \leq 0\} = \int_{\Omega_F} f_{\mathbf{X}}(\mathbf{x})\,\mathrm{d}\mathbf{x} \tag{9}$$

In the case where only second moments and marginal distributions are available, the joint PDF may be approximated using the Nataf model and a transformation of the random variables from the original random variable space to an equivalent independent standard Gaussian space (transformed space) $\mathbf{U} = T(\mathbf{X})$ may be performed (Der Kiureghian & Liu 1986). The probability may then be computed as follows:

$$\Pr(F) = \int_{G(\mathbf{u}) \leq 0} \varphi_{\mathrm{U}}(\mathbf{u})\,\mathrm{d}\mathbf{u} \tag{10}$$

where $\phi_{\mathrm{U}}(\mathbf{u})$ is the standard normal joint PDF and $G(\mathbf{u}) = g[T^{-1}(\mathbf{u})]$.

3.1 *Reliability updating with equality information*

Consider measurements or other observations of the structural or geotechnical system. This information is described by an event Z, which in structural reliability is defined through a domain Ω_Z in the outcome space of the basic random variables $\mathbf{X}$. The conditional probability of F given the information Z is:

$$\Pr(F \mid Z) = \frac{\Pr(F \cap Z)}{\Pr(Z)} = \frac{\int_{\Omega_F \cap \Omega_Z} f_{\mathbf{X}}(\mathbf{x})\,\mathrm{d}\mathbf{x}}{\int_{\Omega_Z} f_{\mathbf{X}}(\mathbf{x})\,\mathrm{d}\mathbf{x}} \tag{11}$$

The domain Ω_Z is defined through a function $h(\mathbf{x})$; if Ω_Z is defined as $\Omega_Z = \{h(\mathbf{x}) \leq 0\}$, the event Z is said to be of inequality-type and if Ω_Z is defined as $\Omega_Z = \{h(\mathbf{x}) = 0\}$, Z is said to be of equality-type. In the latter case, the probability of the event Z is zero and both integrals in Equation 11 will result in zero. To circumvent this problem, a method proposed in (Straub 2010) will be applied, according to which the event Z is replaced by an equivalent inequality-type event Z_e.

To illustrate the methodology, consider the case where Z describes the measurement s_m of a system characteristic $s(\mathbf{x})$, e.g. a deformation measurement of a structure, with corresponding equality-type performance function:

$$h(\mathbf{x}, \varepsilon_m) = s(\mathbf{x}) - s_m + \varepsilon_m \tag{12}$$

where ε_m is a random variable representing the measurement error. Equivalently, the information of the event Z with respect to the random variables $\mathbf{X}$ can be expressed by the likelihood function:

$$L(\mathbf{x}) = f_{\varepsilon_m}[s_m - s(\mathbf{x})] \tag{13}$$

where $f_{\varepsilon m}$ is the PDF of ε_m. It is suggested in (Straub 2010) to utilize this likelihood function to define a new performance function:

$$h_e(\mathbf{x}, u) = u - \Phi^{-1}[cL(\mathbf{x})] \tag{14}$$

where u is the outcome of a standard normal random variable, $\phi^{-1}(.)$ is the inverse of the standard normal cumulative distribution function and c is a positive constant, chosen to ensure that $cL(\mathbf{x}) \leq 1$ for all $\mathbf{x}$. Let $\Omega_{Ze} = \{h_e(\mathbf{x}_+) \leq 0\}$, be the corresponding domain of event Z_e that defines an inequality information, where $\mathbf{x}_+$ is the outcome of $\mathbf{X}_+ = [\mathbf{X}, U]^T$. It is shown in (Straub 2010) that the conditional probability of F given Z is:

$$\Pr(F \mid Z) = \Pr(F \mid Z_e) = \frac{\Pr(F \cap Z_e)}{\Pr(Z_e)} = \frac{\int_{\Omega_F \cap \Omega_{Ze}} f_{\mathbf{X}_+}(\mathbf{x}_+)\,\mathrm{d}\mathbf{x}_+}{\int_{\Omega_{Ze}} f_{\mathbf{X}_+}(\mathbf{x}_+)\,\mathrm{d}\mathbf{x}_+} \tag{15}$$

where $f_{\mathbf{X}+}(\mathbf{x}_+)$ is the joint PDF of $\mathbf{X}_+$. As opposed to the original formulation in Equation 11, both integrals in Equation 15 have finite value and can thus be computed using any structural reliability method, including simulation methods. Here, this formulation will be applied in conjunction with subset simulation for the application to geotechnical reliability updating.

4 SUBSET SIMULATION

The subset simulation is an adaptive simulation method developed by Au & Beck (2001), which is shown to be efficient especially in high dimensions. The method is based on the standard Monte Carlo simulation (MCS) but overcomes its inefficiency in estimating small probabilities, while maintaining its independency on the problem dimensionality. This is achieved by expressing the failure event F as the intersection of M intermediate failure events:

$$F = \bigcap_{i=1}^{M} F_i \tag{16}$$

where $F_1 \supset F_2 \supset \cdots \supset F_M = F$. The probability of failure is estimated by applying the definition of conditional probability, resulting in:

$$\Pr(F) = \Pr(F_1)\prod_{i=2}^{M}\Pr(F_i \mid F_{i-1}) \tag{17}$$

The failure event is defined in the equivalent standard normal space by $G(\mathbf{u}) \leq 0$; each of the intermediate events is defined as $F_i = \{G(\mathbf{u}) \leq G_i\}$, where $G_i > \ldots > G_M = 0$. The values of G_i are chosen adaptively so that the estimates of the conditional probabilities correspond to a chosen value p_0. The probability of F_1 is computed by applying the crude MCS method (Rubinstein 1981). Through MCS, we simulate N samples of $\mathbf{U}$ and take as G_1 the $[(1-p_0)N]$-th largest value among the samples $\{G(\mathbf{u}_k)$: $k = 1,\ldots,N\}$. The samples $\mathbf{u}_i$ for which $G(\mathbf{u}_i) \leq G_1$ are used as starting points for the simulation of samples conditional on F_1, by applying the Metropolis-Hastings algorithm, which is a Markov chain Monte Carlo technique. This procedure is repeated until the maximum level M is reached, for which the threshold $G_M = 0$ is given. The failure probability is then approximated by:

$$\Pr(F) \approx p_0^{\,M-1}\Pr(F_M \mid F_{M-1}) \tag{18}$$

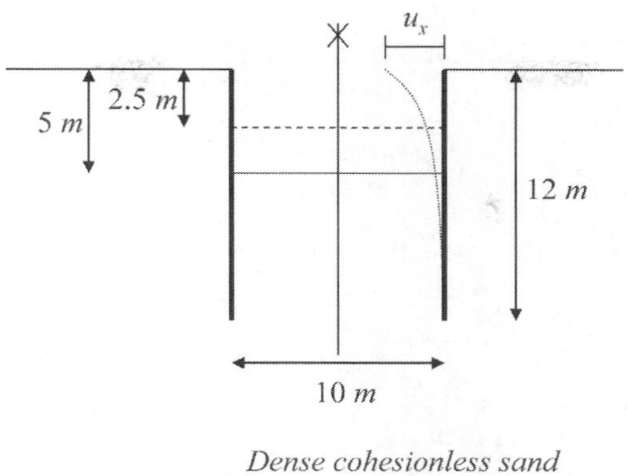

Figure 1. Sheet pile wall in sand.

where the conditional probability $\Pr(F_M \mid F_{M-1})$ is equal to the ratio of the number of samples for which $G(\mathbf{u}) \leq 0$ over the total number of samples N simulated conditional on F_{M-1}.

The choice of p_0 and the number of samples at each step N may in principle be arbitrary. However, the choice of N should be large enough to give an accurate estimation of p_0. In this study, the values $p_0 = 0.1$ and $N = 500$ were chosen, following the initial suggestions given by Au & Beck (2001).

5 APPLICATION

The application considered is a deep excavation in sand with a sheet pile retaining wall. The FE model was built in the SOFiSTiK program and the reliability assessment was performed using a reliability tool, developed as part of the SOFiSTiK software package (Papaioannou et al. 2009).

5.1 *Geotechnical model description*

The site consists of a 5.0 m deep trench with cantilever sheet piles, without anchors or bottom support (Fig. 1), in a homogeneous soil layer of dense cohesionless sand with uncertain spatially varying mechanical properties. The soil is modeled in 2D with plane-strain finite elements. For simplicity, neither groundwater nor external loading is considered. The material model used is an elastoplastic model with a prismatic yield surface according to the Mohr-Coulomb criterion and a non-associated plastic flow. The probability distributions, describing the uncertainty in the material properties of the soil, are shown in Table 1. The spatial variability of the soil is modeled by a second-order homogeneous random field, with the following exponential auto-correlation coefficient function (Vanmarcke 1983):

$$\rho_{XX}(\boldsymbol{\tau}) = \exp\left(-\frac{\tau_x}{l_x} - \frac{\tau_z}{l_z}\right) \tag{19}$$

where $\tau = [\tau_x, \tau_z]^T$ is the vector of absolute distances in the x (horizontal) and z (vertical) directions. The correlation lengths were chosen as $l_x = 20$ m and $l_z = 5$ m. The random field was discretized by the midpoint method using a SFE mesh, consisting of 144 deterministic FE patches. The stochastic discretization resulted in a total of 432 basic random variables. In Figure 2, the stochastic and deterministic FE meshes are shown.

A deterministic design using the conventional method for cantilever sheet pile design in granular soils, which requires equilibrium of the active and passive lateral pressures (e.g. see Tschebotarioff 1951), was performed analytically. A depth of 7.5 m and the sheet pile profile PZC 13 were chosen. The Young's modulus of steel was taken as 210 *GPa*. The pile was modeled using beam elements with an equivalent rectangular cross section that behaves equally to the sheet pile in bending and axial resistance. The interaction between the retaining structure and the surrounding soil was modeled

Table 1. Material properties.

Parameter	Distribution	Mean	COV
Specific weight γ $[kN/m^3]$	Normal	19.0	5%
Young's modulus E $[MPa]$	Lognormal	125.0	25%
Poisson's ratio ν	–	0.35	–
Friction angle ϕ [°]	Beta(0.0,45.0)	35.0	10%
Cohesion c $[MPa]$	–	0.0	–
Dilatancy angle ψ [°]	–	5.0	–

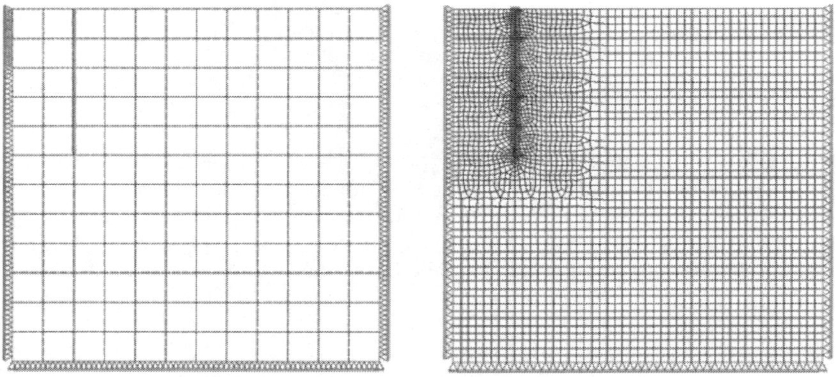

(a) Stochastic finite element mesh (b) Deterministic finite element mesh

Figure 2. Stochastic and deterministic finite element mesh.

using nonlinear interface elements. The properties of these elements are taken from the mean values of the adjacent soil, while the strength properties are reduced by the factor 2/3 (corresponding to a global safety factor of 1.5).

The finite element analysis is performed stepwise. First, the modeling of the in-situ stress state is carried out applying the K_0-procedure, where K_0 is the lateral earth pressure coefficient at rest, computed here using the expression proposed by Jaky (1944) for normally consolidated soils:

$$K_0 = 1 - \sin\varphi \tag{20}$$

Next, the sheet pile is installed by activating the corresponding beam and interface elements. Finally, the excavation is modeled by removing the plane-strain elements corresponding to the trench and applying the necessary loading to establish equilibrium.

5.2 *Limit state functions*

The maximum horizontal u_x displacement occurs at the top of the trench. The failure event F is defined as the event of u_x exceeding a threshold of $u_{x,t} = 10$ cm. Mathematically, this is expressed through the following performance function:

$$g(\mathbf{x}) = u_{x,t} - u_x(\mathbf{x}) \tag{21}$$

We assume that a measurement of the displacement u_x is made at an intermediate excavation step of 2.5 m depth. This information is expressed by an event Z, defined through the following equality-type performance function:

$$h(\mathbf{x}) = u_x(\mathbf{x}) - u_{x,m} + \varepsilon_m \tag{22}$$

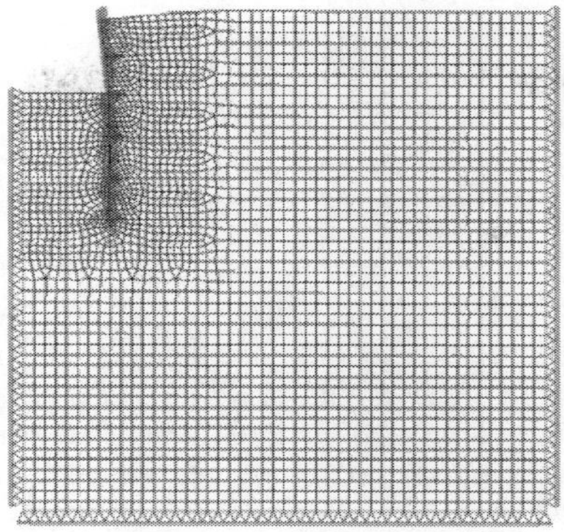

Figure 3. Deformed configuration at the final excavation step.

Table 2. Updated failure probability $\Pr(F) = 1.360 \times 10^{-2}$

	$\Pr(F\|Z)$	
Measurement	$\sigma_{\varepsilon,m} = 2\,mm$	$\sigma_{\varepsilon,m} = 1\,mm$
$u_{x,m} = 10\,mm$	2.185×10^{-1}	3.313×10^{-1}
$u_{x,m} = 5\,mm$	2.091×10^{-2}	3.587×10^{-2}
$u_{x,m} = 2\,mm$	6.743×10^{-3}	1.845×10^{-3}

where ε_m is a zero mean Gaussian random variable, representing the measurement error. The corresponding equivalent inequality performance function is:

$$h_e(\mathbf{x},u) = u - \Phi^{-1}\left[\frac{c}{\sigma_{\varepsilon,m}}\varphi\left(\frac{u_{x,m} - u_x(\mathbf{x})}{\sigma_{\varepsilon,m}}\right)\right] \tag{23}$$

where $\sigma_{\varepsilon,m}$ is the standard deviation of ε_m and the constant is chosen as $c = \sigma_{\varepsilon,m}$.

5.3 Results

In Figure 3, the deformed configuration is depicted at the final excavation stage for the mean values of the random fields.

The reliability analysis is performed by means of subset simulation. Without measurements, the computed failure probability is $\Pr(F) = 1.360 \times 10^{-2}$ with a corresponding reliability index $\beta = 2.209$.

For the estimation of the updated failure probability conditional on the measurement event Z, the integrals in Equation 15 were evaluated with subset simulation. The reliability updating was performed for different values of the assumed measurement $u_{x,m}$, and different values of the standard deviation $\sigma_{\varepsilon,m}$ of the measurement error. The results are summarized in Table 2. For comparison, the (a-priori) expected value of $h(\mathbf{x})$ is computed as 2.57 mm.

Not surprisingly, for measurements higher than the expected value, the updated failure probability is higher than the prior probability. This difference increases with decreasing $\sigma_{\varepsilon,m}$, representing an increase in accuracy of the measurement device. On the other hand, the updated failure probability conditional on a measurement smaller than the expected value is lower than the prior probability, and a decreasing value of $\sigma_{\varepsilon,m}$ increases this difference. It is noted, however, that a measurement that corresponds exactly to the expected value of the deformation would lead to a posterior failure probability that is lower than the prior probability, due to the reduction in uncertainty.

The number of deterministic FE analyses required by the subset simulation ranges between 1900 and 3700. The higher amount of computations is observed in the case where the assumed measurement differs considerably from the expected value (i.e. the case where $u_{x,m} = 10\,mm$). This is due to the small value of the probability $\Pr(Z_e)$ in Equation 15, resulting in a larger number of levels in the corresponding run of the subset simulation algorithm.

6 CONCLUSION

In this paper, a procedure for the reliability updating of geotechnical sites using measurement information is presented. The applied methodology includes the stochastic discretization of the non-Gaussian random fields, representing the spatial variability of the uncertain material parameters. For the reliability updating, a recently proposed method that tackles Bayesian updating with equality information was applied. This approach was combined with the subset simulation, an adaptive Monte Carlo method that is able to handle efficiently reliability problems with a large number of random variables.

The proposed procedure was applied to a nonlinear stochastic finite element model of an excavation with a sheet pile retaining wall. Several different cases of deformation measurements in an intermediate construction stage were considered. The results showed that reasonable estimates of the conditional failure probability may be obtained with a feasible computational cost even when including a large number of basic random variables for representing the soil properties by random fields.

ACKNOWLEDGEMENT

This work is partially supported by SOFiSTiK AG. This support is gratefully acknowledged.

REFERENCES

Au, S.K. & Beck, J.L. 2001. Estimation of small failure probabilities in high dimensions by subset simulation. *Probabilistic Engineering Mechanics* 16(4): 263–277.

Bucher, C. 2009. Asymptotic sampling for high-dimensional reliability analysis. *Probabilistic Engineering Mechanics* 24: 504–510.

Der Kiureghian, A. & Ke, J-B. 1988. The stochastic finite element method in structural reliability. *Probabilistic Engineering Mechanics* 3(2): 83–91.

Der Kiureghian, A. & Liu, P-L. 1986. Structural reliability under incomplete probability information. *Journal of Engineering Mechanics ASCE* 112(1): 85–104.

Der Kiureghian, A. & Zhang, Y. 1999. Space variant finite element reliability analysis. *Comput. Methods Appl. Mech. Engrg.* 168: 173–183.

Jaky, J. 1948. The coefficient of earth pressure at rest. In Hungarian. *J. Soc. Hung. Eng. Arch.*, 355–358.

Katafygiotis, L.S. & Cheung, S.H. 2007. Application of spherical subset simulation method and auxiliary domain method on a benchmark reliability study. *Structural Safety* 29: 194–207.

Papaioannou, I., Heidkamp, H., Düster, A., Rank, E. & Katz, C. 2009. Integration of reliability methods into a commercial finite element software package. In H. Furuta, D.M. Frangopol, M. Shonozuka (eds.), *Proceedings of the 10th International Conference on Structural Safety and Reliability ICOSSAR 2009, Osaka, September, 2009*. London: Taylor & Francis Group.

Rackwitz, R. 2000. Reviewing probabilistic soils modelling. *Computers and Geotechnics* 26: 199–223.

Rubinstein, R.Y. 1981. *Simulation and the Monte Carlo Method*. New York: John Wiley & Sons.

Straub, D. 2010. Reliability updating with equality information, *Probabilistic Engineering Mechanics*, under review.

Sudret, B. & Der Kiureghian, A. 2000. *Stochastic Finite Elements and Reliability: A State-of-the-Art Report*. University of California, Berkeley, 2000 – Technical Report no UCB/SEMM-2000/08.

Tschebotarioff, G.P. 1951. *Soil Mechanics, Foundations and Earth Structures*. New York: M cGraw-Hill.

Vanmarcke, E.H. 1983. *Random Fields: Analysis and Synthesis*. Cambridge, MA: MIT Press.

Zhang, Y. & Der Kiureghian A. 1993. Dynamic response sensitivities of inelastic structures. *Comput. Methods Appl. Mech. Engrg.* 108: 23–36.

Zienkiewicz, O.C. & Taylor, R.L. 2000. *The Finite Element Method. Fifth edition.* Oxford: Butterworth-Heinemann.

Reliability and Optimization of Structural Systems – Straub (ed)
© 2010 Taylor & Francis Group, London, ISBN 978-0-415-88179-1

Optimal design of tall buildings subjected to wind action in the framework of performance based design

F. Petrini, M. Ciampoli & G. Augusti
Sapienza Università di Roma, Department of Structural and Geotechnical Engineering, via Eudossiana, Rome, Italy

ABSTRACT: A probabilistic procedure for the Performance-Based Design of structures subject to wind action is illustrated. The central objective of the procedure is the assessment of the adequacy of the structure through the probabilistic description of a set of decision variables DVs. Each DV is a quantitative measure of a specific structural performance (no collapse, occupant safety, accessibility, full functionality, limited displacements or accelerations, etc.). The structural risk is conventionally measured by the probability of exceeding a relevant value of the corresponding DV; the probability is evaluated by taking into account the Aeolian hazard at the site, the calculated structural response and damage, and the correlation between the attained damage level and the relevant DV. The procedure is applied to an example case: the assessment of the comfort requirement for a high-rise building.

1 INTRODUCTION

During the last decades, the problem of the reliability assessment of structures under wind action has been studied extensively; recently, a probabilistic procedure for the Aeolian risk assessment has been formulated (Augusti and Ciampoli 2008; Ciampoli et al. 2009). The procedure has been developed in the framework of a new branch of Wind Engineering, the so-called Performance-Based Wind Engineering (PBWE), and derived by extending the approach proposed for Performance-Based Seismic Design (PBSD) by the researchers of the Pacific Earthquake Engineering Research Centre (Porter 2003).

The central objective of the procedure is the assessment of the adequacy of the structure through the probabilistic description of a set of decision variables DVs. Each DV is a quantitative measure of a specific structural performance, that can be defined in terms of the interest of the users or the society in general. In Wind Engineering, relevant examples of DVs are the number of lives lost during windstorms, the economic losses resulting from windstorms, the exceeding of a (collapse or serviceability) limit state, the discomfort of the occupants, the length of the out-of-service time, etc.

The starting points of the procedure are the relationships, expressed in probabilistic terms, between the performances specific to the considered construction (no collapse, occupant safety, accessibility, full functionality, limited displacements or accelerations, etc.) and different "intensities", corresponding to different mean return periods of the wind action. With reference to a specific performance, the structural risk is conventionally measured by the probability of exceeding a relevant value of the corresponding DV; this probability is expressed in terms of a Complementary Cumulative Distribution Function (CCDF), and evaluated by taking into account the Aeolian hazard (i.e. the frequency of occurrence of wind actions of specified intensity and characteristics at the site), the calculated structural response and damages, and the correlation between the attained damage level and the relevant DV. The structural design should be optimized by applying a decisional strategy to the risk analyses, with the objective of minimizing the total risk or maximizing an utility function.

In the following, the procedure for PBWE is briefly outlined and applied to an example case: the assessment of the comfort requirements for a high-rise building.

2 THE PROCEDURE FOR PERFORMANCE-BASED WIND ENGINEERING

The procedure for PBWE consists of several steps aimed at: a) defining the Aeolian hazard at the site, in terms of parameters of the wind velocity field; b) defining the models of the interaction phenomena and their relevant parameters; c) analyzing the structural response, mainly in the context of stochastic dynamics; d) defining and evaluating indicators of the structural damage (intended as unacceptable performance), considering performances related to safety and functionality or comfort; e) defining the decisional variables that are appropriate to quantify the performances required for the structure, in terms of consequences of damage (personal damages, restoration costs, costs due to loss or deterioration of service, alterations of users comfort, etc.); f) evaluating the structural risk by a probabilistic characterization of the decision variables; g) optimizing design, that is minimizing risk, by appropriate techniques of decision analysis.

The assessment of the Aeolian hazard requires the use of efficient techniques for modelling wind actions on a slender structure, and the choice of a set of intensity parameters IMs whose stochastic characteristics are sufficient to describe satisfactorily and efficiently the Aeolian hazard at the site. Each IM must be chosen by considering the characteristics of the wind action (mean velocity, turbulence, direction, …), the relevant features of the construction site, the structural properties and the interaction phenomena (wind turbulence, vortex shedding, aeroelastic phenomena, aerodynamic effects).

Modeling the wind-structure interaction phenomena implies the choice and probabilistic characterization of a set of parameters IPs, that allow to take into account the relevant aspects of the interaction between the environment and the structure.

The probabilistic modelling of the structural response requires the choice of the relevant engineering demand parameters EDPs (inter-storey drifts, accelerations and velocities of selected points, stresses and displacements, etc.); the damage evaluation requires the choice (and probabilistic characterization) of the damage parameters DMs, that are able to quantify the structural damage due to wind actions in relation to the considered performances. The choices of any EDP and DM are strongly dependent on the considered structural type and performances. Different parameters can be assumed as DM: they can be defined by one or a combination of relevant EDPs (e.g. the inter-storey drift), or by other parameters, representing, for example, the damage to the partition or façade walls in a building as a function of the inter-storey drift. However, in the latter case it is usually rather difficult to establish sound correlations between the evaluated EDPs and the chosen DMs.

The decision variable DVs, that quantify the performances, must distinguish between low and high performance levels: the former (low performances) imply possible consequences on structural and personal safety (e.g. partial or total collapse, permanent damages); the latter (high performances) affect only serviceability and comfort (e.g. small displacements, limited vibrations, and wind-induced discomfort, also in the area around the structures).

For low performances, the significant DV can be identified with the cost necessary to restore the construction to the undamaged state (or rebuild it in case of collapse); correspondingly, DMs are the damages to be restored, and EDPs are the most significant response parameters for the specific case (peak displacements or accelerations at the building top, overall action at the base, local pressures, etc.). The shortcoming of this definition is the impossibility of including in it the "intangible" losses, i.e. the losses that cannot be measured in "monetary" terms.

High performances are related to the users' comfort/discomfort and, in case of buildings, to inconvenient alterations of the wind field in pedestrian areas around the construction. Appropriate discomfort criteria, that is, statements specifying the maximum acceptable frequencies of occurrence for various degrees of discomfort, can be defined.

In the PBWE procedure adopted in this paper (as in the PEER's approach), the structural risk is defined as the probability of exceeding a threshold level of a relevant DV:

$$\begin{aligned} G(DV) = \int\int\int\int\int & G(DV|DM)\cdot f(DM|EDP)\cdot f(EDP|\mathbf{IM},\mathbf{IP},\mathbf{SP})\cdot \\ & \cdot f(\mathbf{IP}|\mathbf{IM},\mathbf{SP})\cdot f(\mathbf{IM})\cdot f(\mathbf{SP})\cdot dDM\cdot dEDP\cdot d\mathbf{IM}\cdot d\mathbf{IP}\cdot d\mathbf{SP} \end{aligned} \tag{1}$$

where, as already stated: DM is a damage measure; EDP is the corresponding engineering demand parameter, representing the structural response; the basic parameters characterizing the Aeolian hazard are described by the vector **IM** of Intensity Measures; the vector of the Interaction Parameters

IP contains the set of aerodynamic and aeroelastic parameters that allow to take into account the relevant aspect of the interaction between the environment and the structure; **SP** is the vector of the parameters characterizing the structural systems and non environmental actions.

By means of Eq. (1), the problem of risk assessment is disaggregated into the following elements:

- site and structure-specific hazard analyses, that is, the assessment of the probability density functions f(**IM**), f(**SP**) and f(**IP**|**IM**, **SP**);
- structural analysis, aimed at assessing the probability density function of the structural response f(EDP|**IM**, **IP**, **SP**) conditional on the parameters characterizing the wind field and the structural properties;
- damage analysis, that gives the damage probability density function f(DM|EDP) conditional on EDP;
- finally, loss analysis, that is the assessment of G(DV|DM).

If the chosen DV corresponds to a specific EDP, Eq. (1) can be simplified and the exceeding probability is simply given by:

$$\mathrm{G(EDP)} = \int_{-\infty}^{\infty} \mathrm{G(EDP}|\mathbf{IM},\mathbf{IP},\mathbf{SP}) \cdot \mathrm{f}(\mathbf{IP}|\mathbf{IM},\mathbf{SP}) \cdot \mathrm{f}(\mathbf{IM}) \cdot \cdot \mathrm{f}(\mathbf{SP}) \cdot \mathrm{d}\mathbf{IM} \cdot \mathrm{d}\mathbf{IP} \cdot \mathrm{d}\mathbf{SP} \qquad (2)$$

3 MODEL OF THE WIND FIELD

In order to model the wind field, a Cartesian coordinate system (x, y, z), with origin at the ground and the z-axis oriented upward is adopted (Fig. 1). The three components of the wind velocity field $V_x(j)$, $V_y(j)$, $V_z(j)$ at each point j (variation with time is omitted for simplicity) are expressed as the sum of a mean (time-invariant) value $V_m(j)$ and a turbulent component u(j), v(j), w(j) with mean value equal to zero.

Assuming that the mean value of the velocity is not zero only in x direction, the three components are given by:

$$V_x(j) = V_m(j) + u(j);\; V_y(j) = v(j);\; V_z(j) = w(j) \qquad (3)$$

where in general the mean velocity $V_m(j)$ is determined by a database of values recorded at or near the site, and evaluated as the record average over a proper time interval (usually 10 minutes). The variation of the mean velocity V_m with the height z over a horizontal surface of homogeneous roughness can be described by the power law:

$$V_m(z) = V_{10} \cdot \left(\frac{z}{10}\right)^{\alpha} \qquad (4)$$

where V_{10} is the 10-minute mean wind velocity at 10 m height, and α is a site-dependent parameter (taken equal to 0.16 in numerical calculations and corresponding to a roughness length z_0 equal to 0.02).

The turbulent components of the wind velocity have been modeled as zero-mean Gaussian ergodic independent processes. Moreover, in probabilistic calculations, only the random spatial variation with the height z has been taken into account by considering the wind acting on N vertically aligned points. By neglecting the vertical component w of the wind velocity, the turbulent components u and v are completely characterized by the power spectral density matrices $[S]_i$ (i = u, v).

The diagonal terms (auto-spectra) $S_{i_j i_j}(n)$ of $[S]_i$ (j = 1, 2, ..., N) have been expressed in terms of the normalized one-side power spectral densities (Carassale and Solari 2006; Solari and Piccardo 2001):

$$\frac{nS_{u_j u_j}(n)}{\sigma_u^2} = \frac{6.868 \cdot n_u}{\left[1 + 10.302\, n_u^2(z_j)\right]^{5/3}} \qquad (5)$$

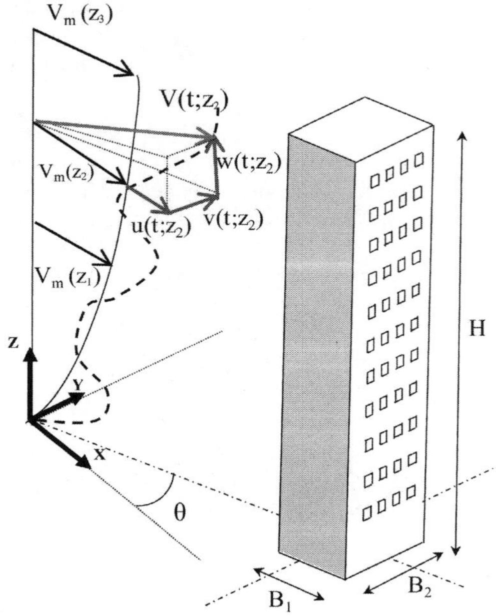

Figure 1. Schematic representation of the elements describing the wind field.

$$\frac{nS_{v_jv_j}(n)}{\sigma_v^2} = \frac{9.434\, n_v}{\left[1 + 14.151\, n_v^2(z_j)\right]^{5/3}} \tag{6}$$

where n is the current frequency (in Hz), z_j is the height (in m) of point j, σ_u^2 and σ_v^2 are the variances of the velocity fluctuations, given by the relationships (Solari and Piccardo 2001):

$$\sigma_u^2 = \left[6 - 1.1\,\mathrm{arctang}\cdot(\ln(z_0)) + 1.75)\right] \cdot u_*^2 \tag{7}$$

$$\frac{\sigma_v}{\sigma_u} = 0.75 \tag{8}$$

u_* is the friction or shear velocity (in m/s), given by: $[(K)^{1/2} \cdot V_{10}]$ where K is a coefficient depending on the roughness length z_0; $n_i(z_j)$ is a non-dimensional parameter given by:

$$n_i(z_j) = \frac{n\, L_i(z_j)}{V_m(z_j)} \tag{9}$$

and the integral scales $L_i(z_j)$ of the turbulent components have been derived for i = u, v according to the procedure given in (ESDU 2001).

The out of diagonal terms (cross-spectra) $S_{i_ji_k}(n)$ of $[S]_i$(j, k = 1,2,…,N) are given by

$$S_{i_ji_k}(n) = \sqrt{S_{i_ji_j}(n) S_{i_ki_k}(n)} \cdot \exp(-f_{jk}(n)) \tag{10}$$

where for vertically aligned points (Di Paola 1998):

$$f_{jk}(n) = \frac{|n| \cdot C_z \cdot |z_j - z_k|}{V_m(z_j) + V_m(z_k)} \tag{11}$$

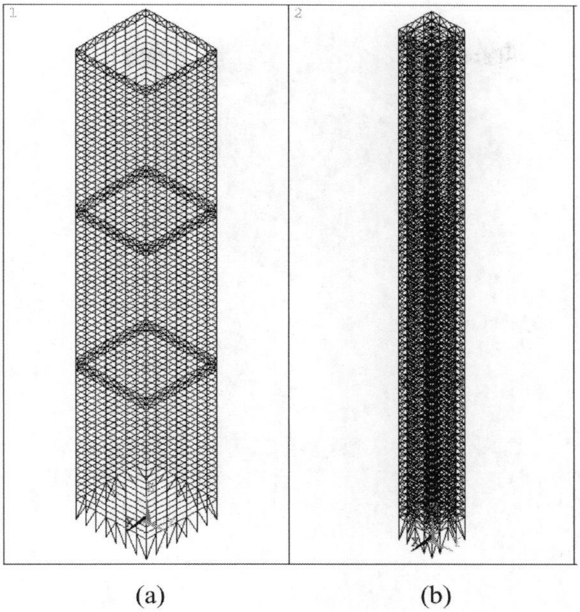

Figure 2. FE model of the structural system: a) the 3d frame on the external perimeter; b) the central core.

and C_z is the decay coefficient, inversely proportional to the spatial correlation of the process.

The spectra of the wind velocity and the global (drag and lift) forces have been evaluated for 74 points ($j = 1, \ldots, 74$), that are located at the height of the master joints of each floor in the FE model. Adopting very rough approximations, the spectra of the global forces have been evaluated by assuming that the force coherence is the same as the velocity coherence, and neglecting the variation of each spectrum on the area of the external surface corresponding to the considered floor.

4 THE EXAMPLE CASE

The examined high-rise steel building has a square plant (side length = 50 m) and a total height of 305 m; the number of floors is 74. The main structural system is composed by a central core (a 3D frame with 16 columns) and a 3D frame on the external perimeter (28 columns). The two substructures are connected at three levels (at the height of 100 m, 200 m and 300 m); the stiffening systems are extended for 3 or 2 floors. The columns have a hollow square section, with dimensions and thickness varying with the height (between 1.20 m and 0.50 m, and 0.06 m and 0.025 m, respectively). The beams are double T steel beams; the beam-column joints are rigid. The bracing system is composed by double T or hollow square struts.

The structural analyses have been carried out using ANSYS V11; the FE structural model is composed by 7592 BEAM4 elements and 2680 LINK8 elements (Figures 2 and 3). The dynamic response has been evaluated by linear modal analysis assuming rigid diaphragms; the damping has been assessed according to Rayleigh's model, by adopting a damping ratio ξ equal to 0.02 for the first and the fourth modes of vibration.

The assessment of the serviceability performance of high-rise buildings under wind action is usually carried out in terms of peak values of lateral displacements or accelerations (Kwok et al. 2009). Excessive lateral displacements are taken into account to prevent damage to façade and non-structural components, such as doors, shutters and elevators; excessive horizontal accelerations may induce discomfort to building occupants. With reference to the considered example case, the comfort requirement has been related to the intensity of the peak acceleration at the top of the building; the human perception thresholds for horizontal (office and apartment) building vibrations have been defined according to the comfort criteria proposed by the Italian Research Council (CNR 2008).

Figure 3. FE model of the bracing system.

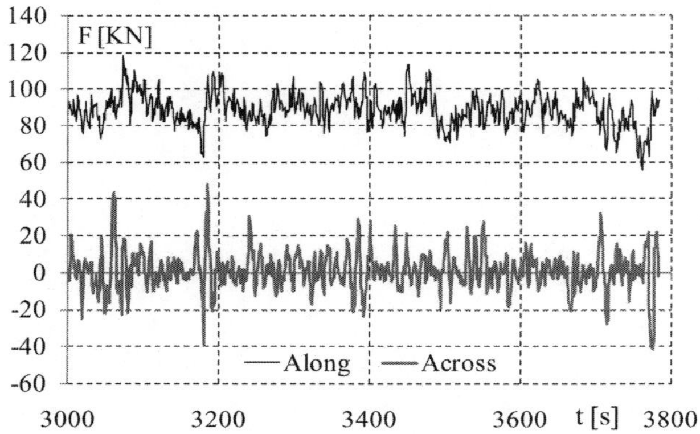

Figure 4. Interval of the time series of the action in along- and across-wind directions, evaluated at the top of the building by scaling the experimental measures.

Dynamic analyses have been carried out both in time and frequency domains. The time-histories of the wind total forces have been obtained by experimental tests on a 1:500 scale model (Spence et al. 2008 a, b). In Figures 4 and 5, samples of the total forces, expressed in full scale, and samples of the absolute accelerations at the top of the building in along- and across-wind directions corresponding to $V_{10} = 20$ m/s, are reported. In frequency domain, the peak values are evaluated as:

$$r^{p}(h) = r_{m} + g \cdot \sigma_{r}(h) \tag{12}$$

where r_m is the mean value of the response parameter (corresponding to the mean wind velocity), σ_r is the standard deviation (evaluated by the power spectral density), and g is the peak factor. The response is described by a Gaussian distribution; therefore, according to (Davenport 1998) the peak factor is equal to:

$$g = \sqrt{2\ln(\eta \cdot T_{wind})} + \frac{0.577}{\sqrt{2\ln(\eta \cdot T_{wind})}} \tag{13}$$

where: η is the cycling rate of the effective frequency of the response, assumed equal to the first natural frequency n_1 of the structure; T_{wind} is the time interval over which the maximum value is evaluated. [In the numerical analyses illustrated in the following, $n_1 = 0.1873$ Hz; $T_{wind} = 3600$ s; therefore, $g = 3.984$.]

The peak values of the along-wind a_D and across-wind a_L accelerations at the top of the building, evaluated by a deterministic analysis in time domain ($V_{10} = 20$ m/s), are compared with the human perception thresholds for horizontal (office and apartment) building vibrations in Fig. 6; as expected, the across-wind component is critical for comfort.

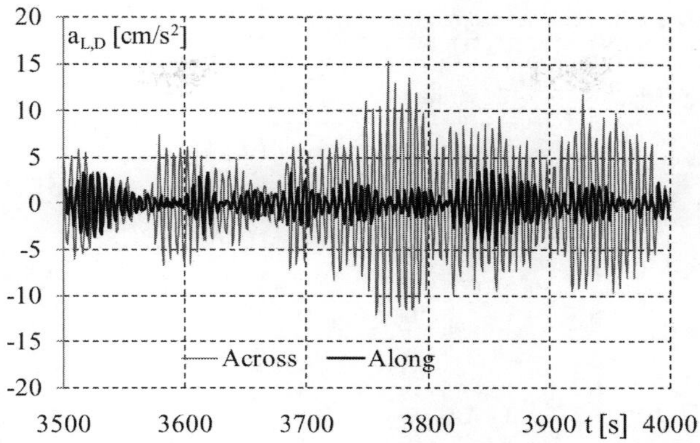

Figure 5. Interval of the time series of the acceleration at the top of the building in along- and across-wind directions evaluated for a mean wind velocity $Vm = 35$ m/s.

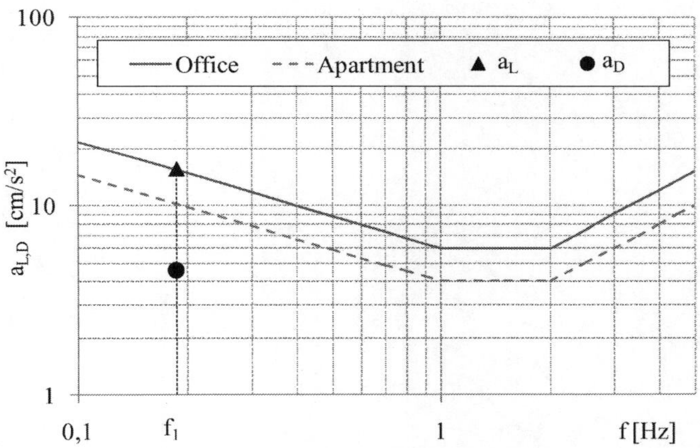

Figure 6. Comparison of the peak values of the along- and across-wind accelerations at the top of the building evaluated by a deterministic analysis with the human perception thresholds for horizontal (office and apartment) building vibrations (CNR 2008).

For risk assessment, the three components of **IM** are: the 10-minute mean wind velocity V_{10} evaluated on an annual basis at 10 m height and characterized by a Weibull distribution; the roughness parameter z_0, characterized by a Lognormal distribution; the direction of the mean wind velocity θ, characterized by a Gaussian distribution. The peak factor g is considered the only interaction parameter IP, and characterized by a truncated Gaussian distribution; the damping ratio ξ is the only random structural property SP, characterized by a Lognormal distribution. The characteristics of the random parameters are synthesized in Table 1.

The probabilities of exceeding the threshold values of the acceleration at the top of the building have been evaluated by Monte Carlo simulation (1000 samples). A first series of sensitivity analyses considers the significance of the input parameters **IM** and IP: in each set of analyses, only one element of **IM** or IP is assumed as random, while the other parameters are deterministic.

In Fig. 7–11, the results of the sensitivity analyses are illustrated, with reference to the peak value a_L of the acceleration in across-wind direction. The plots indicate that: V_{10} is the most relevant parameter; the uncertainty affecting θ is not negligible because it implies a variation of the aerodynamic coefficients, but can be neglected in the examined case because the critical direction can be determined a priori; the uncertainties affecting z_0, g and ξ are not negligible.

In Fig. 12 the results of the analyses carried out considering all random variables (see Table 1) are reported.

Table 1. Characteristics of the considered random variables.

	Stochastic Variable	Description	Distribution type and parameters
IM	V_{10} [m/s]	Mean wind velocity at 10 m height	Weibull $k = 2.02$ (shape parameter) $\sigma^* = 6.2$ (scale parameter)
	θ [deg]	Direction of mean wind velocity	Gaussian $\mu = 22.5$ $\sigma = 30°$
	z_0 [m]	Roughness length	Lognormal $\mu = 0.1$ $\sigma = 0.03$
IP	g	Peak factor	Truncated Gaussian $\mu = 3.9$ $\sigma = 0.3$ $g^- = 1$ $g^+ = 7$
SP	ξ	Structural damping	Lognormal $\mu = 0.02$ CoV = 0.4

μ = mean; σ = standard deviation; g^+, g^- = cut-off values of truncated Gaussian distribution; CoV = coefficient of variation.

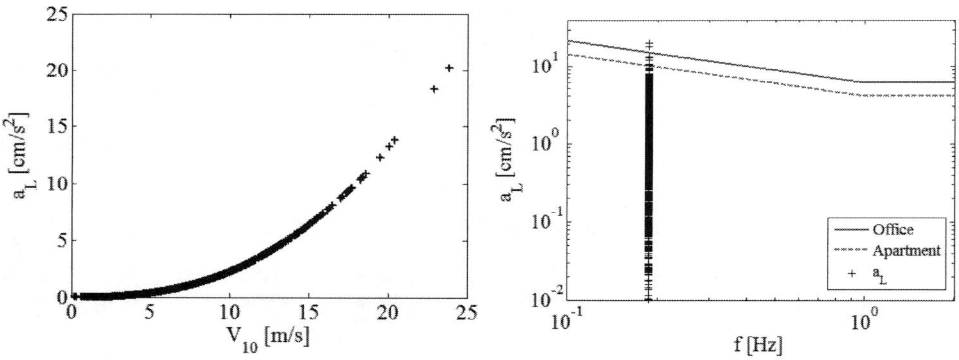

Figure 7. Peak values of a_L (Monte Carlo simulation: only the mean wind velocity V_{10} at 10 m height taken as a random variable): (i) as a function of V_{10}; (ii) compared with the human perception thresholds for horizontal building vibrations.

Finally in Fig. 13 the complementary distribution function $G(a_L)$ is reported, as obtained by considering only V_{10} as random or all random variables. The corresponding annual probabilities of exceeding the human perception thresholds for apartment and office building vibrations are reported in Table 2. While it is not possible to comment about the absolute values of these conventional measures of risk, it is evident that neglecting the random variability of some parameters of the wind velocity field, interaction phenomena and structural behavior, can be seriously misleading, leading to a significant underestimation of risk.

5 CONCLUDING REMARKS

A probabilistic procedure for the Performance-Based Design of structures subject to wind action has been illustrated. The central objective of the procedure is the assessment of the adequacy of

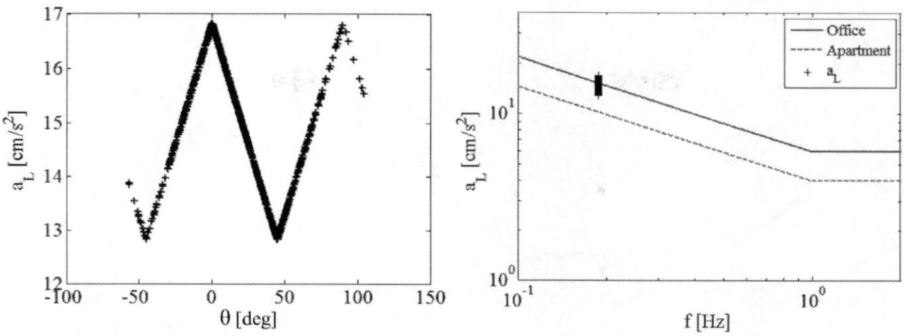

Figure 8. Peak values of a_L (Monte Carlo simulation: only the mean wind direction θ taken as a random variable): (i) as a function of θ; (ii) compared with the human perception thresholds for horizontal (office and apartment) building vibrations.

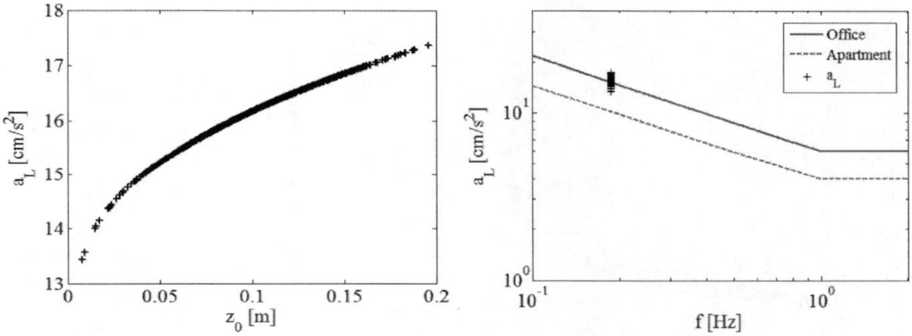

Figure 9. Peak values of a_L (Monte Carlo simulation: only the roughness length z_0 taken as a random variable): (i) as a function of z_0; (ii) compared with the human perception thresholds for horizontal (office and apartment) building vibrations.

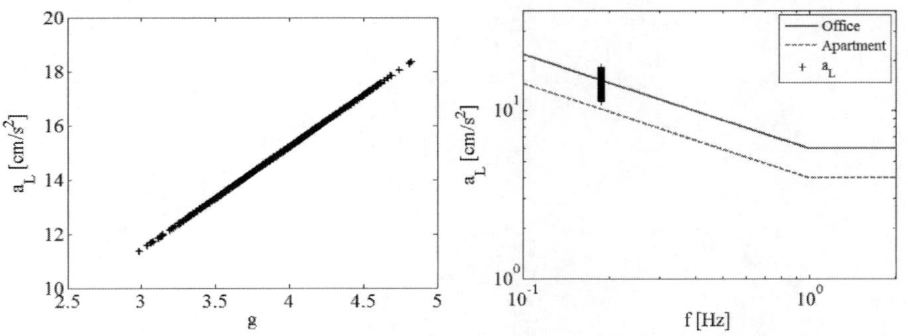

Figure 10. Peak values of a_L (Monte Carlo simulation: only the peak factor g taken as a random variable): (i) as a function of g; (ii) compared with the human perception thresholds for horizontal (office and apartment) building vibrations.

the structure through the probabilistic description of a set of decision variables DVs. The examined structure is a 74 storey steel building; the assumed DV (the peak value of the across-wind acceleration at the top of the building) is a measure of the occupant comfort, that is of the motion perception of the building vibrations due to wind.

The Aeolian risk has been conventionally measured by the probabilities of exceeding relevant values of DV; the probabilities have been evaluated by taking into account the Aeolian hazard at

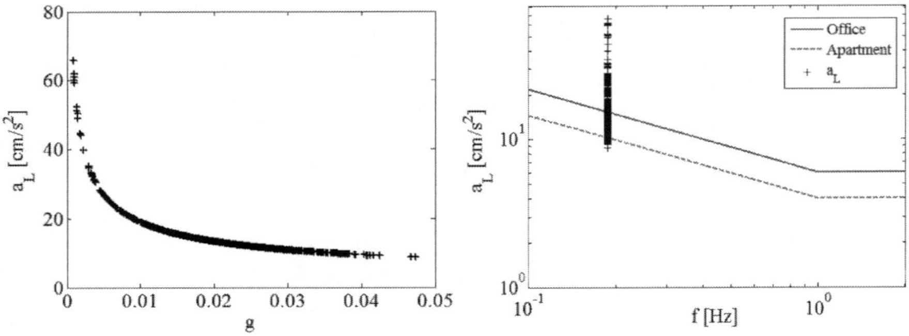

Figure 11. Peak values of a_L (Monte Carlo simulation: only the damping ratio ξ taken as a random variable): (i) as a function of ξ; (ii) compared with the human perception thresholds for horizontal (office and apartment) building vibrations.

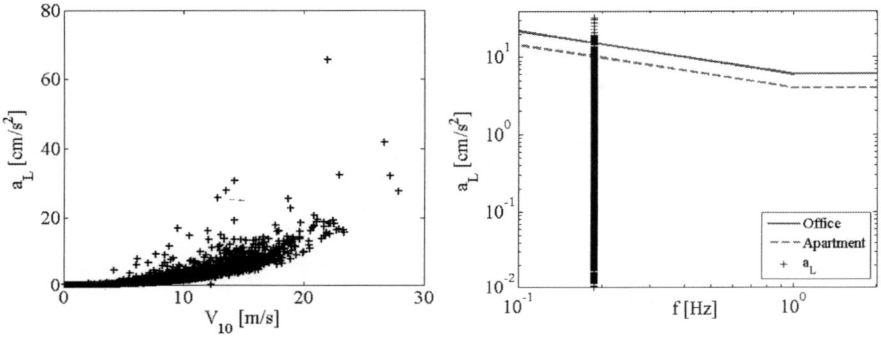

Figure 12. Peak values of a_L (Monte Carlo simulation): (i) as a function of V_{10}; (ii) compared with the human perception thresholds for horizontal building vibrations.

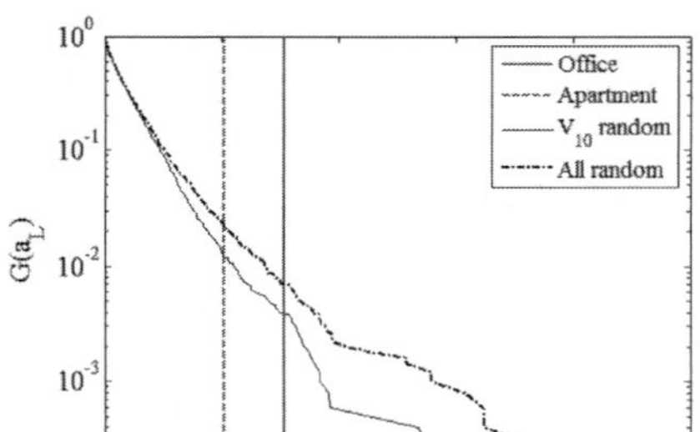

Figure 13. Complementary distribution function of a_L evaluated by considering only V_{10} as a random variable or the effect of all random variables.

the site and the calculated structural response. Probabilistic calculations have been carried out in frequency domain, but the parameters of the wind field have been calibrated on the basis of the results of experimental tests, giving the time-histories of the total forces acting on a 1/500 rigid scale model of the building.

As underlined by Simiu (2009), the risk assessment is essential for developing criteria that may govern the design of structures intended to remain functional when subjected to wind actions, under the constraint that costs were reasonably low, if not optimal. However, the estimation of risk

Table 2. Annual probabilities of exceeding the motion perception thresholds.

Variables considered as random	Office	Apartment
V_{10}	0.0040	0.0124
All (see Tab. 1)	0.0072	0.0226

First natural frequency of the building: $f_1 = 0.18729$ Hz;
Motion perception thresholds (CNR 2008):
15.343 cm/s^2 (for office buildings)
10.228 cm/s^2 (for apartment building)

is typically difficult, as in structural engineering risks are associated with considerable sources of uncertainty; not accounting for uncertainty can result in seriously flawed risk estimates, but accounting for it may require some simplifications. These simplifications concerns the choice of the parameters that are considered as random, their probabilistic description, and also the modeling of the wind velocity field and the total forces adopted for the analyses of the structural response in frequency domain. The authors are well aware of the over-simplified models that have been adopted in numerical calculations; however, these elements could be (and will be in the future) refined, while the main interest of the paper consists in demonstrating that Performance-Based Design of structures under wind actions is feasible and fruitful.

ACKNOWLEDGEMENTS

The research presented in this paper has been developed in the framework of Wi-POD Project (2008–2010), partially financed by the Italian Ministry for Education, University and Research - MIUR.

REFERENCES

ANSYS Inc., http://www.ansys.com/

Augusti, G., Ciampoli, M. (2008). Performance-Based Design in risk assessment and reduction, *Probabilistic Engineering Mechanics*, 23, 496–508.

Carassale, L. and Solari, G. (2006). Monte Carlo simulation of wind velocity field on complex structures, *Journal of Wind Engineering and Industrial Aerodynamics*,. 94 (1) 323–339.

Ciampoli M, Petrini, F. and Augusti G., (2009). A Procedure for the Performance-Based Wind Engineering, *Proc. Tenth international conference on structural safety and reliability (ICOSSAR'09)*, Osaka, Japan, 13–17 September 2009.

CNR (Consiglio Nazionale delle Ricerche) (2008). *CNR-DT 207/2008: Istruzioni per la valutazione delle azioni e degli effetti del vento sulle costruzioni*, Rome, Italy; (in italian).

Davenport, A.G. (1998). Probabilistic methods in wind engineering for long span bridges, *Proc. Int. Symp. on Advances in Bridge Aerodynamics*, May 1998, Copenhagen, Denmark.

Di Paola M. (1998). Digital simulation of wind field velocity, *Journal of Wind Engineering and Industrial Aerodynamics*, vol. 74–76, 91–109.

ESDU (2001). *Report N. 86010: Characteristic of atmospheric turbulence near the ground. Part III: variations in space and time for strong winds (neutral atmosphere)*, http://www.esdu.com.

Kwok, K.C.S., Hitchcock, P.A. and Burton, M.D. (2009). Perception of vibration and occupant comfort in wind-excited tall buildings, *Journal of Wind Engineering and Industrial Aerodynamics*, 97, 368–380.

Petrini, F. (2009). A probabilistic approach to Performance-Based Wind Engineering (PBWE). PhD Dissertation, University of Rome "La Sapienza", Rome, Italy. Available at http://padis.uniroma1.it [Assessed 8 March 2010].

Porter KA (2003). An Overview of PEER's Performance-Based Engineering Methodology, *Proc Ninth International Conference on Applications of Statistics and Probability in Civil Engineering (ICASP9)*. San Francisco, CA, USA; Millpress Rotterdam. Available at: http://keithp.caltech.edu// pubblications.htm [Assessed 8 March 2010].

Simiu, E., (2009). Wind and risk, *Proc. of the Fifth European & African Conference on Wind Engineering (EACWE 5)*, Florence, Italy, July 19–23, 2009.
Solari, G. Piccardo, G. (2001). Probabilistic 3-D turbulence modeling for gust buffeting of structures, *Probabilistic Engineering Mechanics*, (16), 73–86.
Spence S.M.J., Gioffrè M., Gusella V. (2008a). Influence of higher modes on the dynamic response of irregular and regular tall buildings, *Proc. 6th International Colloquium on Bluff Bodies Aerodynamics and Applications (BBAA VI)*, Milano, Italy, July 20–24, 2008.
Spence S.M.J., Gioffrè M., Gusella V. (2008b). Influence of higher modes on the response of irregular tall buildings, *Proc. IABSE Annual Meetings and Congress,* Chicago, USA, September 14–19, 2008.
Stathopoulos, T., (2006). Pedestrian level winds and outdoor human comfort, *Journal of Wind Engineering and Industrial Aerodynamics*, vol. 94, 769–780.
Tamura, Y. (2009). Wind and tall buildings, *Proc. of the Fifth European & African Conference on Wind Engineering (EACWE 5)*, Florence, Italy, July 19–23, 2009.

Reliability and Optimization of Structural Systems – Straub (ed)
© 2010 Taylor & Francis Group, London, ISBN 978-0-415-88179-1

Framework for probabilistic calibration of fatigue design factors for offshore wind turbine support structures

José G. Rangel-Ramírez
Aalborg University, Denmark

John D. Sørensen
Aalborg University and RISØ-DTU, Denmark

ABSTRACT: In this paper is described a reliability-based approach to calibrate Fatigue Design Factors (FDF) for offshore wind turbine support structures. The FDFs are calibrated to a specific minimum reliability level and a particular inspection and maintenance strategy. A similar approach has been used for offshore Oil & Gas (O&G) sub-structures. Based on representative stochastic models and response functions a set of FDF values are calculated which depend on the degree of accuracy of the fatigue analysis and the consequences of fatigue failure of a critical detail. Generally lower FDF values are obtained for offshore wind turbines than for oil & gas structures and reduced FDF values are obtained when inspections are taken into account. Thereby the basis is available for selecting a cost-effective fatigue design for offshore wind turbines substructures. Further, the results can be used as basis for recommendations in standardization.

1 INTRODUCTION

For wind turbine components, fatigue failure represents an important design driver. Some structural components in the substructure of an offshore wind turbine (tower and foundation, e.g. jacket or tripod) are often designed to have a design life where they will not need to be inspected or maintained because they are in hard-to-reach zones in the structure, e.g. the underwater part of the support structure. Unlike these components, structural parts where inspection and maintenance can be carried out "normally", can be 'optimized' in the design, assuming specific maintenance and inspection actions during the design life, e.g. possible for details above the splash and transition zones.

Probabilistic approaches represent a rational way to estimate the fatigue life that is associated with many uncertain processes, e.g. wave and wind load, wake influence in wind farms and influence of control system. The models for the fatigue strength are typically based on experimental tests and modeled by SN-approaches or Fracture Mechanics (FM) methods. These models are highly uncertain due to physical, model and statistical uncertainties. Moreover, the fatigue damage accumulation rules used in design, e.g. the Miner's rule, are very uncertain if is not carefully calibrated together with e.g. the SN curves. Beside of material and load uncertainties, detection and no-detection of fatigue cracks at inspections represent an additional uncertainty that typically is modeled through probability of detection (POD) curves.

In fatigue design, FDF values are used in connection with different fatigue assessment requirements (e.g. damage tolerant/safe life design philosophies) and the consequence of failure (fail- or non-fail-safe components). Offshore codes such as ISO-19902 and NORSOK, use FDF values that take into account the component location (i.e. accessible or not for inspection, change and repair).

In the offshore oil & gas industry reliability-based techniques for planning of inspections have been developed during the last 10-20 years and are used as basis for calibration of fatigue design factors and for cost-optimal planning of inspections, see e.g. Faber et al. (2005).

In this paper application of similar approach is presented for welded details in substructures of offshore wind turbines where the required reliability levels typically are much smaller than for offshore oil & gas structures. Further, the fatigue loading is rather different for offshore wind turbines. Therefore the results from oil & gas industry cannot be used directly, but the basic principles and approaches can be applied for offshore wind turbines. This is described and illustrated in this paper. FDF values are obtained depending on the consequences of fatigue failure and calibrated to a reliability level which is suitable for offshore wind turbines. It is shown how the FDF values can be reduced if inspections are planned to be performed during the design life, and thereby the basis is available for selecting a cost-effective fatigue design with or without inspections (of selected details).

2 CALIBRATION OF FATIGUE DESIGN FACTORS

Calibration of FDF values (or equivalently partial safety factors to be applied to the fatigue load and the fatigue strength) can be performed on a probabilistic basis through the steps recommended by JCSS 2002, see Faber & Sørensen (2003) for general code calibration. Basically, it is assumed that a certain number of inspections has to be performed in the design life and the fatigue reliability for details in wind turbines is obtained assuming no-inspections, using a SN-approach with linear damage accumulation by the Miner's rule. The steps include:

- Definition of code objective: Design of fatigue critical details in offshore wind turbine steel substructures.
- Stochastic modeling of uncertainties: fatigue load (wind, wave and influence of wakes in wind farms), calculation of fatigue stress effects (stress ranges), SN-curves and model uncertainty (Miner's rule).
- Target reliability level: this should for different classes of fatigue critical details be selected such that the same general reliability level is obtained as for other ultimate failure modes, but taking into account the consequences of failure, see Straub & Faber (2005). In special cases it can be argued that the reliability level should be increased/decreased if the marginal cost of improving the reliability is low/high, see e.g. JCSS (2002).
- Identification of typical fatigue critical details: these have to be selected and used in the calibration to secure that the FDF values can be used in the whole range of typical applications.
- Determination of FDF values: FDF values are calibrated such that the reliability level for the different typical fatigue critical details are as close as possible to the target reliability level.
- The calibrated FDF values have to be verified for use in practical design.

An extension of this calibration approach is to include the effect of performing inspections during the design life. This is achieved by considering a FM model (see appendix B) that is calibrated to the SN-approach and including the reliability of possible inspection methods though POD-curves (see section 4 and table 1).

3 PROBABILISTIC MODELING

The underlying assumptions of the probabilistic models are briefly described in this section. The turbulence intensity represents an important aspect to consider due to its effects on the fatigue life for an offshore wind turbine. It can be taken into account by specifying a characteristic influence function where the fluctuation of the (10-minutes) wind speed can be related with internal forces (stress ranges) for a specific component, sector or detail (hot spots) of the OWT support structure.

When an OWT is within a wind farm (IWF) the model should be modified to include additional turbulence due to wakes from surrounding turbines. To incorporate this additional IWF turbulence the following code-based (IEC-61400) model is used:

$$\sigma=\left[\left(1-N_w \cdot p_w\right)\sigma_0^m + N_w \cdot p_w \cdot \sigma_w^m \right]^{1/m} \qquad (1)$$

In equation (1), N_w is for number of wakes affecting the wind turbine, p_w is the probability of wake condition, m is the Wöhler exponent, σ_0 is the turbulence standard deviation under free flow conditions and σ_w is the maximum wake turbulence standard deviation. It is assumed that the standard deviation of a considered response is proportional to the standard deviation of turbulence, see Frandsen (2007).

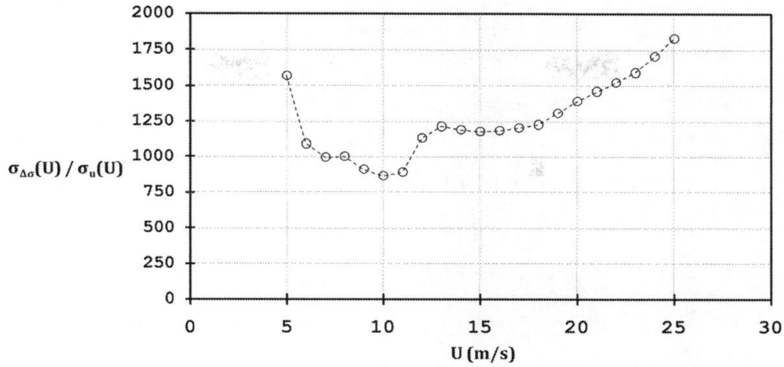

Figure 1. Influence coefficient function ($\sigma_{\Delta\sigma}(U)/\sigma_u(U)$) – pitch controlled wind turbine.

In the design equation, it is assumed that the response of the structure can be modeled by a narrow-banded stationary Gaussian process and that the failure occurs as a result of the accumulated effect of stress cycles for the given mean wind speed distribution. With these assumptions the frequency of maxima in the response (typical cycles per year, ν) are proportional to the time interval t (years).

If inspections are performed during the design life a reliability-based approach can as mentioned above be used to determine the times for the inspections such that the reliability during the whole design life is larger than a minimum reliability level. The reliability of the inspection technique (e.g. visual inspection, Eddy current or MPI) is modeled by a probability of detection (POD) curve; see table 1 and appendix B. Further, it is assumed that all detected cracks are perfectly repaired by grinding, welding or replacement.

In the fatigue failure limit state equation, the uncertainties related with the wind, stress concentration factor and Miner rule (see appendix A) are taken into account. In case of IWF, the model in equation (1) is integrated in the design and limit state equations.

4 EXAMPLES

A probabilistic calibration of FDF for OWT support structure is presented in this section and illustrative results are presented. A representative influence coefficient function is used where the stress ranges are related to the wind load (wind load is assumed to be dominating). These functions could be obtained in different manners such as by simulations and/or by measurements.

If measurements or monitoring data are available indirect information about the OWT performance can be obtained. Data from e.g. strain gauges at specific hot spots can provide direct information about stresses at the critical detail.

In the following example is used an influence coefficient function, $\alpha_{\Delta\sigma}$ (U) which is representative for fatigue load effects in a critical welded detail in a steel support structure. The influence function is shown in figure 1. Due to the effect of the control system, the internal stresses will not vary linearly, and thereby the influence function will vary non-linearly as illustrated in figure 1.

4.1 *Example cases*

FDF values are obtained for four different cases for welded steel details. These will be the combinations of linear (L) and bilinear (BL) SN-curve with a wind turbine placed within a wind farm (IWF) and stand alone/single (S).

4.2 *Stochastic models*

An offshore wind turbine with steel jacket/lattice support structure is considered. An expected design life T_L equal to 20 years is assumed. It is noted that for a linear SN-curve, FDF is related to the partial safety factors on load, γ_f and fatigue resistance, γ_m by FDF $= (\gamma_f \cdot \gamma_m)^m$. In tables 1 and 2 representative stochastic models, equations and parameters are shown (see explanation of parameters

Table 1. Stochastic model.

Variable	Distribution	Parameters	Comment
$f_U(U)$	$W(\alpha, \beta_U)$	$\alpha = 2.3, \beta_U = 10.0$ m/s	Mean wind speed
$f_{\Delta\sigma D}(\cdot)$	$W(\alpha_{\Delta\sigma D}, \beta_{\Delta\sigma D})$	$\alpha_{\Delta\sigma_D} = 0.8$	Stress ranges
$f_{\sigma u}(\cdot)$	$LN(\mu, \sigma)$	$\mu = I_{ref} \cdot (0.75 \cdot U + 3.6)$, $\sigma = 1.4$ m/s $\cdot I_{ref}$.	Mean turbulence
POD(x)	$P_o \cdot (1 - \exp(-a/\lambda))$	$P_O = 1.0$, $\lambda = 2.67$ mm	POD curve
$N_1(s)$	$K_1 \cdot s^{-m1}$	$s \geq \Delta\sigma_D$	SN curve linear
$N_2(s)$	$K_2 \cdot s^{-m2}$	$s < \Delta\sigma_D$	SN curve bi-linear

D: Deterministic, N: Normal, LN: Lognormal, W: Weibull

Table 2. Stochastic variables.

Variable	Distr	E[·]	Standard deviation	Comment
Δ	N	1.0	(1) 0.0, (2) 0.2	Damage accumulation Wind
X_W	LN	1.0	V_W	
X_{SCF}	LN	1.0	V_{SCF}	Stress concentration factor Wake
X_{wake}	LN	1.0	(1) 0.10, (2) 0.15	
m_1	D	3.0	–	SN-curve. Wöhler exponent (linear)
m_2	D	5.0	–	SN-curve. Wöhler exponent (bi-linear)
$\Delta\sigma_D$	D	71 MPa	–	Constant amplitude fatigue limit
Log K_1	N	Determined from $\Delta\sigma_D$	0.20	Material parameter
Log K_2	N	Determined from $\Delta\sigma_D$	0.25	Material parameter
N_W	D	5/–	–	In-wind farm/single OWT
N	D	5×10^7	–	Fatigue cycles per year
U_{in}–U_{out}	D	5–25 m/s	–	Cut in – out velocities
p_W	D	0.06/0.0	–	In-wind farm/single OWT
d_j	D	4.0	–	Normalized distance of OWT
Ln C_C	N	μ_{lnCc} (fitted)	0.77	Crack growth ratio
N_I	W	$\mu_O = T_{init} \cdot \nu$	$0.35 \cdot \mu_O$	T_{init} (fitted), Initiation Time
Y	LN	1.0	0.10	Shape factor
a_C	D	50 mm	–	Critical crack size
a_O	D	0.4 mm	–	Initial crack size
f_{cr}	D	4.0	–	Crack length/depth ratio
thickness	D	50 mm	–	Thickness
m	D	3.0	–	Material parameter

Log K_1 and Log K_2 are assumed fully correlated. Ln C_C and N_I are correlated with correlation coefficient $\rho_{LnCc,NI} = -0.5$

in appendix A and B), based on uncertainty models used for offshore oil & gas installations, see Faber et al. (2005) and for offshore wind turbines, see Rangel-Ramirez, J.G. & Sørensen (2008) and Sørensen et al. (2007).

4.3 *Results*

In tables 3 and 4 results are shown for one typical fatigue critical detail with the influence coefficient function in figure 1 and without inspections. The minimum target annual reliability index is chosen to $\Delta\beta_{min} = 3.1$, as used in Tarp-Johansen et al. (2002) and Tarp-Johansen (2003). In figure 2 the reliability indices as function of time are shown for IWF-L, $V_W = 0.1$, $V_{SCF} = 0.05$, Vwake-1 and $V_{\Delta-1}$. Both FDF values and fatigue design lives TF = FDF • TL are shown in tables 3 and 4.

The results in table 3 and 4 show that the FDF values are generally smaller than those used for fatigue design for oil & gas installations, but are dependent on especially the uncertainty of the stress range calculation (SCF). It is important to take into consideration the probabilistic modeling of the quality of inspection methods. When this quality varies the probability of detection will influence the reliability analysis and thereby the calibration of FDF values. The effects of variation

Table 3. FDF and T_F (in years) for minimum annual reliability index ($\Delta\beta = 3.1$) and no-inspections during design life.

V_W	V_{SCF} FDF	0.05 IWF-L	S-L	IWF-BL	S-BL
0.10	$V_{wake-2}, V_{\Delta-2}$	2.00	1.40	2.00	1.60
	$V_{wake-2}, V_{\Delta-1}$	1.85	1.30	3.90	1.50
	$V_{wake-1}, V_{\Delta-2}$	1.65	1.40	1.95	1.60
	$V_{wake-1}, V_{\Delta-1}$	1.55	1.30	1.85	1.50

V_W	V_{SCF} T_F	0.05 IWF-L	S-L	IWF-BL	S-BL
0.10	$V_{wake-2}, V_{\Delta-2}$	40	28	40	32
	$V_{wake-2}, V_{\Delta-1}$	37	26	78	30
	$V_{wake-1}, V_{\Delta-2}$	33	28	39	32
	$V_{wake-1}, V_{\Delta-1}$	31	26	37	30

Table 4. FDF and T_F (in years) for minimum annual reliability index ($\Delta\beta = 3.1$) and no-inspections during design life.

V_W	V_{SCF} FDF	0.1 IWF-L	S-L	IWF-BL	S-BL
0.10	$V_{wake-2}, V_{\Delta-2}$	2.15	1.60	2.40	1.90
	$V_{wake-2}, V_{\Delta-1}$	2.00	1.45	5.65	4.50
	$V_{wake-1}, V_{\Delta-2}$	1.85	1.60	2.35	1.90
	$V_{wake-1}, V_{\Delta-1}$	1.70	1.45	5.60	4.50

V_W	V_{SCF} T_F	0.1 IWF-L	S-L	IWF-BL	S-BL
0.10	$V_{wake-2}, V_{\Delta-2}$	43	32	48	38
	$V_{wake-2}, V_{\Delta-1}$	40	29	113	90
	$V_{wake-1}, V_{\Delta-2}$	37	32	47	38
	$V_{wake-1}, V_{\Delta-1}$	34	29	112	90

in the detectable crack depth modeled by an Exponential distribution with mean value is shown in more details in tables 5 to 7. The FDF values are calculated for equidistant inspection times for a critical crack depth of 40 mm.

Table 5 show the results in the case with a linear SN curve and one single wind turbine. If no inspections are carried out then the required FDF value is 1.9. As expected more inspections and better inspection quality (lower λ) result in lower FDF values. If the mean value of the inspection quality is around 10 mm then the influence of the inspection disappears if there is only one inspection. Due to the larger uncertainty associated with the turbulence in wind farm, the required FDF values are larger than a single wind turbine, see table 6.

When a bilinear SN-curve is used in design then smaller values of the design parameter (z) is obtained compared to the case with a linear SN-curve. On the other hand, the larger uncertainty of the lower part of the SN-curve implies that larger FDF values have to be applied, see results in table 7.

The above results illustrate the potential of the reliability-based approach for calibration of FDF values or equivalently partial safety factors for fatigue design – and for including the effect of possible inspections during the life time of the wind turbine. However, more results are needed considering a range of different fatigue critical details, material thicknesses, SN-curves, number

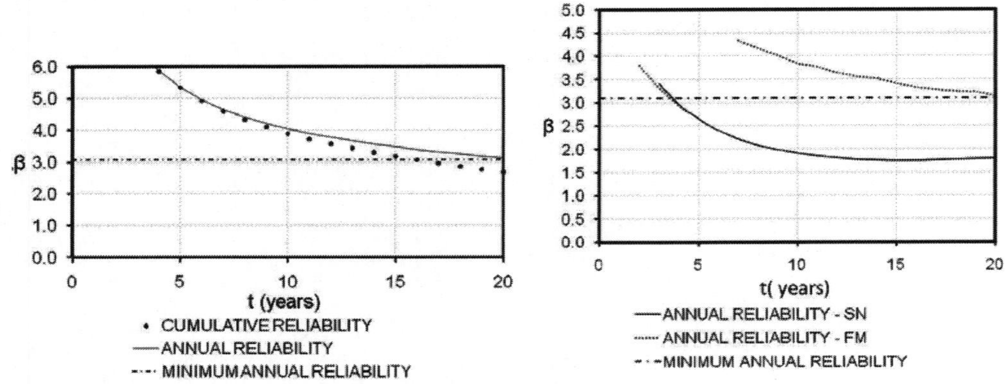

(a) (b)

Figure 2. (a) Annual reliability indices of SN and FM approach for IWF-L, $V_W = 0.1$, $V_{SCF} = 0.05$, V_{wake-1} and $V_{\Delta-1}$ without inspections during design life. (b) Reliability indices of SN and FM approach for IWF-L, $V_W = 0.1$, $V_{SCF} = 0.05$, V_{wake-1} and $V_{\Delta-1}$ with one-inspection during design life.

Table 5. FDF for minimum annual reliability index = 3.1 and linear SN-curve with a single wind turbine. If no inspections FDF = 1.9.

λ (mm)	2	4	6	8	10
1 inspection: year 10	1.5	1.6	1.8	1.8	1.9
2 inspections: year 7 and 14	1.3	1.5	1.7	1.7	1.8
3 inspections: year 5, 10 and 15	1.0	1.4	1.5	1.6	1.7

Table 6. FDF for minimum annual reliability index = 3.1 and inear SN-curve within wind farm location. If no inspections FDF = 2.1.

λ (mm)	2	4	6	8	10
1 inspection: Year 10	1.7	1.8	2.0	2.1	2.1
2 inspections: Year 7 and 14	1.4	1.7	1.9	1.9	2.0
3 inspections: Year 5, 10 and 15	1.1	1.6	1.7	1.8	1.9

Table 7. FDF for minimum annual reliability index = 3.1 and bilinear SN-curve with a single wind turbine. If no inspections FDF = 3.2.

λ (mm)	2	4	6	8	10
1 inspection: Year 10	2.3	2.5	2.9	3.0	3.1
2 inspections: Year 7 and 14	1.8	2.3	2.6	2.7	2.8
3 inspections: Year 5, 10 and 15	1.4	2.1	2.3	2.6	2.7

of inspections, inspection techniques and repair strategies. Further, the sensitivity of the stochastic models should be investigated, especially for Δ, X_W and X_{SCF}. Finally, for some details where the consequence of failure is less severe, a smaller value of β_{min} can be accepted.

5 CONCLUSION

A methodology for probabilistic calibration of fatigue design factors for offshore wind turbine support structure is described. Representative stochastic models and influence coefficient functions are presented. For calibration examples single wind turbines and wind turbines within a wind farm are considered for a linear and a bi-linear SN-curve for welded steel details. Representative FDF values are obtained without and with one inspection during the design life.

Using the reliability level used for calibration of partial safety factors for ultimate limit states in the IEC 61400-1 standard lower FDF values are obtained than used for civil engineering structures and offshore oil and gas steel platforms. If an inspection is performed during the design life then the FDF values can be decreased even more. However, the decrease in cost of the substructure due to the lower FDF values should be compared to the cost of the inspection and possible repair if a crack is detected. The influence of the quality of inspection on the FDF is shown. It is noted that the results presented are illustrative – more examples (influence coefficients modeling different types of critical details) and different types of inspections should be considered before recommendations for standardization can be formulated. This framework of reliability-based calibration of FDF values can also be used as a tool to select for a given wind turbine (wind farm) a cost-effective fatigue design considering initial design cost and service life costs.

ACKNOWLEDGEMENT

The financial supports from the Mexican National Council of Science and Technology (CONACYT) and the Integrated Project "UpWind" supported by the EU sixth Framework Program, grant no. 019945 are greatly appreciated.

REFERENCES

Faber, M.H., Sørensen, J.D., Tychsen, J. & Straub, D. (2005) Field Implementation of RBI for Jacket Structures. *Journal of Offshore Mechanics and Arctic Engineering*, Vol. 127, pp. 220–226.

Faber, M.H. & Sørensen, J.D. (2003) Reliability Based Code Calibration – The JCSS Approach. *Proc. ICASP'9 2003 conf. San Francisco*, pp. 927–935.

Frandsen, S.T. (2007) Turbulence and turbulence generated structural loading in wind turbine clusters. *Risø-R-1100*(EN). Risø National Laboratory, Roskilde, Denmark. January 2007.

IEC-61400-1. *International Electrotechnical Committe-Wind turbines, part 1: Design requirements*. Third edition 2005–08.

ISO 19902 (2007) *Petroleum and natural gas industries – Fixed steel offshore structures*.

JCSS. *Joint Committee of Structural Safety-Probabilistic Model Code*. (2002) http://www.jcss.ethz.ch. 2002.

NORSOK, (2004). *Design of Steel Structures N-001*. Rev. 2004.

Rangel-Ramirez, J.G. & Sørensen, J.D. (2008) Optimal Risk-Based Inspection Planning for Offshore Wind Turbines. *International Journal of Steel Structures*. Volume 8(4), pp. 295–303.

Sørensen, J.D., Frandsen, S. & Tarp-Johansen, N.J. (2007) Fatigue reliability and effective turbulence models in wind farms. *Applications of statistics and Probability in Civil Engineering – Kanda*, Takada & Furuta (Eds).

Straub, D. & Faber (2005) M.H. Risk Based Acceptance Criteria for joints subject to Fatigue Deterioration. *Journal of Offshore Mechanics and Artic Engineering*. Vol. 127. pp. 150–157.

Tarp-Johansen N.J., Madsen P.H., Frandsen S. (2002) *Partial Safety Factors for Extreme Load Effects*. Risø National Laboratory, Roskilde, Denmark, Report No.: Risø-R-1319(EN).

Tarp-Johansen N.J. (2003) *Examples of Fatigue Lifetime and Reliability Evaluation of Larger Wind Turbine Components*. Risø National Laboratory, Roskilde, Denmark. Report No.: Risø-R-1418(EN).

APPENDICES – PROBABILISTIC MODELS

Appendix A. SN-assessment of reliability

Design equation for free flow condition

Based on the assumptions mentioned in section 3 the design equation for a single/alone OWT can be written:

$$G(z)=1-\frac{\nu\cdot FDF\cdot T_L}{K_c}\int_{U_{in}}^{U_{out}} D(m;\sigma_{\Delta\sigma}(U,z))\cdot f_U(U)\,dU = 0 \tag{2}$$

where ν is the number of stress cycles per year, FDF is the fatigue design factor ($T_F = T_L \cdot FDF$ and $FDF = (\gamma_f \cdot \gamma_m)^m$ for a linear SN-curve using load and material partial safety factors γ_f and γ_m, T_L is the design life time, K_c is characteristic value of K (material parameter in SN-curve), U_{in} and U_{out} are the cut-in and – out wind speeds, m is the Wöhler exponent in SN-curve, $f_U(U)$ is the probability density function of mean wind speed U, z is the design parameter and D represents the expected value of fatigue for the all stress ranges given a mean wind speed U and standard deviation $\sigma_{\Delta\sigma}$ and considering medium turbulence characteristics (IEC-61400). The expected value of Δ_σ can be obtained for linear and bilinear SN-curve with the following formulae:

$$D_L = \int_0^{\infty} s^m \cdot f_{\Delta\sigma}(s|\sigma_{\Delta\sigma}(U,z))\, ds \tag{3}$$

$$D_{BL} = \int_0^{\Delta\sigma_D} s^{m_1} \cdot f_{\Delta\sigma}(s|\sigma_{\Delta\sigma}(U,z))\, ds + \int_{\Delta\sigma_D}^{\infty} s^{m_2} \cdot f_{\Delta\sigma}(s|\sigma_{\Delta\sigma}(U,z))\, ds \tag{4}$$

where $f_{\Delta\sigma}(s|\sigma_{\Delta\sigma})$ represents the probability density function for stress ranges given a standard deviation $\sigma_{\Delta\sigma}$ at the mean wind speed U. $\sigma_{\Delta\sigma}$ and ν can be obtained by cycle counting methods, e.g. Rainflow counting and $\sigma_{\Delta\sigma}$ is written:

$$\sigma_{\Delta\sigma} = \alpha_{\Delta\sigma}(U) \cdot \frac{\sigma_u(U)}{z} \tag{5}$$

$$\sigma_u = I_{ref} \cdot (0.75 \cdot U + b), \quad b = 5.6 m/s \tag{6}$$

In equation (5), $\alpha_{\Delta\sigma}$ is the influence function referring to an specific detail (hot spot) or sector in the OWT and is a function of the wind speed. σ_U is the (normal) turbulence standard deviation (IEC-) considering class B.

Design equation for in-wind farm location

For wind farm location the turbulence model is integrated at the model, resulting in the following design equation:

$$G(z) = 1 - \frac{v \cdot FDF \cdot T_L}{K_c} \cdot \int_{U_{in}}^{U_{out}} \left[\begin{matrix} (1 - N_W \cdot p_W) \cdot D(m; \sigma_{\Delta\sigma}(U,z)) \\ + p_W \cdot \sum_{i=1}^{N_W} D(m; \sigma_{\Delta\sigma,j}(U,z,j)) \end{matrix} \right] \cdot f_U(U)\, dU = 0 \tag{7}$$

where $\sigma_{\Delta\sigma,j}$ is the standard deviation of stress ranges considering wake condition from j-neighboring wind turbine and is defined as follows:

$$\sigma_{\Delta\sigma,j} = \alpha_{\Delta\sigma}(U) \cdot \frac{\sigma_{u,j}(U,j)}{z} \tag{8}$$

$$\sigma_{u,j} = \sqrt{\frac{0.9 \cdot U^2}{\left(1.5 + 0.3 \cdot d_j \sqrt{U/c}\right)^2} + \sigma_u^2} \tag{9}$$

d_j is the distance between OWT normalized by the rotor diameter of the neighboring wind turbine *j* and *c* is the constant equal to 1 m/s.

Limit state equation for free flow condition

The limit state equation in free flow conditions is written:

$$g(t)=\Delta-\frac{v\cdot t}{K}\int_{U_{in}}^{U_{out}}\int_{0}^{\infty}(X_w\cdot X_{SCF})^m\cdot\begin{pmatrix}D(m;\sigma_{\Delta\sigma}(U,z))\cdot\\ f_{\sigma_u}(\sigma_u|U)\cdot f_U(U)\end{pmatrix}\,d\sigma_u\,dU \tag{10}$$

where Δ, X_w and X_{SCF} represent the uncertainties related with the Miner rule of damage accumulation, wind load effects and local stress concentration and analysis. t is the reference time and $f_{\sigma u}$ is the probability density function for free flow turbulence that is modeled as LogNormal distributed with a representative mean turbulence equal to:

$$I_{ref}\cdot(0.75\cdot U+3.6) \tag{11}$$

and standard deviation equal to 1.4 m/s times the I_{ref}.

Limit state equation for in-wind farm location

To assess the reliability in wind farm location the following limit state equation is used:

$$g(t)=\Delta-\frac{v\cdot t}{K}(X_{wake}\cdot X_{SCF})^m\cdot\int_{U_{in}}^{U_{out}}\int_0^{\infty}\begin{bmatrix}\left(1-N_w\cdot p_w\right)\cdot D(m;\sigma_{\Delta\sigma}(U,z))\\ +p_w\cdot\sum_{j=1}^{N_w}D(m;\sigma_{\Delta\sigma,j}(U,z,j))\end{bmatrix}\cdot f_{\sigma_u}(\sigma_u|U)\cdot f_U(U)\,d\sigma_u\,dU \tag{12}$$

where X_{wake} models the uncertainty related to the wake turbulence conditions coming from surrounding wind turbines. When equation (8) is used with this model, equation (9) will change in order to take into account the wind turbulence with the additional turbulence coming from the wake condition. The formula becomes:

$$\sigma_{u,j}=\sqrt{\frac{X_w\cdot U^2}{\left(1.5+0.3\cdot d_j\sqrt{U/c}\right)^2}+\sigma_u^2} \tag{13}$$

Appendix B. FM-assessment of reliability

Crack model

For the assessment of the FM-fatigue life, a one dimensional representative crack growth model is used with crack length *c* related with the crack depth *a* through a constant factor f_{cr}. The following (Paris-Erdogan) model is used:

$$\frac{da}{dN}=C_a\cdot(\Delta K_a)^m\,,\qquad a(N_0)=a_0\ ,\ c(f_{cr}\cdot a_0)=c_0 \tag{14}$$

where C_a is a material parameter (crack growth rate), ΔK_a is the stress intensity factor range in a stress cycle. a_0 and c_0 represent the initial dimensions of the crack. The fatigue life is represented by two stages: fatigue initiation life (N_I) and fatigue propagation life (N_P). Adding these periods is obtained the number of stress cycles to fatigue failure. The stress intensity factor is written:

$$\Delta K_a=Y\cdot\Delta\sigma_e\cdot\sqrt{a\cdot\pi} \tag{15}$$

Y is a stochastic variable taking into account the uncertainty of the geometry function and $\Delta_{\sigma e}$ is the equivalent stress range that can be calculated as follow for free flow condition:

$$\Delta\sigma_e = X_w \cdot X_{SCF} \cdot \left[\int_{U_{in}}^{U_{out}} \int_0^{\infty} \begin{pmatrix} D(m;\sigma_{\Delta\sigma}(U,z)) \cdot \\ f_{\sigma_u}(\sigma_u|U) \cdot f_U(U) \end{pmatrix} d\sigma_u \, dU \right]^{1/m} \tag{16}$$

In case of wind farm location:

$$\Delta\sigma_e = X_{wake} \cdot X_{SCF} \cdot \left[\int_{U_{in}}^{U_{out}} \int_0^{\infty} \begin{bmatrix} (1-N_w \cdot p_w) \cdot D(m;\sigma_{\Delta\sigma}(U,z)) \\ +p_w \cdot \sum_{j=1}^{N_w} D(m;\sigma_{\Delta\sigma,j}(U,z,j)) \end{bmatrix} \cdot f_{\sigma_u}(\sigma_u|U) \cdot f_U(U) \, d\sigma_u \, dU \right]^{1/m} \tag{17}$$

Limit state equation

The limit state equation used in the FM-approach is associated with failure that occurs when the crack a(t) at the considered detail exceeds the critical size a_c of the crack (e.g. thickness). The limit state equation is thus written:

$$g(t) = a_c - a(t) \tag{18}$$

The uncertainty related to a chosen inspection technique is modeled by a probability of detection (POD) curve (see table 1).

Reliability and Optimization of Structural Systems – Straub (ed)
© 2010 Taylor & Francis Group, London, ISBN 978-0-415-88179-1

Multiobjective non-linear random vibration analysis for performance-based earthquake engineering

Salvatore Sessa
Dipartimento per le Tecnologie, Università di Napoli "Parthenope", Napoli, Italy

ABSTRACT: A new multiobjective linearization method for nonlinear random vibration analysis is presented. The strategy employs the Tail-Equivalent Linearization Method (TELM) which is a non-parametric linearization algorithm for multi-DOFs nonlinear systems. Due to the definition of conditioned probability, the joint tail probability of a multi-response structural system can be written as the product of a first marginal probability and some lower order conditioned tail probabilities. The algorithm decomposes the joint probability into a conveient form so that the conditioned probabilities are arranged consequently. In this case, each probability but the first is conditioned to previously computed responses. Then, TELM is recursively applied in order to define a set of interconnected linearized systems, each one defined in terms of its impulse response function. The definition of the base excitation by its cross-covariance or its power spectral density leads to the first marginal tail probability and to the power spectral density of the corresponding response. Afterwards, the interconnected linearized system is used to compute the tail probability and the power spectral density of each response in function of the previously analyzed responses' statistics. The computed joint probability can be used in random vibration analysis in order to get various statistics of the nonlinear response, such as the mean level-crossing rate and the joint first-passage probability. This work analyzes a series system, however, the procedure can be easily extended to the general case. Also, numerical applications illustrate the features of the method and comparison with results obtained by Monte Carlo simulations demonstrate its accuracy, in particular for high response thresholds.

1 INTRODUCTION

In performance-based earthquake engineering, it is important to properly consider non-linearities since failure usually occurs in the non-linear range of structural behavior. Equivalent linearization is one of the most powerful techniques in Random Vibrations because of its versatility in application to MDOF Finite Element Models. Several criteria have been proposed in years, in particular, one efficient non-parametric algorithm is the Tail-Equivalent Linearization Method (TELM) (Fujimura and Der Kiureghian 2007). It is able to predict non-Gaussian response distributions; also, non-stationary analysis can be performed (Der Kiureghian and Fujimura 2009), and it can even account for asymmetries providing non-symmetric response PDFs (Sessa and Der Kiureghian 2009).

However, dependence on a specific response affetcs equivalent linearization. TELM defines probability distributions of a single specific response only, thus, it is a single-objective analysis. Common structures usually require a multi-objective analysis: buildings' structural safety conditions often require that *each* inter-storey drift remains below a safety threshold. Thus, structural RV analysis can be considered as the investigation of the global reliability of a stochastic system which may be defined by a set of mutually related components both in series (such as in the inter-storey drift case) and parallel subsets.

Efficient algorithm have been developed for this purpose. Joint First-Passage Probability of interconnected systems (Song and Der Kiureghian 2006) leads to reasonably accurate estimates, however, the joint-PDF of the responses of interest is required. Matrix-based System Reliability (MSR) and Linear Programming (LP) (Song and Der Kiureghian 2003) efficiently compute reliability for any general system; the procedure is not dependent on the number of components and it is able to define the narrowest possible bounds for the available informations. However, because of its wide capability, Linear Programming can be demanding because it must be able to address any possible system configuration.

A procedure for multi-objective random vibration analysis is presented in this work. Its purpose is to define a discretized form of a non-Gaussian responses joint-PDF. This formulation is very fruitful because, eventually, the computed joint-PDF can be employed either for stationary or non-stationary analysis applying classical stochastic methods regardless of dimensions and parameters of the structural system. The underlying philosophy consists in defining the statistics of a "multi-objective" linearized system, whose equivalence condition is defined in therms of joint-tail probability of the system components.

Briefly, the joint-tail probability of a structural system is written as the product of conditioned probabilities of each response. Then, an equivalent linear system (TELS) is computed by TELM for each response. The TELS are defined in terms of a collection of Impulse Response Functions (IRF), each corresponding to a specified threshold.

Note that the IRF is defined as the evolution in time of a chosen response, given by a generical impulsive action. Thus, the proposed strategy will compute two different kinds of IRFs: a ground motion–response relationship, if only the marginal distribution of the specified response is needed, or a response-*i*–response-*j* relationship, if the needed distribution of the response *j* is conditioned on the response *i*.

Once the input excitation process statistics have been defined, it is possible to evaluate all the marginal distributions of the non-conditioned responses and also their power spectral densities which will be employed in order to get the statistics of the conditioned responses.

In Section 2 a review of the classic formulation of TELM is presented. Afterwards, the decomposition of the multiobjective problem is shown in Section 3 and then, the conditioned tail probabilities are computed in Section 4. Finally, numerical applications and comparisons with the results of a Monte Carlo simulation are provided in Section 5.

2 TELM REVIEW

Let $F(t)$ be a random process defined as the response of a linear filter excited by a white noise $W(\tau)$:

$$F(t) = \int_0^t h_f(t-\tau) W(\tau)\, d\tau \tag{1}$$

where $h_f(\tau)$ is the Impulse Response Function (IRF) of the linear filter. The random process $F(t)$ can be expressed in discretized form as a random pulse train (Der Kiureghian 2000) as:

$$F(t) = \sum_{i=1}^{n} s_i(t) u_i = \mathbf{s}^T(t)\, \mathbf{u} \tag{2}$$

where $\mathbf{u}$ is a vector of standard normal random variables and $\mathbf{s}(t)$ is a deterministic vector depending on the base-excitation's covariance.

A multi-degrees-of-freedom (MDOF) nonlinear system can be defined by its equation of motion:

$$\mathbf{M}\ddot{\mathbf{U}} + \mathbf{C}\dot{\mathbf{U}} + \mathbf{R}\left(\mathbf{U}\dot{\mathbf{U}}\right) = \mathbf{P}F(t) \tag{3}$$

where $\mathbf{U}$ denotes the displacements vector, $\mathbf{M}$, $\mathbf{C}$ and $\mathbf{P}$, respectively, the mass, damping and loads matrixes and $\mathbf{R}$ is the nonlinear restoring force. A generic response of interest $X(t)$ can be defined as nonlinear function of nodal displacements, velocities and accelerations. Our interest is in determining the tail probability $\Pr[x \le X(t_n)]$ for a specified threshold x at time t_n. Defining a limit state function such as:

$$G(x, t_n, \mathbf{u}) = x - X(t) \tag{4}$$

TELM (Fujimura and Der Kiureghian 2007) is able to perform the First-Order Reliability Method (FORM) which evaluates the tail probability and the "design point" as the solution to the costrainted optimization problem:

$$\mathbf{u}^\star(x, t_n) = \text{arg min}\left[\|\mathbf{u}\| \,|G(x, t_n, \mathbf{u}) = 0\right]. \tag{5}$$

Note that $\mathbf{u}^*$ is the closest point of the limit state surface 4 to the origin in the standard-normal space and corresponds to its maximum likelihood. Furthermore, FORM linearizes the limit state function and gets the first-order approximation of its tail-probability.

Now, the generic response of a linear system can be expressed in terms of its IRF:

$$X_L(t) = \int_0^t h(t-\tau) F(\tau)\, d\tau \tag{6}$$

or, in discretized form:

$$X_L(t) = \int_0^t h(t-\tau) \sum_{i=1}^{n} s_i(\tau) u_i \, d\tau = \mathbf{a}(t)\,\mathbf{u} \tag{7}$$

where the deterministic vector $\mathbf{a}(t)$ has elements:

$$a_i(t) = \int_0^t h(t-\tau) s_i(\tau)\, d\tau. \tag{8}$$

The limit state function 4 for the response 6 describes an hyperplane in the standard normal space, due to the linearity of the system. Geometrical considerations yield:

$$\mathbf{a}(t_n) = \frac{x}{\|\mathbf{u}^\star(x,t_n)\|} \frac{\mathbf{u}^\star(x,t_n)^T}{\|\mathbf{u}^\star(x,t_n)\|}. \tag{9}$$

The TELM equivalent condition requires that the tail probability of the nonlinear and linearized system must be the same, then, both systems must lead to the same design point. Once the nonlinear design point is computed by FORM, it is possible to get the linera IRF by the 9 and the 8. The Tail-equivalent Linearized System (TELS), then, is fully defined by a *collection* of IRFs, each one depending on a different threshold x_i.

3 MULTIOBJECTIVE RANDOM VIBRATION ANALYSIS

TELM is very powerful because it can be performed regardless of the number of DOFs and of the constitutive models (as long as they can be smoothed). Its only requirements are that the nonlinear system must reach stationarity before the considered time t_n and the Limit State Function's first derivative must be continuous in the standard normal space. While the first requirement can be easily achieved, the latter one could be harder to satisfy. In fact, most of the common problems require to evaluate the first-excursion probability with respect a safety threshold of either displacements or internal forces. While for some structures safety can be expressed in terms of a single response (such as for tanks, bridges and electrical substations), in many cases the global safety depends on a set of responses of interest. For example, the safety condition of a building requires that *each* inter-storey drift remains smaller than the safety threshold during the seismic motion. In that case, the global limit state function and the tail probability can be written as:

$$G(\mathbf{x},t_n) = \min\left[x_1 - X_1(t_n), x_2 - X_2(t_n), \ldots, x_m - X_m(t_n)\right] \tag{10}$$

$$\Pr\left[G(\mathbf{x},t_n) \le 0\right] = \Pr\left[x_1 \le X_1 \cup x_2 \le X_2 \cup \ldots \cup x_m \le X_m\right] = 1 - \Pr\left[\bigcap_{i=1}^{m} x_i \ge X_i\right]. \tag{11}$$

Note that in the 11 the dependence on t_n of the X_i has been omitted for simplicity. TELM is not able to get the 11 because the first derivative of the 10 is usually non-continuous in the standard normal space. Furthermore, the LSF itself cannot be easily approximated by an hyperplane.

This drawback can be overcome by decomposing the joint-probability into the product of lower-order conditioned probabilities:

$$\Pr\left[G(\mathbf{x},t_n) \ge 0\right] = \Pr\left[x_1 \ge X_1\right] \Pr\left[x_2 \ge X_2 | X_1\right] \ldots \Pr\left[x_m \ge X_m | X_1 \ldots X_{m-1}\right]. \tag{12}$$

The proposed stategy consists in computing the conditioned probability of the 16 by TELM and then to get the joint-tail probability in order to perform nonlinear random vibration analysis. Note that the Equations 10, 11 and 16 refer to a series system, *i.e.* the global chrisis occurs if any of the responses crosses its safety threshold. In case of parallel or hybrid systems, the Equation 16 can be easily modified by computing the complementary-conditioned tail probabilities.

4 THE CONDITIONED PROBABILITIES

Each term of the right side of the 16 can be computed separately. The first one is a marginal probability which can be computed by the usual formulation of TELM. In particular, let's consider a set of threshold of interest $\mathbf{x} = [x_1\, x_2\, \ldots x_m]^T$, for each one of them, TELM evaluates the correponding IRF in discretized form: $h_k^{1,l}$, where the indexes 1, k and l designate respectively the response, the time step and the threshold. The Fourier transform of the IRF is the Frequency Response Function (FRF) $H^{1,l}(\omega)$, corresponding to the lth threshold, which leads to the variance of the response:

$$\sigma_{1,l}^2 = 2\int_0^\infty \left\|H^{1,l}(\omega)\right\|^2 \Phi_{FF}(\omega)\, d\omega \tag{13}$$

where Φ_{FF} is the power spectral density of the base excitation. Then, the tail probability is:

$$\Pr\left[x_1^l \geq X_1\right] = \Phi\left[\frac{x_1^l}{\sigma_{1,l}}\right] \tag{14}$$

where $\Phi[\,\cdot\,]$ denotes the Standard Normal Cumulative Distribution Function. Also, the power spectral density of the displacements and of the acceleration is:

$$\Phi_{X_1X_1}^l(\omega) = \left\|H^{1,l}(\omega)\right\|^2 \Phi_{FF}(\omega) \tag{15}$$

note that the tail probability and the power spectral density are referred only to the threshold x^l. The power spectral density of the derivatives of X_1 can be easily obtained from the 15 as shown in (Lutes and Sarkani 2004). Performing a similar procedure, also the conditioned tail probabilities can be computed:

$$\Pr\left[x_2^l \geq X_2 | x_1^k = X_1\right] = \Phi\left[\frac{x_2^l}{\sigma_{2|1,l|k}}\right] \tag{16}$$

where:

$$\sigma_{2|1,l|k}^2 = 2\int_0^\infty \left\|H^{2|1,l}(\omega)\right\|^2 \Phi_{\ddot{X}_1\ddot{X}_1}^k(\omega)\, d\omega. \tag{17}$$

The frequency response function $H^{2|1,l}(\omega)$ it the response X_2 for a steady-state action in X_1. Its evaluation will be described in the following. Note that the 17 can be written only if the response X_2 given by an excitation in X_1 is uncoupled with the base excitation $F(t)$. This hypothesis is usually right for common buildings as long as the responses of interest are either the interstorey drifts or the floor displacements. In fact, if the first floor displacement is assigned, then the second floor drift can be considered independent of the base excitation. Thus, TELM can be performed in order to get the IRF $h_k^{2|1,l|k}$ where the index l denotes the lth threshold for X_2 and k denotes the kth threshold of X_1. For each threshold value of X_1, then, TELM defines a set of IRFs each corresponding to a threshold value of X_2. This procedure can be applied recursively to each response of interest in order to get all the conditioned tail probabilities; its generalized formulation is:

$$\Pr\left[x_j^l \geq X_j | x_i^k = X_i\right] = \Phi\left[\frac{x_j^l}{\sigma_{j|i,l|k}}\right]; \quad \sigma_{j|i,l|k}^2 = 2\int_0^\infty \left\|H^{j|i,l}(\omega)\right\|^2 \Phi_{\ddot{X}_i\ddot{X}_i}^k(\omega)\, d\omega. \tag{18}$$

The conditioned IRFs and FRFs $h_k^{j|i,l}$ and $H^{j|i,l}(\omega)$ can be easily computed by TELM, by using the jth limit state function and an excitation at the ith response. In case of interstorey drift responses, the ith excitation would be a storey acceleration.

Note that the input excitation of TELM should be defined with respect the power spectral density $\Phi_{\ddot{X}_i\ddot{X}_i}(\omega)$: first, the auto-covariance $\Gamma_{\ddot{X}_i\ddot{X}_i}$ is computed by taking the Fourier Transform of the PSD, then, the covariance matrix $\mathbf{G}$ can be built as discretized form of the auto-covariance (where each row-column corresponds to a specific time-step) which leads to the matrix $\mathbf{S}$:

$$\mathbf{S}\mathbf{S}^T = \mathbf{G}; \quad \mathbf{S} = [\mathbf{s}(t_1)\, \mathbf{s}(t_2) \ldots \mathbf{s}(t_n)] \tag{19}$$

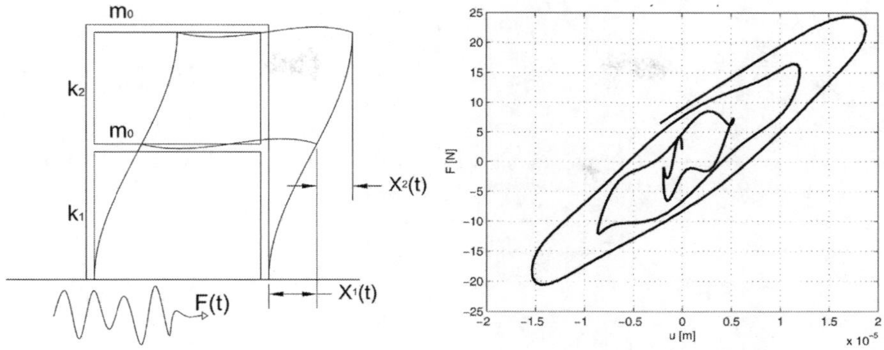

(a) Structural Model (b) Bouc–Wen hysteresis loop

Figure 1.

Table 1. Mechanical parameters of the structural models

Model	$k_1\left[\frac{N}{m}\right]$	$k_1\left[\frac{N}{m}\right]$	$m[kg]$
2DOF-1	$7.5\cdot10^7$	$7.0\cdot10^7$	$3.0\cdot10^4$
2DOF-2	$4.0\cdot10^7$	$2.0\cdot10^7$	$3.0\cdot10^4$

which defines the excitation process by the Equation 2. However, this procedure could be demanding because the linearized system would depend on each specific threshold of the previous responses and on the base excitation. It has been shown in (Fujimura and Der Kiureghian 2007) that the linearized system does not depend on the excitation scale and that in common practice it is possible to evaluate the TELS with respect a White Noise excitation regardless of its real power spectral density. The so-called "White Noise approximation" leads to reasonably approximated linearized systems. Thus, the set of IRFs of each response of interest will be computed for each threshold of the response itself and regardless of the statistics of the other responses.

5 NUMERICAL APPLICATION

In order to test the proposed procedure, a numerical application has been developed. The chosen structural models are two degrees-of-freedom nonlinear buildings, whose geometric scheme is shown in Figure 5; also, mechanical parameters are summarized in Table 1. Note that the first-floor displacement has been set as the first response of interest $X_1(t)$ and the drift between the first and the second floot has been set as the second response $X_2(t)$. Nonlinearities are modeled by Bouc-Wen constitutive models (Bouc 1963), (Wen 1976), (Wen 1980) with $\alpha = 0.05$, $A_0 = n = 1$ and $\beta = \gamma = 35$. A sample hysteretic loop of the Bouc-Wen model is shown in Figure 5. The material's stiffness shown in Table 1 is the tangent of the Bouc-Wen loop at zero displacement.

The first performance of TELM computes the IRFs of $X_1(t_n)$ for a set of 20 thresholds with step $\Delta x = 1$ cm; some of the computed IRFs are shown in Figure 2. Then, a random vibration analysis is performed in order to get the statistics and the power spectral density of the displacement X_1. The base excitation has been defined as an unfiltered white noise. The analysis has been performed for the white noise cutoff standard deviation $\sigma_{WN} = 0.50$ g. The displacement power spectral densities are shown in Figure 3.

Then, TELM has been performed again; the input excitation is the first displacement X_1 and the Limit State Function has been defined in terms of X_2. The computed IRFs are shown in Figure 4; note that each IRF is the evolution in time of the interstorey drift given by an impulsive action of the first floor displacement, as explained in the previous section.

The computed IRFs lead to the evaluation of the conditioned tail probabilities of X_2 given X_1. The conditioned response statistics are easily computed by evaluating the FRFs as the Fourier transform

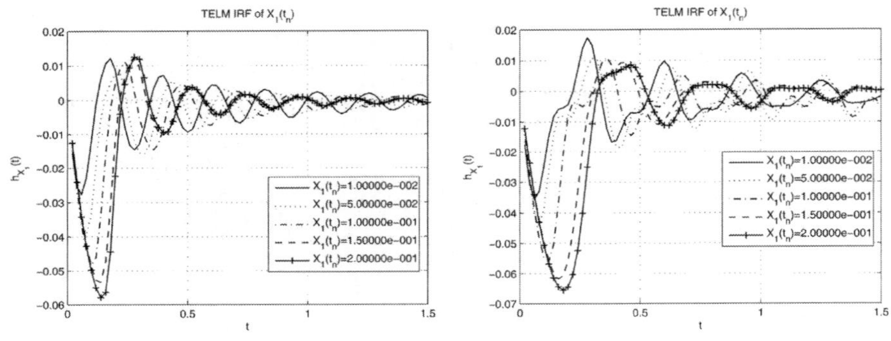

(a) Model 2DOF–1 (b) Model 2DOF–2

Figure 2. Impulse Response Functions – $X_1(t)$.

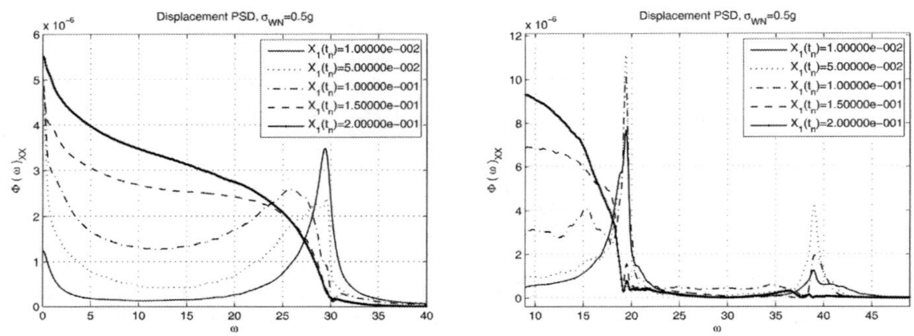

(a) Model 2DOF–1 (b) Model 2DOF–2

Figure 3. Power spectral densities of $X_1 - \sigma_{WN} = 0.50$ g.

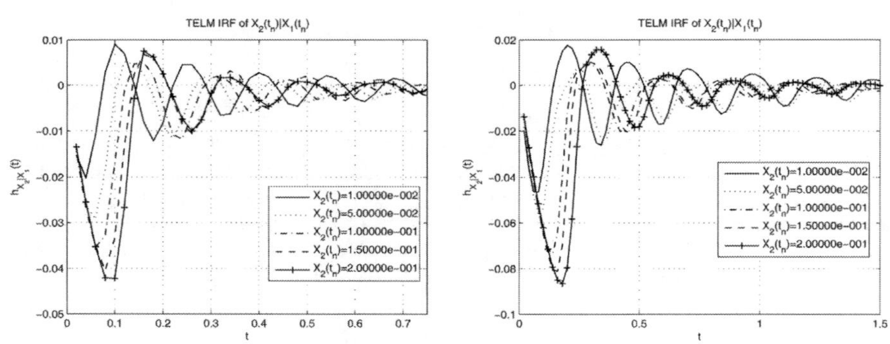

(a) Model 2DOF–1 (b) Model 2DOF–2

Figure 4. Impulse Response Functions – $X_2(t)$

of the IRFs, and then to combine them with the power spectral densities of the displacement X_1. The conditioned PDFs of X_2 given X_1 are shown in Figure 5.

Finally, it is possible to employ the 16 in order to get the joint distribution of the responses. In Figure 6 the complementary-joint-PDF are shown, *i.e.* the probability that at least one of the responses is greater than the corresponding threshold.

In order to check the approximation of the proposed strategy, a Monte Carlo simulation has been performed. The response statistics obtained with 10,000 samplings have been compared

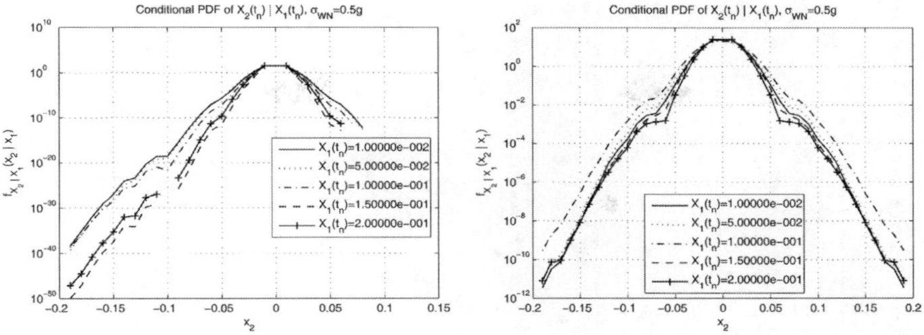

(a) Model 2DOF–1 (b) Model 2DOF–2

Figure 5. PDF of $X_2(t)$ given $X_1 - \sigma_{WN} = 0.5$ g.

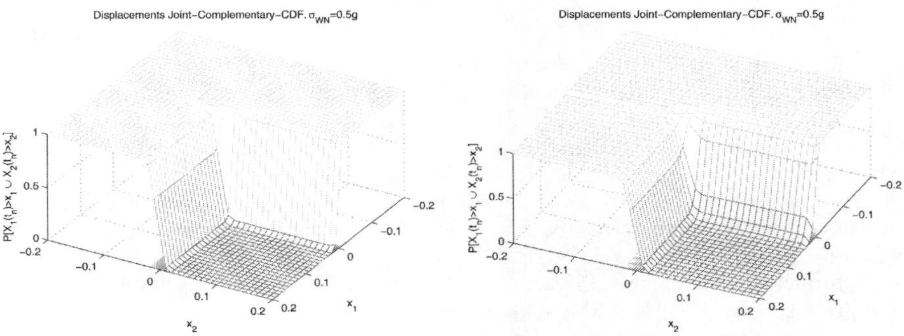

(a) Model 2DOF–1 (b) Model 2DOF–2

Figure 6. Complementary Joint-CDF – $\sigma_{WN} = 0.50$ g.

with the results of TELM. In particular, Figure 7 shows the difference of the complementary-joint-PDFs with respect the thresholds x_1 and x_2, specifically, in both cases, the computed error is not negligible for small thresholds. This expected behavior is due a drawback of TELM: the linearized system depends on the dimensions of the hysteresis loops. For this reason, the nonlinear relationship between forces and displacements is close to be linear at small thresholds, then, the response statistics are quite close to be Gaussian. However, it has been shown in (Fujimura 2006) that this drawback is overcomed by the evaluation of the extreme-response statistics: the error of the response statistics at t_n does not affects the first-passage probability. Also, in common practice, the structural safety is usually computed for higher values of the threshold, where the error between TELM and the Monte Carlo simulation is negligible.

6 CONCLUSIONS

An extension of the Tail Equivalent Linearization Method is presented. Its goal is to compute the joint tail probability of structural systems whose reliability depends on two or more responses of interest. The proposed procedure can be performed regardless of the system dimensions, the only requirements are the existence and uniqueness of TELS. In this paper, just the series system case has been investigated, however, the algorithm can be easily extended to the general case of parallel and hybrid systems. Furthermore, numerical applications have been provided; in order to be able to show the joint probabilities, only the two-degrees of freedom has been presented. In any case, a generalized procedure can be applied also to multi-dimensional systems.

The proposed algorithm leads to reasonably accurate results which can be employed in order to get the level-crossing rate and the Joint First-Passage Probability of interconnected systems

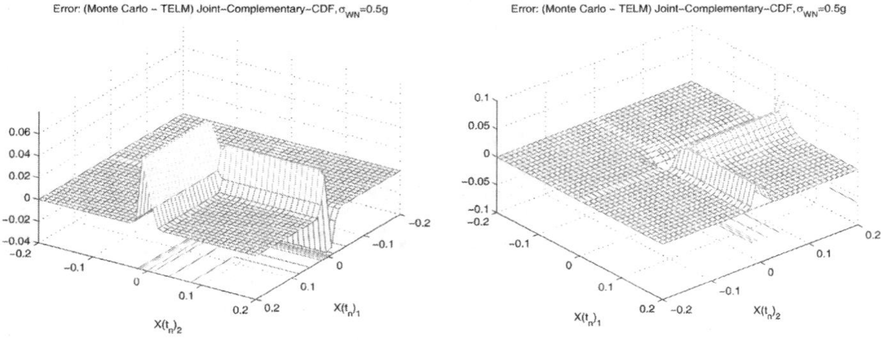

(a) Model 2DOF–1 (b) Model 2DOF–2

Figure 7. Monte Carlo – TELM error – $\sigma_{WN} = 0.50$ g.

(Song and Der Kiureghian 2006); it appears particularly convenient for buildings. Also, the TELM algorithm has been implemented in OpenSees, a framework for structural analysis provided by the Pacific Earthquake Engineering Research Center and it is available for scientific purposes.

However, further work can be developed. In fact, the *white noise approximation* affects the computation of the conditioned Impulse Response Functions (IRFs); the error is negligible for the provided applications, however, it could be higher for complex systems and narrow band excitations. A further development would define the conditioned IRFs with respect conveniently filtered white noises in order to get the structural filtering.

Also, a further development is the implementation of multi-support excitations: the provided structure presents two responses whose the second one can be expressed in function of the first response and regardless of the base excitation. If the structure has coupled responses, then TELM should take into account of the multiple dependence by using multi-support excitations.

REFERENCES

Bouc, R. (1963). Forced vibrations of mechanical systems with hysteresis. In A. Publishing (Ed.), *Proceedings of the 4th conference on nonlinear oscillation*, Prague, pp. 315. Academia Publishing.

Der Kiureghian, A. (2000). The geometry of random vibrations and solutions by form and sorm. *Probabilistic Engineering Mechanics 15*(1), 81–90.

Der Kiureghian, A. & K. Fujimura (2009). Nonlinear stochastic dynamic analysis for performance-based earthquake engineering. *Earthquake Engineering & Structural Dynamics 38*(5), 719–738.

Fujimura, K. (2006). *Tail Equivalent Linearization Method for Nonlinear Random Vibration*. PhD dissertation, University of California–Berkeley, Department of Civil and Environmental Engineering.

Fujimura, K. & A. Der Kiureghian (2007). Tail-equivalent linearization method for nonlinear random vibration. *Probabilistic Engineering Mechanics 22*(1), 63–76.

Lutes, L. D. & S. Sarkani (2004). *Random Vibrations–Analysis of Structural and Mechanical Systems* (2004 ed.). Oxford, UK: Elsevier.

Sessa, S. & A. Der Kiureghian (2009, September). Application of the tail–equivalent linearization method for stochastic dynamic analysis with asymmetric hysteresis. In *Proceedings of the 10th International Conference on Structural Safety And Reliability–Osaka*.

Song, J. & A. Der Kiureghian (2003). Bounds on system reliability by linear programming. *Journal of Engineering Mechanics 129*(6), 627–636.

Song, J. & A. Der Kiureghian (2006). Joint first-passage probability and reliability of systems under stochastic excitation. *Journal of Engineering Mechanics 132*(1), 65–77.

Wen, Y. (1976). Method for random vibration of hysteretic systems. *Journal of Engineering Mechanics Division, ASCE 102*(2), 249–163.

Wen, Y. (1980). Equivalent linearization for hysteretic systems under random excitations. *Journal of Applied Mechanics ASME 47*(3), 150–154.

Reliability and Optimization of Structural Systems – Straub (ed)
© 2010 Taylor & Francis Group, London, ISBN 978-0-415-88179-1

Fuzzy reliability of concrete bridge elements with respect to the reinforcement chloride corrosion

R. Sieniawska, P. Sniady & St. Zukowski
Wroclaw University of Technology, Wroclaw, Poland

ABSTRACT: The complex problem of assessing the reliability of a reinforced concrete structure exposed to the reinforcement chloride corrosion taking into account two type of uncertainty, namely the randomness and fuzziness of structure as well as load and corrosion parameters is considered. An original algorithm for evaluating the reliability of beam and beam-plate reinforced concrete bridges is presented. Many arbitrary simplifications are assumed and for this reason the limit condition is described by a fuzzy number and the ion-propagation time by a fuzzy random variable. The most important element of the elaborated algorithm is considering the problem in two separate time periods: in the period of the ion propagation through the concrete cover and in the period after the initiation of the reinforcement corrosion. Such approach leads to an integral expression for calculating the probability of failure. The theoretical considerations are illustrated by a numerical example.

1 INTRODUCTION

An important factor which causes the load capacity degradation of reinforced concrete bridge structures during their exploitation is the concrete and steel corrosion. One of the essential reasons of corrosion is the presence of chlorides. Chloride corrosion in reinforced structures is a result of chloride ion propagation through the cover layer after the critical chloride concentration on the reinforcement surface is reached. The phenomenon of the chloride ion propagation through the concrete cover is very complex and random by nature due to changing environmental conditions and concrete inhomogeneity. From the reliability point of view it is very important to determine the corrosion initiation time. Empirical investigations are very difficult, therefore in many papers the corrosion modelling based on Fick's law of diffusion was used to derive an expression for corrosion initiation time. Thoft-Christensen (1998) showed by means of the simulation technique that the corrosion initiation time can be modelled by the Weibull distribution while Frangopol & Enright (1999) proposed the lognormal distribution for the corrosion initiation time modelling.

Many reliability assessment methods which take into account deterioration effects due to chloride corrosion have been developed. Val and Melchers (1998) presented a method for reliability assessment of RC frame structures based on the combination of the nonlinear finite element structural model and the first order method for reliability estimation. The papers of Thoft-Christensen et al. (1997), Thoft-Christensen (1998, 2000), Thoft-Christensen et al. (1997) present stochastic modelling of reinforced concrete bridges and other structures deterioration based on the detailed modelling of corrosion initiation and cracking. A new service lifetime definition based on the evolution of the corrosion crack width was proposed by Thoft-Christensen (2000), Thoft-Christensen et al. (1997). The optimal planning of measurements of the chloride content in reinforced concrete structures using the FORM-analysis was considered by Sörensen & Engelund (1996). They used Bayesian statistics to obtain predicted distributions of the main uncertain quantities describing the transport of chloride. In the paper of Frangopol & Enright (1999) the reliability assessment of deteriorating reinforced concrete highway bridges under aggressive conditions was presented. The Monte Carlo simulation, in which the strength degradation parameters were estimated on the basis of expected corrosion initiation and propagation rates, was applied. For estimating the time-variant system failure probability of an existing reinforced concrete bridge the important sampling method was used. Enright & Frangopol (1999) investigated a similar problem. The authors proposed a wide variety of models for estimating the reliability of highway bridges and considered the influence of resistance degradation and post-failure load redistribution. A spatial time-dependent reliability model developed for a RC beam for shear and the flexural limit state was considered by Stewart (2009). The couple reliability problem of biodeterioration, cloride ingress and cracking for reinforced concrete structures was solved by Basiga-Arteaga et al. (1998). Original algorithm for evaluating the reliability of bridge based on integral formulae was presented by Sniady et al. (2004).

Val & Stewart (2009) provide a brief description of the framework for reliability-based assessment deterioration due to corrosion reinforced concrete structures.

The complex problems of assessing the reliability of a concrete structure with the respect to the reinforcement chloride corrosion involve two kinds of uncertainty: randomness and fuzziness. The uncertainty related to fuzziness comes, among others, from the interaction of the construction with a variable environment and from the simplifications of the theoretical model used to describe the problem. The concept of fuzzy random variables were introduced by Kwarkernaak (1978), Puri and Ralescu (1986) to describe fuzzy random quantities. In the paper by Szeliga & Witkowski (1997) the assessment of the structural reliability using the fuzzy sets theory was presented. The problem of safety assessment of structures in view of fuzzy randomness has been considered in the books (Moller & Beer 2004, Moller & Reuter 2007).

In the presented paper an original algorithm for evaluating the reliability of beam and beam-plate type reinforced concrete bridges, taking into account chloride corrosion of the reinforcement, fuzzy random nature of parameters describing the corrosion process as well as geometrical parameters of the cross-section is presented. Presented procedure is an extension of that described by Sniady et al. (2004). In the considered problem many arbitrary simplifications are assumed and for this reason the limit condition is described by a fuzzy number and the ion propagation time by a fuzzy random variable.

The most important element of the elaborated algorithm is considering the problem in two separate time periods: in the period of the ion propagation through the concrete cover and in the period after the initiation of the reinforcement corrosion. Such approach leads to an integral expression for calculating the probability of failure. If estimating the reliability of a bridge treated as a system the correlation between failures of two different cross-sections are determined, where the fact that the failures can take place in the two different phases of corroded beams has been taken into account.

2 ASSUMPTIONS

During the structure capacity degradation process due to chloride corrosion the following phases could be distinguished: the initial phase, the phase of chloride ion propagation and, finally, the reinforcement corrosion phase.

The initial phase begins when the structure is established and ends at some moment, when the concentration of the chlorides at the concrete surface reaches some value C_0. The next phase is the propagation of chloride ions from the outside surface of the beam through the concrete cover towards the reinforcement bars. At the time when the chloride ion concentration on the reinforcement bar surface reaches the critical value C_{cr}, the process of corrosion of the reinforcing steel starts, which causes the gradual decrease of the usable cross-section of the reinforcement. With increasing corrosion the rust products increase volume and cause tensile stresses. These stresses cause the cracks in the concrete cover. The process of cracks size increase is accelerated by the corrosion of the structure reinforcement and, vice versa, the crack size influences the corrosion of steel bars. The next phase of the reinforced concrete beams degradation due to chloride corrosion is the spalling of the concrete cover. Pantazopoulou & Papoulia (2001) formulated an analytical model to demonstrate the mechanical consequences of corrosion-product build-up around the bar. This model was used for estimating the service life of RC structures. All phases and steps of structure deterioration due to chloride corrosion were described in detail by Thoft- Christensen (1998, 2001).

In the literature the following models for the reinforcing steel corrosion are considered: a uniform corrosion, a non-uniform corrosion and an in-depth (local, pitting) corrosion (Val & Melchers

1998). The uniform corrosion process takes place almost equally within the whole perimeter of the bar. The non-uniform one takes place within almost whole perimeter of the bar but not equally. The two first cases are characteristically accompanied by visible dark dyed water stains spanned over large areas of the cover surface; they can be also treated as general corrosion. The local corrosion has a restricted character and proceeds from one side of the bar. The course of the local corrosion phenomenon is faster than that for the uniform one but the scope of it is significantly smaller. The corrosion causes the steel volume increase (swelling) and, finally, leads to the damage of the girder.

Let $C(x, t)$ denote the chloride ion concentration in the concrete at the moment $t$in the distance x from the surface. We assume that chlorides appear in the girder's environment and the ion flow proceeds from the girder surface (higher concentration) towards reinforcement lower concentration. Additionally, we assume that the chloride ions do not disperse in other directions. The process of chloride ion flow through the cover is described by the diffusion equation, so called second Fick's equation

$$D_c \cdot \frac{\partial^2 C(x,t)}{\partial x^2} = \frac{\partial C(x,t)}{\partial t} \tag{1}$$

where D_c is a diffusion coefficient.

The solution of the equation (1) has the form

$$C(x,t) = C_0\left[1 - erf\left(\frac{x}{2 \cdot \sqrt{D_c \cdot t}}\right)\right] \tag{2}$$

where $C_0 = C(0, t_0)$ is the concentration of the chlorides at the concrete surface, and $erf(\cdot)$ denotes the error function.

The corrosion of reinforcement bars begins when the chloride ion concentration, after defeating the concrete cover of the thickness d, reaches the critical value C_{cr} at the reinforcement. The moment t_J in which the corrosion starts, so called the initiation time for the corrosion, may be shown as proposed by Thoft-Christensen (1998)

$$t_J = \frac{d^2}{4 \cdot D_c} \cdot \left[erf^{-1}\left(\frac{C_0 - C_{cr}}{C_0 - C_i}\right)\right]^{-2} \tag{3}$$

where C_i is the initial chloride concentration at the reinforcement surface. As stressed in the work of Vu & Steward (2000, 2001), the second Fick's equation works fine in the case of wet concrete. In other situations, the course of the penetration process not always conforms the equation. The chloride propagation process is complex; it is connected with the propagation of water and oxygen hence it is a multiphase and non-linear phenomenon. Nevertheless, the Fick's equation is often applied in structure reliability assessment because of two reasons:

- it is easy to solve the diffusion equation (1) both for the function $C(x, t)$ describing the chloride concentration change and for the value t_J describing the amount of time until the chloride concentration at the reinforcement surface becomes critical,
- it is relatively easy to determine the C_0 constant being the chloride concentration at the concrete surface as well as (from the empirical data) the diffusion coefficient D_c.

For further considerations we also make simplification of the process of the chloride ion flow through the cover and apply the Fick's equation. The chloride ion propagation is a fuzzy random process. Let us assume, as in the work by Vu & Steward (2001), that the diffusion coefficient D_c, the chloride concentration at the concrete surface C_0 and the critical concentration of the chloride at the reinforcement surface C_{cr} are random Gaussian variables. We also assume, that the cover thickness d is described by a random variable. The initial chloride concentration at the reinforcement surface C_i may be assumed as a deterministic value equal to zero or equal to a small fraction of the average value of the surface concentration C_0, i.e. $C_i = \chi$ E[C_0] where $0 \leq \chi << 1$. The chloride concentrations C_0, C_{cr}, C_i are identified as the percentage of the concrete mass.

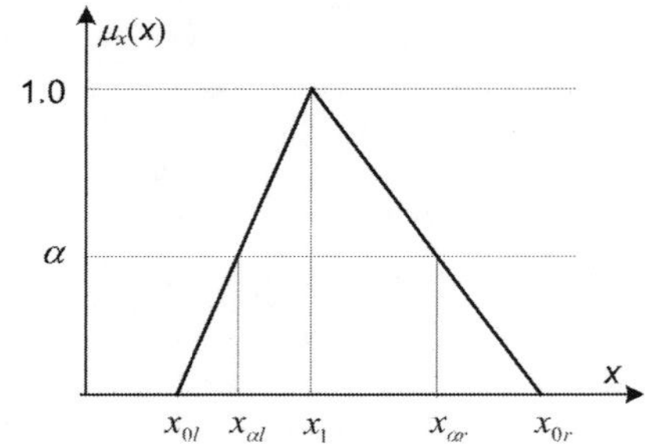

Figure 1. Triangular membership $\mu_X(x)$.

The process of steel corrosion can be initiated by the critical concentration of chlorides at the reinforcement surface formed not only due to diffusion but also as a result of cracks and damages in a concrete cover. As mentioned earlier, when the chloride ion concentration at the surface surrounding the reinforcement bars reaches the critical level C_{cr}, the corrosion begins.

Thoft-Christensen (1998) showed by means of the simulation technique and using the formula (3) that the corrosion initiation time t_J can be modelled by a Weibull distribution while Frangopol & Enright (1999) proposed the lognormal distribution for the corrosion initiation time modelling. Taking into account the uncertainity connected with initiation time of the corrosion and the results of the papers Thoft-Christensen (1998) and Frangopol & Enright (1999) it is assumed that the probability density function of the initiation time t_J is fuzzy and given by equation

$$f_{t_J\alpha}(t) = a_\alpha f_{1\alpha}(t) + (1 - a_\alpha) f_{2\alpha}(t), \tag{4}$$

where $f_{1\alpha}(t)$ is fuzzy Weibull distribution, $f_{0\alpha}(t)$ is fuzzy lognormal distribution and $0 \le a_\alpha \le 1$ is fuzzy number (Moller & Beer 2004) and $0 \le \alpha \le 1$ is crisp number.

In the paper it is assumed that fuzzy numbers x_α have triangular membership function $\mu_X(x)$ (Fig. 1) given by the formulae

$$\mu_X(x_\alpha) = \begin{cases} \alpha = 0 & \text{for } x_\alpha \le x_{0l}, \\ \alpha = \dfrac{x_{\alpha l} - x_{0l}}{x_1 - x_{0l}} & \text{for } x_{0l} < x_{\alpha l} \le x_1, \\ \alpha = \dfrac{x_{0r} - x_{\alpha r}}{x_{0r} - x_1} & \text{for } x_1 < x_{\alpha r} < x_{0r}, \\ \alpha = 0 & \text{for } x_{0r} \le x_\alpha. \end{cases} \tag{5}$$

Further, it is assumed that the mean values $E[X_\alpha]$ of the fuzzy random variables be fuzzy numbers with triangular membership function, as shown in Fig. 1.
In this case we have

$$E[X_\alpha] = [E(X_{\alpha l}), E(X_{\alpha r})] = [(1-\alpha)E(X_{0l}) + \alpha E(X_1), (1-\alpha)E(X_{0r}) + \alpha E(X_1)], \tag{6}$$

It is also assumed that the variation coefficients ν_X are constant and do not depend on the level of fuzziness. Therefore the standard deviation of the fuzzy random variables is a fuzzy number as well as the function of the expected value $\sigma_{X\alpha} = \nu_X \cdot E[X_\alpha]$.

We will compute the radius of the pit $r(t)$ (Fig. 2) in the same way as Val & Melchers (1998) from the formula

$$r(t) = C_{corr} i_{corr} (t - t_{J\alpha}) R + r_\alpha(t), \tag{7}$$

where C_{corr} is a corrosion coefficient of the steel, i_{corr} is a rate of corrosion, R is a constant and $r_\alpha(t)$ is a fuzzy function in which some uncertainty is taking into account among other uncertainty of the parameters C_{corr} and i_{corr}.

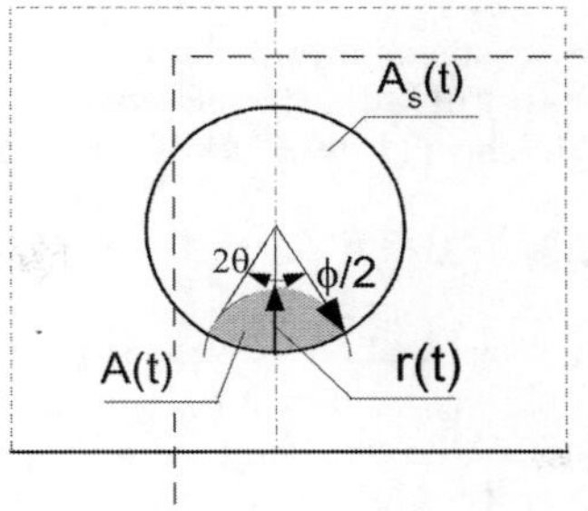

Figure 2. Model of corrosion.

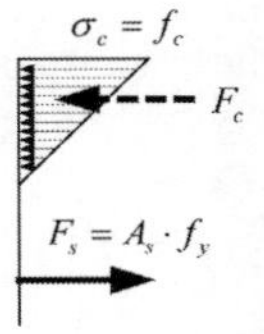

Figure 3. Stress distribution in a beam.

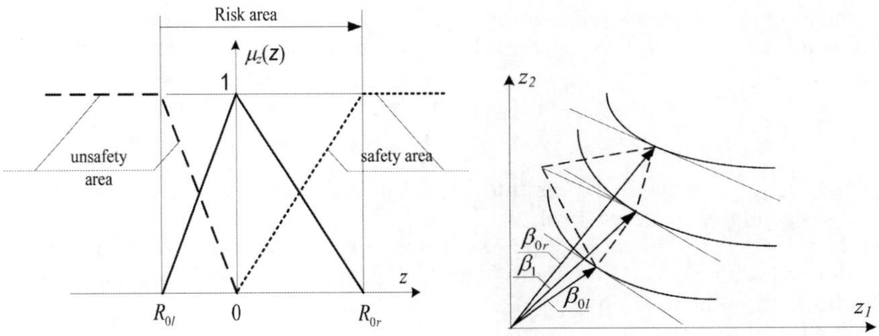

Figure 4. Safety, unsafety and risk area.

The value $i_{corr} = 1\,\mu A/\text{cm}^2$ corresponds to the decrease of the steel thickness equal to $11.6\,\mu m$/year. It is assumed that the corrosion coefficient and the rate of corrosion are random variables.

3 BEAM GIRDER RELIABILITY

While determining the load capacity $M_C(\cdot)$ of the reinforced concrete we will assume that the compressive stresses in the concrete have a triangular distribution (Fig. 3) and the tensile stresses in steel reach the plasticity limit.

We will assume the limit condition for the cross-section in the form

$$Z_{R\alpha}(\ldots, z_i, \ldots) = M_{C\alpha} - M_{S\alpha} = R_\alpha \tag{8}$$

where $Z_{R\alpha}(\cdot)$ denotes the fuzzy load bearing capacity reserve (the difference between the load bearing capacity of the cross-section $M_{C\alpha}$ and the bending moment $M_{S\alpha}$ in this cross-section due to the acting load), z_i stands for a set of parameters defining the load and the load bearing capacity of the reinforced concrete cross-section with regard to reinforcement corrosion. The fuzzy risk R_α is a fuzzy number with a membership function $\mu_R(x)$. The fuzzy risk, fuzzy safety and fuzzy unsafety areas are shown in the Figure 4 (Szeliga & Witkowski 1997).

As the reliability measure we will assume the probability of no failure $(1-p_{f\alpha})$ and the reliability index β_α connected with the probability of failure by the relationship shown, among others, in works by Ditlevsen & Madsen (1996), Melchers (1999), Nowak & Collins (2000)

$$p_{f\alpha} = p_{f\alpha}(t) = P\{Z_\alpha < R_\alpha\} = \Phi(-\beta_\alpha(t)) \tag{9}$$

We assume that the reinforcement cross-section as well as the load bearing capacity of the cross-section do not change in the time until $t_{J\alpha}$ and then $t_{J\alpha}$ is a function of time dependent corrosion model applied. Taking into account that the corrosion initiation time $t_{J\alpha}$ is a fuzzy random variable, the probability that the corrosion process has already started $(t > t_{J\alpha})$ is equal to the cumulative distribution function (c.d.f.) $F_{tJ\alpha}(t)$ of the variable $t_{J\alpha}$ and the probability that the corrosion process has not already started $(t < t_{J\alpha})$ is equal to $1 - F_{tJ\alpha}(t)$.

The probability that the fuzzy random variable $t_{J\alpha}$ is within the interval $t_{J\alpha} \in (\tau, \tau + d\tau)$ equals $f_{tJ\alpha}(\tau)\mathrm{d}\tau$, where $f_{tJ\alpha}(\tau) = \mathrm{d}F_{tJ\alpha}(\tau)/\mathrm{d}\tau$ denotes the probability density function (p.d.f.) of the fuzzy random variable $t_{J\alpha}$. Assuming that the probability $p_{fo\alpha}$ of the load bearing capacity reserve $(Z_R < R_\alpha)$ for $t \le t_{J\alpha}$ and the probability $p_{fcorr\alpha}(t - t_J)$ of the load bearing capacity reserve for $t > t_{J\alpha}$ are known, the probability of failure could be determined by the formula

$$p_{f\alpha}(t) = p_{f0\alpha} \cdot [1 - F_{t_J\alpha}(t)] + \int_0^t p_{f\,corr\alpha}(t-\tau) \cdot f_{t_J\alpha}(\tau) d\tau \tag{10}$$

for any t and random corrosion initiation time $t_{j\alpha}$..

The probabilities $p_{fo\alpha}(t - t_J)$, $p_{fcorr\alpha}(t - t_J)$ may be obtained from the relationship

$$p_{f\alpha} = \int \cdots \int_{Z_\alpha(\cdot) \le R_\alpha} \cdots \int f_{Z_\alpha}(z_1 \ldots, z_i, \ldots) \cdot dz \tag{11}$$

where $f_Z(z_1, \ldots, z_i, \ldots)$ denotes the multidimensional load bearing capacity reserve function for the cases $t \le t_{J\alpha}$ and $t > t_{J\alpha}$, respectively.

The probability density function $f_{tJ\alpha}(t)$ of the fuzzy random variable $t_{J\alpha}$ is an boundary distribution of the multidimensional probability density function of parameters d, D_c, C_0, C_{cr}, C_i and $t_{J\alpha}$ what follows from the relationship (3).

The probabilities $p_{f0\alpha}$, $p_{fcorr\alpha}(t - t_J)$ and the cumulative distribution function $F_{tJ\alpha}(t)$ could be also determined from the relationships

$$p_{f0\alpha} = \Phi(-\beta_{0\alpha}), \tag{12}$$

$$p_{f\,corr\alpha}\left(t - t_{j\alpha}\right) = \Phi(-\beta_{corr\alpha}\left(t - t_{J\alpha}\right)), \tag{13}$$

$$F_{t_{J\alpha}}(t) = P\{t_{J\alpha} \le t\} = 1 - P\{t - t_{J\alpha} < 0\} = 1 - P\{Z_{t_{J\alpha}} < 0\} = 1 - \Phi(-\beta_{t_{J\alpha}}(t)), \tag{14}$$

after the reliability indices $\beta_{0\alpha}$, $\beta_{corr\alpha}(t - t_{J\alpha})$, and $\beta_{tJ\alpha}$ have been computed. These indices may be determined numerically by means of the FORM (First Order Reliability Method) or SORM (Second Order Reliability Method) methods (Ditlevsen & Madsen 1996, Melchers 1999, Nowak & Collins 2000).

The reliability indices $\beta_{0\alpha}$ and $\beta_{corr\alpha}(t - t_{J\alpha})$ are being determined for crossing the limit condition (8), for the case in which the corrosion has not started yet $(t \le t_{J\alpha})$ and for the case of running corrosion $(t > t_{J\alpha})$, respectively.

Using the *α–level* optimization procedure (Moller & Beer 2004) for an arbitrary $\alpha = \alpha_k \in [0,1]$ or the max-min procedure in the extension principle (Zadeh 1965), the membership function of the failure probability $p_{f\alpha}(t)$ and reliability index $\beta_\alpha(t)$ can be found.

4 BRIDGE RELIABILITY

The values of $p_{f\alpha}(t)$ and $\beta_\alpha(t)$ obtained for a given cross-section of the beam are the basis for determining the reliability of the whole bridge treated as a system. In general serial, parallel and mixed-type reliability systems are distinguished. Actually, the failure in one cross-section (including statically-indeterminate structures) does not necessarily lead to the failure of the whole structure. However, as a rule, it requires the bridge reconstruction. Moreover, assuming that the failure, in sense of crossing the $Z_\alpha = R_\alpha$ condition, brings about the necessity of reconstruction, each structure can also be treated as a serial reliability system.

Let us assume that the structure has n cross-sections called dangerous in which the above mentioned failure can occur. In our discussion we will skip the influence of decreasing load bearing capacities of the individual beams cross-sections due to corrosion on the redistribution of load transmitted by beams. Let us denote the reliability index and the probability of failure in the i-th dangerous cross-section by $\beta_{i\alpha}(t)$, $p_{f\alpha}(t)$, $(i = 1,2,\ldots, n)$, respectively, in which

$$p_{f_i\alpha}(t) = \Phi(-\beta_{i\alpha}(t)) \tag{15}$$

and

$$\beta_{i\alpha}(t) = -\Phi^{-1}(p_{f_i\alpha}(t)) \tag{16}$$

For simplification, we assume that the "failures" in individual cross-sections are mutually independent. Then, the reliability of the bridge is treated as the reliability of the serial system consisting of $n$elements. Hence, the failure probability of the bridge is shown by the formula

$$p_{f\alpha}(t) = 1 - \prod_{i=1}^{n}[1 - \Phi(-\beta_{i\alpha}(t))] \tag{17}$$

Therefore, the reliability index for the bridge is equal to

$$\beta_\alpha(t) = -\Phi^{-1}(p_{f\alpha}(t)) \tag{18}$$

Applying the solutions obtained for non-correlated load bearing capacity reserves $Z_\alpha(\ldots, z_i,\ldots) = M_{R\alpha} - M_{S\alpha}$ for different $i = 1,2,\ldots,n$ to the case of correlated ones leads to approximate results. For more detailed study of this question, let us denote the event that $Z_{Ri\alpha}(t) \leq R_{i\alpha}$ by $U_{i\alpha}(t)$. The probability of failure is then equal to

$$p_{f\alpha} = P\{U_{1\alpha} \cup U_{2\alpha} \cup \ldots \cup U_{n\alpha}\} \tag{19}$$

Because the real events $U_i(t)$ are mutually correlated for different i, the failure probability $p_{f\alpha}$ is difficult to calculate exactly it could be only estimated. Below, we will show the evaluation by Cornell & Ditlevsen described, among others, by Ditlevsen & Madsen (1996), Melchers (1999), Nowak & Collins (2000). The Cornell's evaluation results in

$$\max_{i=1}^{n} \Phi(-\beta_{i\alpha}(t)) \leq p_{f\alpha}(t) \leq 1 - \prod_{i=1}^{n}[1 - \Phi(-\beta_{i\alpha}(t))] \tag{20}$$

Note, that the right-hand side of the formula (20) is identical with the one in the formula (17). Hence, the assumption about independence of events leads to the upper estimation.

The evaluation of Ditlevsen is more accurate and has the form

$$\begin{aligned}\Phi(-\beta_{1\alpha}(t)) + \sum_{i=2}^{n} \max\{\Phi(-\beta_{i\alpha}(t)) - \sum_{j=1}^{i-1} \Phi_2(\beta_{i\alpha}(t), \beta_{j\alpha}(t), \rho_{ij\alpha}), 0\} \leq p_{f\alpha}(t) \leq \\ \leq \sum_{i=1}^{n} \Phi(-\beta_{i\alpha}(t)) - \sum_{i=2}^{n} \max\{\Phi_2(\beta_{i\alpha}(t), \beta_{j\alpha}(t), \rho_{ij\alpha})\}\end{aligned} \tag{21}$$

where the symbol $\Phi_2(x, y, \rho_{ij\alpha})$ denotes the two-dimensional cumulative distribution function of the normal distribution of the average (0,0), variance (1,1) and correlation coefficient $\rho_{ij\alpha}$.

The correlation coefficient of two events Ui and Uj is presented by the relationship

$$\rho_{ij\alpha} = \frac{\text{cov}(U_{i\alpha}, U_{j\alpha})}{\sigma_{U_i\alpha}(t) \cdot \sigma_{U_j\alpha}(t)} \tag{22}$$

Table 1. Calculation parameters.

Parameter	$E[x]_{0l}$	$E[x]_1$	$E[x]_{0r}$	ν_x	Distribution
C_{cr}	0.27 %	0.3 %	0.33 %	0.1	normal
C_0	0.585 %	0.65 %	0.715 %	0.05	normal
D_c	25 mm^2/year	25 mm^2/year	25 mm^2/year	0.2	normal
C_{corr}	2.09 mm^3/μA/year	2.32 mm^3/μA/year	2.55 mm^3/μA/year	0.01	normal
i_{corr}	0.027 μA/mm^2	0.03 μA/mm^2	0.033 μA/mm^2	0.01	normal
Φ_0		16 mm		0.02	normal
f_y	369000 kN/m^2	410000 kN/m^2	451000 kN/m^2	0.02	normal
f_c	19980 kN/m^2	22200 kN/m^2	24420 kN/m^2	0.02	normal
b	0.45 m	0.50 m	0.55 m	0.02	normal
h	0.63 m	0.70 m	0.77 m	0.02	normal
Ms		400 kN m		0.01	normal
t_J		10 years		0.3	Weibull or lognormal

The event $U_{1\alpha}$ in the formula (22) denotes the event characterised by the greatest failure probability $P\{U_{l\alpha}\} = \Phi(-\beta_{l\alpha}(t)) = \max P\{U_{i\alpha}\} = \max \Phi(-\beta_{i\alpha}(t))$.

While determining the standard deviation and the covariance function in the formula (22) it should be taken into consideration that the event U_i might occur when the reinforcement corrosion has not started yet ($t < t_J$) or the corrosion has already been running ($t \geq t_J$). Therefore, the quantities $\sigma^2_{Ui}(t)$ and $\text{cov}(U_i,, U_j)$ will be determined from the following relationships

$$\sigma^2_{U_i\alpha}(t) = \sigma^2_{U_i\alpha}(t \mid t \leq t_{Ji\alpha}) \cdot \left[1 - F_{t_{Ji\alpha}}(t)\right] + \sigma^2_{U_i\alpha}(t \mid t > t_{Ji\alpha}) \cdot F_{t_{Ji\alpha}}(t) \quad (23)$$

$$\begin{aligned} \text{cov}[U_{i\alpha}(t), U_{j\alpha}(t)] = {} & \text{cov}[U_{i\alpha}(t), U_{j\alpha}(t) \mid t \leq t_{Ji\alpha}, t \leq t_{Jj\alpha}] \cdot \left[1 - F_{t_{Ji\alpha}}(t)\right] \cdot \left[1 - F_{t_{Jj\alpha}}(t)\right] + \\ & + \text{cov}[U_{i\alpha}(t), U_{j\alpha}(t) \mid t > t_{Ji\alpha}, t \leq t_{Jj\alpha}] \cdot F_{t_{Ji\alpha}}(t) \cdot \left[1 - F_{t_{Jj\alpha}}(t)\right] + \\ & + \text{cov}[U_{i\alpha}(t), U_{j\alpha}(t) \mid t \leq t_{Ji\alpha}, t > t_{Jj\alpha}] \cdot \left[1 - F_{t_{Ji\alpha}}(t)\right] \cdot F_{t_{Jj\alpha}}(t) + \\ & + \text{cov}[U_{i\alpha}(t), U_{j\alpha}(t) \mid t > t_{Ji\alpha}, t > t_{Jj\alpha}] \cdot F_{t_{Ji\alpha}}(t) \cdot F_{t_{Jj\alpha}}(t) \end{aligned} \quad (24)$$

In above formulas $\sigma^2_{Ui}(t\ t \leq t_{Ji}, t \leq t_{Jj}) = const$ and $\text{cov}[U_i(t), U_j(t)\ t \leq t_{Ji}, t \leq t_{Jj}] = const.$

5 EXAMPLE

The procedure presented has been used for the reliability analysis of the reinforced concrete two-beam bridge. The bridge span is equal to 20 m, the beams have rectangular cross-sections of the dimensions $b \times h$ and are reinforced by 12 bars. Applying the limit condition (8) and using the Comrel software the reliability indices have been calculated for fuzzy random parameters of the structure, load and corrosion as in the table 1.

The results obtained are presented in Figures 5 and 6. In Figure 5 the reliability indices as functions of the structure exploitation time and the level of fuzziness are plotted and in the Figure 6 the membership function of fuzziness of the reliability index calculated for the time ($E[t_{J0}] + 50$) years is shown.

6 SUMMARY AND CONCLUDING REMARKS

A numerical-analytical algorithm for reliability estimation of a structure exposed to chloride corrosion taking into account the two types of uncertainty, namely the randomness and fuzziness of

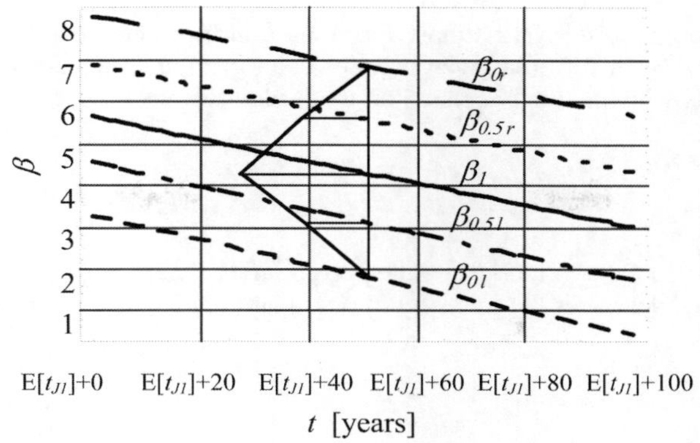

Figure 5. Safety index as a function of time and the level of fuzziness.

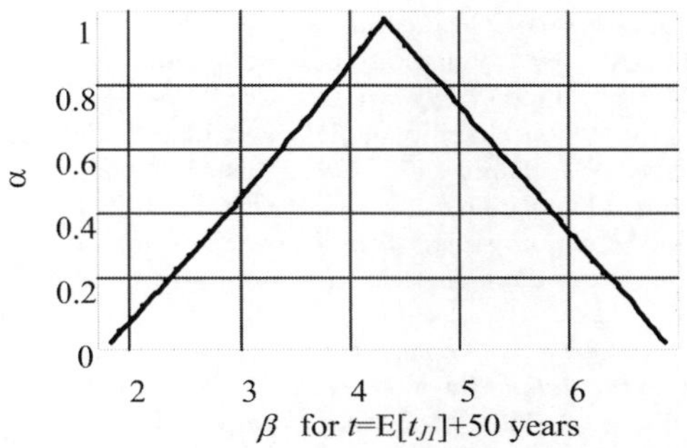

Figure 6. Membership of fuzziness of the reliability index.

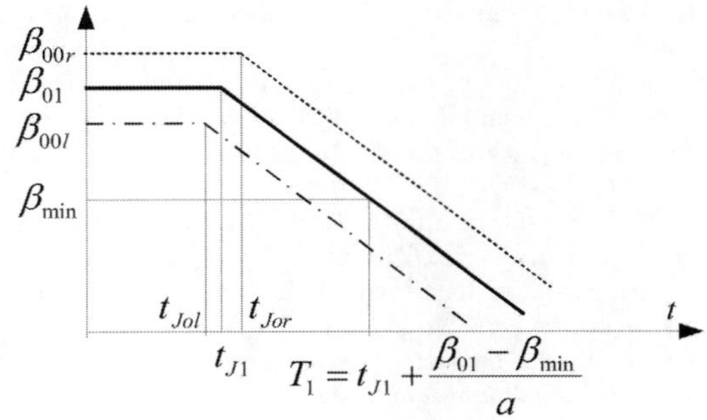

Figure 7. Time of safe exploitation.

structure as well as the load and corrosion parameters has been elaborated. The probability of failure has been computed as the superposition of the failure probability without taking into account the corrosion process and the failure probability with the corrosion process taken into account, where the corrosion initiation time is deterministic. For computation of probability distributions of the corrosion initiation time t_J, the load bearing capacity M_R, the load M_S and other quantities required for determining the reliability, an original procedure based on joint probability distributions has been applied. The solution presented could be applied in the standardization works and in the setting-up of the bridge reconstruction schedules, as well as in forecasting the time of the safe exploitation of a given bridge. This issue should be taken into account from the bridge maintenance point of view. The algorithm presented and the randomly made analysis in this work and mostly in Sniady et al. (2004) allow, after wide-range calculations for different parameters and exploitations conditions of the bridge, to derive a simple formula describing the change in time of the bridge reliability index $\beta(t)$. Such formula given by expression (25) and displayed in the Figure 7 may turn out to be helpful when making decision on the bridge repair.

$$\beta_{\alpha}(t) = \begin{cases} \beta_{0\alpha} & \text{for } t \le t_{J\alpha}, \\ \beta_{0\alpha}\text{-a}(t - t_{J\alpha}) & \text{for } t > t_{J\alpha}. \end{cases} \tag{25}$$

In the formula above the safety index $\beta_{0\alpha}$ is the design safety index, $\beta_{\min}$ is the assumed limit value and the parameter "a" describes the intensity of degradation.

REFERENCES

Bastida-Arteaga, E., Sanchez-Silva, M., Chateauneuf, A. & Ribas Silva, M. 2008. Coupled reliability model of biodeterioration, chloride ingress and cracking for reinforced concrete structures, *Structural Safety*100: 110–129.

Ditlevsen, O. & Madsen, H.O. 1996. *Structural Reliability Methods*. John Wiley & Sons.

Enright, M.P. & Frangopol, D.M. 1999. Reliability-based condition assessment of deteriorating concrete bridges considering load redistribution, *Structural Safety* 21: 159–195.

Frangopol, D.M. & Enright, M.P. 1999. Reliability of highway bridges under aggressive conditions, In Spanos (ed.), *Computational Stochastic Mechanics*: 173–179. Balkema: Rotterdam.

Kwarkernaak, K. 1978. Fuzzy random variables (I). *Inform. Sci.* 15: 1–29.

Melchers, R.E. 1999. *Structural Reliability Analysis and Prediction.* John Wiley & Sons.

Moller, B. & Beer, M. 2004. *Fuzzy Randomness. Uncertainty in Civil Engineering and Computational Mechanics*, Springer.

Moller, B. & Reuter, U. 2007. *Uncertainty Forecasting in Engineering*, Springer.

Nowak, A.S. & Collins, K.R. 2000. *Reliability of Structures*, Mc Graw Hill.

Pantazopoulou, S.J. & Papoulia, K.D. 2001. Modelling Cover-Cracking due to Reinforcement Corrosion in RC Structures, *Journal of Engineering Mechanics* 127(4): 342–351.

Puri, M.L. & Ralescu, D.A. 1986. Fuzzy random variables, *J. Math. Anal. Appl.* 114: 409–422.

Sorensen, J.D. & Engelund, S. 1996. Reliability-based planning of chloride measurements. In Frangopol D. M., Corotis R. B. & Rackwitz R. (eds), *Reliability and Optimization of Structural System, proc. of IFIP WG7.5*: 313–320. Pergamon.

Stewart, M.G. 2009, Mechanical behaviour of pitting corrosion of flexural and shear reinforcement and its effect on structural reliability of corroding RC beams, *Structural Safety* 31: 19–30.

Szeliga, E. & Witkowski, M. 1997. Assessment of structural reliability using the fuzzy sets theory, *Archives of Civil Engineering* XLIII (4): 327–352.

Śniady, P., Sieniawska, R. & Żukowski, S. 2004. Reliability of a concrete bridge with respect to the reinforcement chloride corrosion, *Archives of Civil Engineering* 3: 381–399.

Thoft-Christensen, P. 1998. Assessment of the service lifetime of existing reinforced concrete bridges. In Shinozuka & Wen (eds), *Structural Safety and Reliability*: 121–127. Rotterdam: Balkema.

Thoft-Christensen, P. 2000. Modelling of the Deterioration of Reinforced Concrete Structures, In A. Nowak & M. Szerszen (eds), *Reliability and Optimization of Structural System, proc. of IFIP WG7.5*: 15–26. The University of Ann Arbor: Michigan.

Thoft-Christensen, P. 2001. Corrosion crack based assessment of the life-cycle reliability of concrete structures, In Corotis et al., Swets & Zeitlinger (eds), *Structural Safety and Reliability*: 1–7.

Thoft-Christensen, P., Jensen, F.M., Middleton, C.R. & Blackmore, A. 1997. Assessment of the reliability of concrete slab bridges, In Frangopol D.M., Corotis R.B., Rackwitz R. (eds), *Reliability and Optimization of Structural Systems*:, 321–328. Pergamon.

Val, D. & Melchers, R.E. 1998. Reliability analysis of deteriorating reinforced concrete frame structures, In Shinozuka & Wen (eds), *Structural Safety and Reliability*: 105–112. Balkema: Rotterdam.

Val, V.D. & Stewart, M.G. 2009. Reliability assessment of ageing reinforced concrete structures – current situation and futures challenges, *Structural Engineering International* 2: 211–219.

Vu, K.A.T. & Stewart, M.G. 2000. Structural reliability of concrete bridges including improved chloride-inducted corrosion models, *Structural Safety* 22: 313–333.

Vu, K.A.T. & Stewart, M.G. 2001. Cracking and spalling reliability analysis considering spatial variability for reinforced concrete structures. In Corotis et al, Swets & Zeitlinger (eds), *Structural Safety and Reliability*: 1–8.

Zadeh, L.A. 1965. Fuzzy sets. *Information and Control* 8: 338–353.

Reliability and Optimization of Structural Systems – Straub (ed)
© 2010 Taylor & Francis Group, London, ISBN 978-0-415-88179-1

Dynamic response optimization of tall buildings subject to extreme wind loads

Seymour M.J. Spence & Massimiliano Gioffrè
Department of Civil and Environmental Engineering, University of Perugia, Italy

ABSTRACT: In this paper an alternative approach for the dynamic response optimization of wind excited tall buildings is proposed. Instead of optimizing the structure in terms of the response time histories, it is proposed to optimize under probabilistic constraints defined by the second order statistics of the response. Each constraint is formulated considering the mean hourly wind speed that causes a peak response level with specified mean recurrence interval. This mean hourly wind speed is calculated in terms of the response through a directional vulnerability model that rigorously accounts for the directional aerodynamics and climatological information. Finally the proposed method is illustrated on a full size 2D frame analyzed in a fully 3D extreme wind climate.

1 INTRODUCTION

The design of super tall buildings is in general governed by serviceability limit states rather than those associated with the ultimate response and, in particular, those limiting the inter-story drift will in general govern the design compared to the member capacity requirements (Chan et al. 1995). Numerous methods have been proposed for the drift optimization of tall buildings (Baker 1990, Chan et al. 1995, Kargahil et al. 2006, Kargahil and Anderson 2006, Kim et al. 2007). However, common to all these methods is the necessity to define a certain number of idealized Equivalent Static Wind Loads (ESWLs) (Chen and Kareem 2005a). These loads will depend on the dynamic response of a critical load effect, usually the base bending moments, which is then distributed over the buildings height (Chen and Kareem 2005a, Chen and Kareem 2005b). The response level achieved by applying these equivalent static loads is calibrated by considering the base bending moments generated by non directional mean hourly wind speeds with specified Mean Recurrence Intervals. It assumed that by blowing these mean wind speeds, responses with identical MRIs will be achieved. While the dependency of the static loads on the dynamic characteristics of the structure during minimum weight drift optimization has been the focus of recent studies (Chan et al. 2010), the importance of correctly estimating the mean wind speeds has been neglected. Also an extreme wind climate is in general highly directional (Simiu and Scanlan 1996), as is the aerodynamic response of typical tall buildings. The rigorous combination of these effects in the analysis and optimization of tall buildings is lacking.

In this paper a technique is proposed for member size optimization subject to serviceability constraints on lateral drift without the need to define any idealized ESWLs. Instead the problem is formulated directly in the time domain as a probabilistic dynamic response optimization problem. The mean wind speeds that cause a particular probabilistic response to assume a level with specified MRI are individually estimated. This is achieved by rigorously combining the directional aerodynamic climatological information through a vulnerability model based on a bi-dimensional extreme value distribution of the extreme wind climate. The implicit probabilistic structural optimization problem is then replaced by a sequence of explicit sub-problems based on the Principle of Virtual Work (PLV). The convergence of the sub-problems includes updating of the directional vulnerability model therefore ensuring consistent mean hourly speeds at the optimum. Quick and steady convergence is seen demonstrating the applicability of the proposed method.

2 THE OPTIMIZATION PROBLEM

A tall building will in general be composed of a certain number of floors, $j = 1, \ldots, N$, and members, $i = 1, \ldots, M$, which make up the structural system. Within this system there will be a number of column lines, $k = 1, \ldots, K$, for which the inter-story drift ratio d, with specific MRI, y_d, is to be controlled for a number of wind directions $\alpha = 1, \ldots, D$ in two orthogonal directions $(s = x, y)$. Within a member size optimization, the objective will be to minimize the weight of the structure, W, in terms of a design variable vector $\mathbf{x} = (x_1, \ldots, x_n)^T$ while controlling the inter-story drift ratio, d, and member level capacity ratios b. The response of these last will also be defined in terms of an acceptable response level, b_y, for a specified MRI y_b. Under these conditions the following dynamic response optimization problem may be defined:

$$\text{Find} \qquad \mathbf{x} = (x_1, ..., x_n)^T \tag{1}$$

$$\text{to minimize} \qquad W = f(\mathbf{x}) \tag{2}$$

subject to:

$$\underbrace{d_{jks\alpha}(\mathbf{x}, \ddot{\mathbf{z}}, \dot{\mathbf{z}}, \mathbf{z}, t)}_{\bar{V}_H^{d_y}(\mathbf{x},\ddot{\mathbf{z}},\dot{\mathbf{z}},\mathbf{z},\alpha)} - d_{jk}^{U,y_b} \le 0 \qquad (j = 1, ..., N)(k = 1, ..., K)(s = x, y)(\alpha = 1, ..., D) \tag{3}$$

$$\underbrace{b_{ip\alpha}(\mathbf{x}, \ddot{\mathbf{z}}, \dot{\mathbf{z}}, \mathbf{z}, t)}_{\bar{V}_H^{b_y}(\mathbf{x},\ddot{\mathbf{z}},\dot{\mathbf{z}},\mathbf{z},\alpha)} - b_p^{U,y_b} \le 0 \qquad (i = 1, ..., M)(p = 1, ..., P)(\alpha = 1, ..., D) \tag{4}$$

$$\mathbf{M}\ddot{\mathbf{z}}(t) + \mathbf{C}\dot{\mathbf{z}}(t) + \mathbf{K}\mathbf{z}(t) - \mathbf{f}(t) = 0 \tag{5}$$

where f; $d_{jks\alpha}$; d_{jk}^{U,y_d}; $b_{ip\alpha}$; b_p^{U,y_b}; $\mathbf{f}(t)$; $\ddot{\mathbf{z}}(t)$; $\dot{\mathbf{z}}(t)$; $\mathbf{z}(t)$; $\bar{V}_H^{d_y}(\mathbf{x}, \ddot{\mathbf{z}}, \dot{\mathbf{z}}, \mathbf{z}, \alpha)$; $\bar{V}_H^{b_y}(\mathbf{x}, \ddot{\mathbf{z}}, \dot{\mathbf{z}}, \mathbf{z}, \alpha)$; H; $\mathbf{M}$; $\mathbf{C}$; $\mathbf{K}$; P; and t are the objective function, the inter-storey drift ratios, the associated upper limits, the capacity ratios, their corresponding upper limits, the external load vector, the acceleration vector, the velocity vector, the displacement vector, the mean hourly wind speed at the building top that causes the drift response $d_{jks\alpha}$ to take the value d_y with MRI y_d, the corresponding speed for $b_{ip\alpha}$; the building height, the mass matrix, the damping matrix, the stiffness matrix, the number of capacity constraints, and the time, respectively.

This represents a classical dynamic response optimization problem (Kang et al. 2006) with one important difference. The implicit dependence of the time dependent constraints on the mean hourly wind speeds at the top of the building, $\bar{V}_H^{r_y}(\mathbf{x}, \ddot{\mathbf{z}}, \dot{\mathbf{z}}, \mathbf{z}, \alpha)$, causing a peak in the generic time history response r with specified MRI.

As outlined in equations (3–4) a classical dynamic response optimization attempts to limit the peak of the response time histories. In this paper an alternative approach based on probabilistically describing the peak and therefore limiting this value is adopted. In particular the probabilistic peak is expressed in terms of the first two statistical moments and the peak factor. Under these conditions the constraints in (3–4) are replaced by:

$$\underbrace{d_{jks\alpha}^{p}(\mu_d(\mathbf{x}, \ddot{\mathbf{z}}, ...), \sigma_d(\mathbf{x}, \ddot{\mathbf{z}}, ...), g_{\%}d(\mathbf{x}, \ddot{\mathbf{z}}, ...))}_{\bar{V}_H^{d_y}(\mathbf{x},\ddot{\mathbf{z}},\dot{\mathbf{z}},\mathbf{z},\alpha)} - d_{jk}^{U,y_d} \le 0 \qquad (j = 1, ..., N)(k = 1, ..., K)(s = x, y)(\alpha = 1, ..., D) \tag{6}$$

$$\underbrace{b_{ip\alpha}^{p}(\mu_b(\mathbf{x}, \ddot{\mathbf{z}}, ...), \sigma_b(\mathbf{x}, \ddot{\mathbf{z}}, ...), g_{\%}b(\mathbf{x}, \ddot{\mathbf{z}}, ...))}_{\bar{V}_H^{b_y}(\mathbf{x},\ddot{\mathbf{z}},\dot{\mathbf{z}},\mathbf{z},\alpha)} - b_p^{U,y_b} \le 0 \qquad (i = 1, ..., M)(p = 1, ..., P)(\alpha = 1, ..., D) \tag{7}$$

where μ indicates the mean value, σ the standard deviation while $g_{\%}$ the peak factor with specified probability of non-exceedance.

The first step in resolving this probabilisitic response optimization problem is the estimation of $\bar{V}_H^{d_y}(\mathbf{x}, \ddot{\mathbf{z}}, \dot{\mathbf{z}}, \mathbf{z}, \alpha)$ and $\bar{V}_H^{b_y}(\mathbf{x}, \ddot{\mathbf{z}}, \dot{\mathbf{z}}, \mathbf{z}, \alpha)$.

3 DIRECTIONAL VULNERABILITY MODEL

One of the limitations in the models used for the analysis and optimization of tall buildings is the adoption of directionless wind models or simplified approaches for attempting the account for the inherently directional characteristics of extreme wind climates. In this paper a fully directional description of the wind hazard is ensured by adopting the wind model proposed in (Grigoriu 2006, Grigoriu 2009). This model describes the wind hazard in a particular site by a set of independent wind vectors $\mathbf{V}(k)$, $k = 1, 2, \ldots, K$ containing the maximum sectorial mean wind speeds at a given height above ground recorded during the Kth wind storm. These vectors can be used to fit an extreme joint probability density of the hazard.

Once the extreme wind climate has been defined, the 1-dimensional domain of mean wind speeds at the building top $\bar{V}_H^{r_y}(\mathbf{x}, \ddot{\mathbf{z}}, \dot{\mathbf{z}}, \mathbf{z}, \alpha)$ that will cause the response r to take the value r_y with MRI y_r can be estimated from:

$$P(r \leq r_y) = \left[\int_0^{\bar{V}_H^{r_y}(\mathbf{x},\ddot{\mathbf{z}},\dot{\mathbf{z}},\mathbf{z},\alpha)} \int_0^{2\pi} p(\bar{V}_H, \alpha) d\bar{V}_H d\alpha \right]^{\lambda} = 1 - \frac{1}{y_r} \tag{8}$$

where $p(\bar{V}_H, \alpha)$ is the joint probability density function of the directional extreme wind climate describe after appropriate scaling to ensure consistent averaging times and height above ground H (Simiu and Scanlan 1996, Simiu and Miyata 2006) while λ is the average number of extreme wind events in an epoch of a year.

The resolution of the double integral of (8) requires the knowledge of the vulnerability surface associated with r. That is the value of $r^{peak}(\mu_r(\mathbf{x}, \ddot{\mathbf{z}}, \ldots), \sigma_r(\mathbf{x}, \ddot{\mathbf{z}}, \ldots), g_{\%r}(\mathbf{x}, \ddot{\mathbf{z}}, \ldots))$ in terms the various hazard intensities. In the case of wind excited buildings the response processes can be considered stationary (Simiu and Scanlan 1996). Therefore the peak generic constrained response, r^{peak} may be estimated from:

$$r^{peak}(\mu_r(\mathbf{x}, \ddot{\mathbf{z}}, \ldots) = \mu_r(\mu_r(\mathbf{x}, \ddot{\mathbf{z}}, \ldots) + g_{\%r}(\mu_r(\mathbf{x}, \ddot{\mathbf{z}}, \ldots)\sigma_r(\mu_r(\mathbf{x}, \ddot{\mathbf{z}}, \ldots) \tag{9}$$

where μ_r and σ_r are the mean value and standard deviation r respectively while $g_{\%r}$ is the peak value with specified probability of non-exceedance % (Davenport 1964). It should be observed that any non-Gaussian features of the response can be estimated form incorporating an appropriate peak factor (Grigoriu 1984, Gioffrè et al. 2000, Gioffrè and Gusella 2007).

The estimation of $\bar{V}_H^{r_y}(\mathbf{x}, \ddot{\mathbf{z}}, \dot{\mathbf{z}}, \mathbf{z}, \alpha)$ may then be obtained by intersecting the vulnerability surface with a horizontal plane passing through the response level r_y, Fig. 1(a). This intersection then defines the domain under which $p(\bar{V}_H, \alpha)$ must be integrated Fig. 1(b).

4 APPROXIMATE PROBABILISTIC OPTIMIZATION

The computational burden associated with updating the directional vulnerability model is considerable. This implies the need for optimization algorithms with high convergence rates such as those experienced with gradient based algorithms (Arora 2004). These however require gradient evaluation, which considering the implicit relationship that exists between $\bar{V}_H^{r_y}$ and the second order statistics together with the implicit non linear relationship between these last and the design variables (Jensen 2000) will cause the gradient evaluation, or design sensitivity analysis, to become excessively cumbersome. To overcome this, the concept of defining a sequence of explicit sub-problems can be adopted. In this paper a high quality explicit sub-problem is defined through the Principle of Virtual Work (PVW).

4.1 *Drift constraints*

Assume that the dynamic equilibrium of a multistory system may be adequately described with rigid floor diaphragms. Then by considering $\bar{V}_H^{d_y}(\mathbf{x}_0, \ddot{\mathbf{z}}, \dot{\mathbf{z}}, \mathbf{z}, \alpha)$ to be constant for small changes in

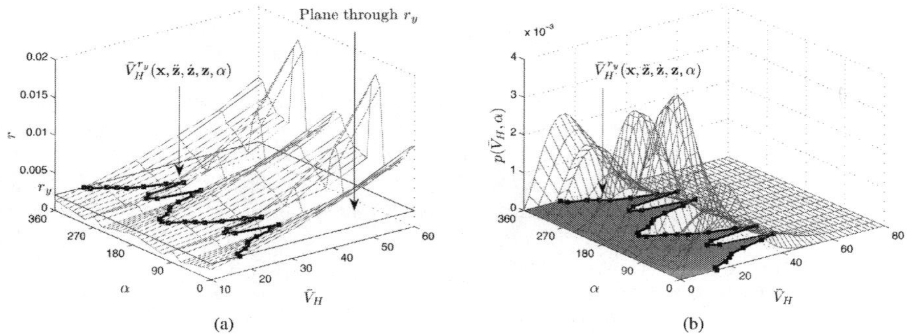

Figure 1. (a) Vulnerability surface of r and the 1-dimaensional domain for response level r_y, (b) Extreme joint probability density function, $p(\bar{V}H, \alpha)$, and the integration domain.

the design variables around the current design point $\mathbf{x}_0$, the inter-story drift ratio may be expressed through the PVW as:

$$\hat{d}_j(\mathbf{x}_0, \ddot{\mathbf{z}}, ..., t) = \sum_{i=1}^{M} \int_0^{L_i} \left(\frac{f_{1i}(\mathbf{x}_0)}{A_i} \hat{N}_i(\mathbf{x}_0, \ddot{\mathbf{z}}, ..., t) + \frac{f_{2i}(\mathbf{x}_0)}{A_{Yi}} \hat{V}_{i,Y}(\mathbf{x}_0, \ddot{\mathbf{z}}, ..., t) + \frac{f_{3i}(\mathbf{x}_0)}{I_{Xi}} \hat{M}_{i,X}(\mathbf{x}_0, \ddot{\mathbf{z}}, ..., t) \right) dl \tag{10}$$

where the functions f_{1i}, f_{2i} and f_{3i} are given by:

$$f_{1i}(\mathbf{x}_0) = \frac{n_{ji}(\mathbf{x}_0) - n_{j-1i}(\mathbf{x}_0)}{Eh_j}$$

$$f_{2i}(\mathbf{x}_0) = \frac{v_{jYi}(\mathbf{x}_0) - v_{j-1Yi}(\mathbf{x}_0)}{Gh_j} \tag{11}$$

$$f_{3i}(\mathbf{x}_0) = \frac{m_{jXi}(\mathbf{x}_0) - m_{j-1Xi}(\mathbf{x}_0)}{Eh_j}$$

where the hat indicates the dependency on a specified wind direction and so associated mean hourly wind speed evaluated in $\mathbf{x}_0$ while the quantities m_{jXi}, v_{jYi}, and n_{ji} are the various internal forces in the local reference system (X, Y) of member i due to a unit force acting through the center of mass of floor j. $\hat{N}_i(\mathbf{x}_0, \ddot{\mathbf{z}}, \ldots, t)$, $\hat{V}_{i,Y}(\mathbf{x}_0, \ddot{\mathbf{z}}, \ldots, t)$, and $\hat{M}_{i,X}(\mathbf{x}_0, \ddot{\mathbf{z}}, \ldots, t)$ are the time varying internal forces while A_i and A_{Yi} are the axial and shear areas of member i while I_{Xi} is the flexural moments of inertia. E and G are the axial and shear material moduli while L_i is the length of the ith member. It should be observed that equations (10–11) are written considering a 2D structure for the sake of simplicity. The extension to the third dimension is immediate.

In the case of structural systems consisting of mono-dimensional elements this maybe written as:

$$\hat{d}_j(\mathbf{x}_0, \ddot{\mathbf{z}}, \dot{\mathbf{z}}, \mathbf{z}, t) = \sum_{i=1}^{M} \sum_{l=1}^{3} P_{li}(x_i) \hat{G}_{li}(\mathbf{x}_0, \ddot{\mathbf{z}}, \dot{\mathbf{z}}, \mathbf{z}, t) \tag{12}$$

where $1/A_i = P_{1i}(x_i)$, $1/A_{Yi} = P_{2i}(x_i)$ and $1/I_{Xi} = P_{3i}(x_i)$ are explicit link functions while the functions $\hat{G}_{1i}(\mathbf{x}_0, \ddot{\mathbf{z}}, \dot{\mathbf{z}}, \mathbf{z}, t)$ are given by:

$$\hat{G}_{1i}(\mathbf{x}_0, \ddot{\mathbf{z}}, ..., t) = C_{1i}(\mathbf{x}_0) \hat{N}_i(\mathbf{x}_0, \ddot{\mathbf{z}}, ..., t)$$

$$\hat{G}_{2i}(\mathbf{x}_0, \ddot{\mathbf{z}}, ..., t) = C_{2i}(\mathbf{x}_0) \hat{V}_{i,Y}(\mathbf{x}_0, \ddot{\mathbf{z}}, ..., t) \tag{13}$$

$$\hat{G}_{3i}(\mathbf{x}_0, \ddot{\mathbf{z}}, ..., t) = C_{3i}^{(1)}(\mathbf{x}_0) \hat{M}_{i,X}^{(1)}(\mathbf{x}_0, \ddot{\mathbf{z}}, ..., t) + C_{3i}^{(2)}(\mathbf{x}_0) \hat{M}_{i,X}^{(2)}(\mathbf{x}_0, \ddot{\mathbf{z}}, ..., t)$$

where $C_{1i} = L_i f_1(\mathbf{x}_0)$, $C_{2i} = L_i f_3(\mathbf{x}_0)$, $C_{3i}^{(1)} = \dfrac{L_i}{6}(2f_4^{(2)}(\mathbf{x}_0) - 7f_4^{(1)}(\mathbf{x}_0))$ and $C_{3i}^{(2)} = \dfrac{L_i}{6}(f_4^{(2)}(\mathbf{x}_0) -4f_4^{(1)}(\mathbf{x}_0))$ are constants in which the apexes (1) and (2) indicate that the internal forces have be evaluated in the initial and final cross sections of the ith member. No apex simply implies that the internal forces are constant along the member.

It can be seen that the functions $\hat{G}_{1i}(\mathbf{x}_0, \ddot{\mathbf{z}}, \dot{\mathbf{z}}, \mathbf{z}, t)$ depend on the virtual static and dynamic internal force distribution. Therefore by assuming that both these quantities are relatively insensitive for small changes in the design variables, equation (12) represents an approximate explicit representation of the inter-story drift ratio around the current design point:

$$\hat{d}_j(\mathbf{x}_0, \ddot{\mathbf{z}}, \dot{\mathbf{z}}, \mathbf{z}, t) \approx \sum_{i=1}^{M} \sum_{l=1}^{3} P_{li}(x_i)\hat{G}_{li}(\mathbf{x}_0, t) \tag{14}$$

By taking advantage of the stationary nature of the inter-story drift response, and assuming that the peak factor is also insensitive to small changes in the design variables, the following approximate explicit expression for the probabilistic peak is obtained:

$$\hat{d}_j^{peak}(\mu_d(\mathbf{x}_0, \ddot{\mathbf{z}}, ...), \sigma_d(\mathbf{x}_0, \ddot{\mathbf{z}}, ...), g_{\%d}(\mathbf{x}_0, \ddot{\mathbf{z}}, ...))$$

$$\approx \sum_{i=1}^{M} \sum_{l=1}^{3} P_{li}(x_i) \underbrace{\mu_{\hat{G}_{1i}}}_{j} + g_{\%d} \sqrt{\sum_{i=1}^{M} \sum_{w=1}^{M} \sum_{l=1}^{3} \sum_{m=1}^{3} P_{li}(x_i) P_{mw}(x_i) \underbrace{C_{\hat{G}_{li}\hat{G}_{mw}}}_{j}} \tag{15}$$

where $C_{\hat{G}_{li}\hat{G}_{mw}}$ are the zero lag covariance coefficients between $\hat{G}_{li}$ and $\hat{G}_{mw}$.

4.2 *Capacity constraints*

The member level capacity ratios for a typical 2D steel framed structure are of the general form:

$$\hat{b}_{ip}(\mathbf{x}_0, \ddot{\mathbf{z}}, ..., t) = \frac{|\hat{N}_{ip}(\mathbf{x}_0, \ddot{\mathbf{z}}, ..., t)|}{\alpha \phi N_{ni}(x_i)} + \beta \left(\frac{|\hat{M}_{ip,X}(\mathbf{x}_0, \ddot{\mathbf{z}}, ..., t)|}{\phi_b M_{ni,X}(x_i)} \right) \tag{16}$$

subject to the condtions:

$$\begin{cases} \dfrac{|\hat{N}_{ip}(\mathbf{x}_0, \ddot{\mathbf{z}}, ..., t)|}{\phi N_{ni}(x_i)} \geq \tilde{N} & \Rightarrow \begin{cases} \alpha = \alpha_1 \\ \beta = \beta_1 \end{cases} \\ \dfrac{|\hat{N}_{ip}(\mathbf{x}_0, \ddot{\mathbf{z}}, ..., t)|}{\phi N_{ni}(\mathbf{x}_i)} < \tilde{N} & \Rightarrow \begin{cases} \alpha = \alpha_2 \\ \beta = \beta_2 \end{cases} \\ \hat{N}_{ip}(\mathbf{x}_0, \ddot{\mathbf{z}}, ..., t) \geq 0 & \Rightarrow \quad \phi = \phi_t \\ \hat{N}_{ip}(\mathbf{x}_0, \ddot{\mathbf{z}}, ..., t) < 0 & \Rightarrow \quad \phi = \phi_c \end{cases} \tag{17}$$

where N_{ni} and $M_{ni,X}$ are the nominal axial and flexural capacities of member i, ϕ and ϕ_b are axial and flexural resistance factors, $\hat{N}_{ip}(\mathbf{x}_0, \ddot{\mathbf{z}}, \ldots, t)$, and $\hat{M}_{ip,X}(\mathbf{x}_0, \ddot{\mathbf{z}}, \ldots, t)$ are the total internal forces due to specified factored load combination p, $\tilde{N}$ represents the axial force ratio for which the bending strength is reduced while α_1, α_2, β_1 and β_2 are the coefficients that implement this reduction. Equation (16) represents a 2D normalized strength domain for member i. In particular each section, s, of the member will have a representative point depending on whether the member is in compression, c, or traction, t. Indeed ϕ depends on whether the member is subject to local buckling. Again it should be observed that the extension to the third dimension is immediate.

Equation (16) represents a discontinuous function due to the conditions in (17). Therefore physically meaningful probabilistic results cannot be obtained by directly analyzing (16). This difficulty can be overcome by seeing the conditions in (17) from a probabilistic point of view. Therefore substituting $\hat{N}_{ip}(\mathbf{x}_0, \ddot{\mathbf{z}}, \ldots, t)$ with $N_{ip}^{\pm} = \mu_N(\mathbf{x}_0, \ddot{\mathbf{z}}, \ldots) + g_{\%N}^{\pm}\sigma_N(\mathbf{x}_0, \ddot{\mathbf{z}}, \ldots)$. In this way the conditions can be seen as an indicator of which peak of the four continuous capacity functions $\hat{b}_{ip}^{peak(id)}$ for

$id = 1t, 2t, 1c, 2c$ defined for the various values of α, β and ϕ is governing. Where given a set of values for α, β and ϕ the peak $\hat{b}_{ip}^{(id)}$ may be obtained from:

$$\hat{b}_{ip}^{peak(id)}(\mu_b(\mathbf{x}_0, \ddot{\mathbf{z}}, ...), \sigma_b(\mathbf{x}_0, \ddot{\mathbf{z}}, ...), g_{\%b}(\mathbf{x}_0, \ddot{\mathbf{z}}, ...)) = \sum_{i=1}^{2} I_i(x_i)|\mu_{\hat{IF}}| + g_{\%b}\sqrt{\sum_{j=1}^{2}\sum_{k=1}^{2} I_j(x_i)I_k(x_i)|C_{\hat{IF}_j\hat{IF}_k}|} \tag{18}$$

where $\hat{IF}_1 = \hat{N}_{ip}(\mathbf{x}_0, \ddot{\mathbf{z}}, \ldots, t)$ and $\hat{IF}_2 = \hat{M}_{ip,X}(\mathbf{x}_0, \ddot{\mathbf{z}}, \ldots, t)$ while $C_{\hat{IF}_j\hat{IF}_k}$ is the zero lag covariance coefficient between the internal forces $\hat{IF}_j$ and $\hat{IF}_k$ with $id = 1t, 2t, 1c, 2c$ function identification index and I_j for $j = 1, 2$ given by:

$$I_1 = \frac{1}{\alpha\phi N_{ni}(P_{1i}(x_i))}$$
$$I_2 = \frac{1}{\phi_b M_{ni,X}(P_{3i}(x_i))} \tag{19}$$

Once the governing capacity ratio has been identified a zero order treatment of the capacity constraints can be employed by defining a vector of probabilistic side constraints $\mathbf{x}_i^L = (x_1^L, x_2^L, \ldots)$. Each value of which may be obtained by placing the governing capacity ratio, written in the explicit form of (18) while assuming constant internal forces, equal to one.

4.3 *Explicit sub-problem*

From the expressions derived so far the following approximate explicit sub-problem can be defined around the current design point:

$$\text{Find} \quad \mathbf{x} = (x_1, x_2, ..., x_M) \tag{20}$$

$$\text{to minimize} \quad W(\mathbf{x}) = \sum_{i=1}^{M} \rho \frac{L_i}{P_{i1}(x_i)} \tag{21}$$

subject to:

$$\underbrace{d_{j\alpha}(\mathbf{x})}_{\text{peak}_{\%},d_y} = \sum_{i=1}^{M}\sum_{l=1}^{3} P_{il}(x_i)\underbrace{\bar{G}_{1i}}_{j\alpha,d_y} + g_{\%}\sqrt{\sum_{i=1}^{M}\sum_{w=1}^{M}\sum_{l=1}^{3}\sum_{m=1}^{3} P_{il}(x_i)P_{wm}(x_i)\underbrace{C_{G_{li}G_{wm}}}_{j\alpha,d_y}} - d_j^{U,d_y} \leq 0 \tag{22}$$

$(j = 1, ..., N)(\alpha = 1, ..., D)$

$$\max\Big(\underbrace{x_i^L}_{\alpha,b_y}, \breve{x}_i^L\Big) \leq x_i \leq x_i^U$$
$$(\alpha = 1, ..., D)(i = 1, ..., M) \tag{23}$$

Each sub-problem can be solved extremely efficiently by any gradient based optimization algorithm. In particular in this study an Optimality Criteria (OC) algorithm was adopted. The optimization procedure, including vulnerability updating, is shown in Figure 2 where DC stands for Design Cycle and represents the convergence of each sub-problem.

5 DESIGN EXAMPLE

To illustrate the proposed method a 2D frame is analyzed and optimized in a extreme 3D wind environment (Fig. 3). The frame is taken from a fully 3D building on which wind tunnel tests were carried out (Spence et al. 2008). The analysis and optimization is carried out considering 1/3rd of

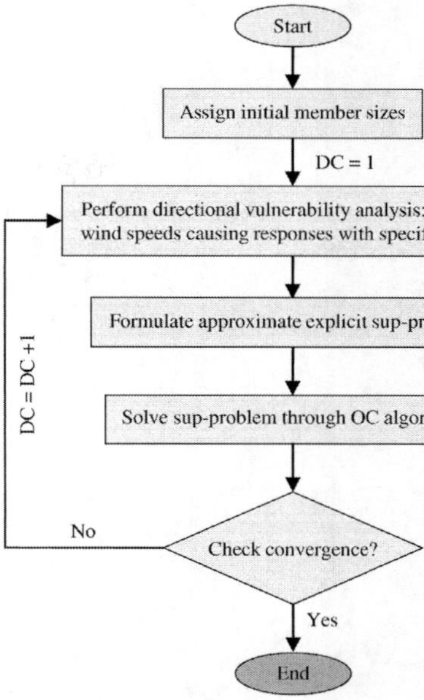

Figure 2. Flowchart of proposed dynamic response optimization algorithm.

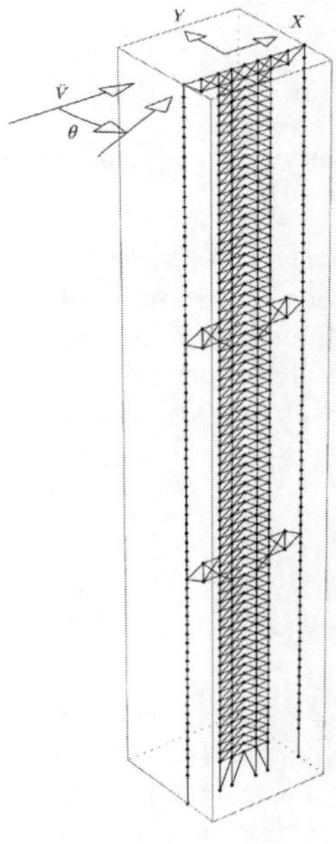

Figure 3. Planar frame as set in the 3D wind environment.

the wind load acting parallel to the frames plane. The beams and diagonals are American AISC (AISC 2001) standard wide-flange W sections. The initial beams size are W24 × 176 while the diagonals are W14 × 370. The columns and outriggers are steel box sections of flange thickness 1/20th of the mid line diameter.

The wind environment comes from simulated hurricane databases of the National Institute of Standards and Technology (NIST). The building is considered in the Miami area of Florida. Therefore milepost 1500 was considered.

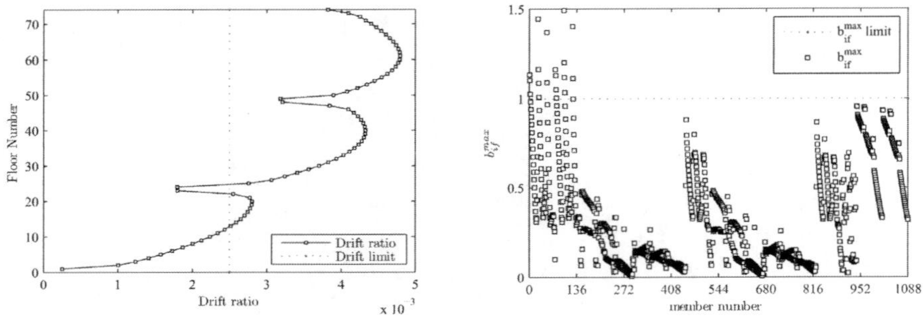

(a) (b)

Figure 4. (a) Initial 50 year drift performance, (b) Initial 50 year factored member capacity performance.

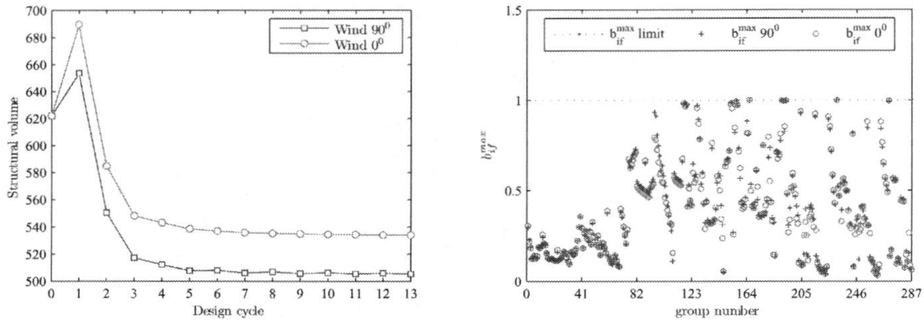

(a) (b)

Figure 5. (a) Objective function design history, (b) Optimized capacity ratio performance.

Figure 4 shows the initial performance of the frame in terms of the inter-story drift ratios and member capacity ratios. As can be seen the frame severely exceeds the 1/400 limit imposed on the drift ratio. All analysis considers the response processes to be Gaussian in nature. To calibrate the optimization algorithm the explicit link functions must be defined. For members with standard AISC sections the cross sectional area are chosen as the design variables. Therefore the functions can be expressed as (Chan et al. 1995):

$$\begin{cases} P_{1i}(x_i) = 1/A_i \\ P_{2i}(x_i) = C_{A_X}/A_i + \grave{C}_{A_X} \\ P_{3i}(x_i) = C_{I_X}/A_i + \grave{C}_{I_X} \end{cases} \tag{24}$$

During the optimization the beams and columns may vary within the groups W24 and W14 respectively. The columns and outriggers are defined as steel box sections with a flange thickness 1/20th of the mid line diameter. For these section the flange to diameter ratio is kept constant throughout the optimization. This allows the functions $P_{1i}(x_i)$, $P_{2i}(x_i)$ and $P_{3i}(x_i)$ to be written as:

$$\begin{cases} P_{1i}(x_i) = 5/D_i^2 \\ P_{2i}(x_i) = 19/2D_i^2 \\ P_{3i}(x_i) = 30/D_i^4 \end{cases} \tag{25}$$

where D is the mid line diameter. The lower limit was fixed at $D_i = 0.3$ m with an upper limit of $D_i = 1.8$ m. The design variables were gathered so as to ensure symmetry of the frame with respect to the central vertical axis. The beams and columns were also grouped two levels at a time while the outriggers were designed individually. The number of design variables was 287.

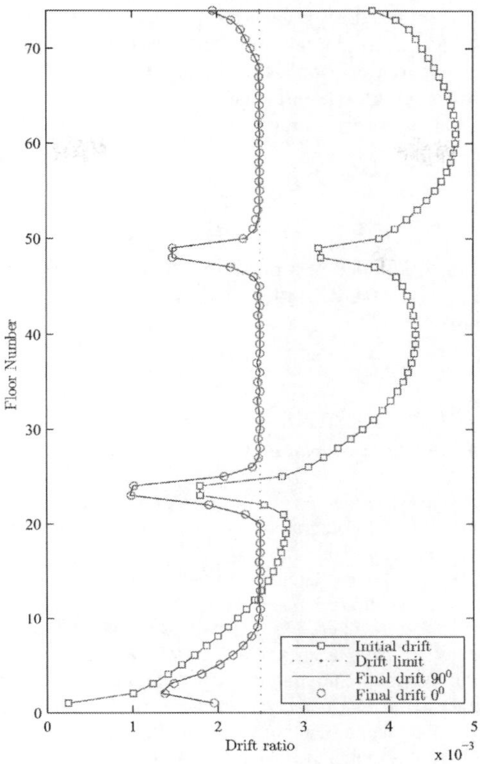

Figure 6. Optimized drift ratio performance.

5.1 *Results*

Figure 6 shows the optimized inter-story drift ratio. In particular two wind directions were considered. Where the difference between the two directions is only interesting from algorithm performance point of view as the directional vulnerability model ensures that drift performance for all wind directions. The high quality of the approximate sub-problem is clearly visible from the number of active or near active constraints at the optimum design. From figure 5(a) the rapid convergence of the proposed probabilistic dynamic response algorithm again shows the quality of the sub-problems. The difference in the objective function by optimizing down different wind directions is due to the numerical resolution of the vulnerability curves. Indeed the optimum design coincides for both wind directions. This is clearly visible from figures (6-5(b)) where both the inter-story drift and capacity ratios tend to be identical. It is interesting to observe from Figure 5(b) how few member capacity constraints are active at the optimum.

6 CONCLUSIONS

In this paper an alternative approach to the dynamic response optimization of wind excited tall buildings was presented. The method is based on replacing the time-dependent constraints with probabilistically equivalent constraints. In particular the stochastic response functions are described through their first two statistical moments and a potentially non-Gaussian peak factor. The critical response level for each constrained function is estimated by a vulnerability model that rigorously combines the directional aerodynamic and climatological information yielding the mean hourly wind speeds that will cause a peak response with specified MRI. An optimization procedure is presented based on defining a sequence of explicit sup-problems in terms of the response statistics. The sup-problems are defined by explicitly expressing the second order statistics of the drift ratios in terms design variables through the PVW. The capacity ratios are dealt with by defining probabilistic side constraints on the design variables. The sup-problems can then be solved by any conventional optimization algorithm. At each design iteration the vulnerability model is updated. The applicability of the proposed algorithm is demonstrated on a full scale 2D frame analyzed in an extreme 3D wind environment. Quick and steady convergence is seen confirming the quality of the approximate sup-problems.

ACKNOWLEDGEMENTS

The researched presented in this paper was partially financed by the Italian Ministry for Education, University and Research – MIUR, under the WI-POD project (PRIN 2007) and is greatly acknowledged by the authors.

REFERENCES

AISC (2001). *Manuel of steel construction: load and resistance factor design, 3rd Edition*. AISC, American Institute of Steel Construction.

Arora, J. S. (2004). *Introduction to optimum design – 2nd ed.* San Diego: Elsevier Inc.

Baker, W. F. (1990). Sizing techniques for lateral systems in multi-story steel buildings. In *Proceedings, 4th World Congress on Tall Buildings: 2000 and Beyond*, pp. 545–553. Hong Kong.

Chan, C., M. Huang, and K. Kwok (2010). Integrated wind load analysis and stiffness optimization of tall buildings with 3d modes. *Engineering Structures 32*(5), 1252–1261.

Chan, C. M., D. E. Grierson, and A. N. Sherbourne (1995). Automatic optimal design of tall steel building frameworks. *Journal of Structural Engineering, ASCE 121*(5), 838–847.

Chen, X. and A. Kareem (2005a). Coupled dynamic analysis and equivalent static wind loads on buildings with three-dimensional modes. *Journal of Structural Engineering 131*(7), 1071–1082.

Chen, X. and A. Kareem (2005b). Dynamic wind effects on buildings with 3d coupled modes: Application of high frequency force balance measurements. *Journal of Engineering Mechanics 131*(11), 1115–1125.

Davenport, A. G. (1964). Note on the distribution of the largest value of a random function with application to gust loading. In 28 (Ed.), *Proceedings, Institution of Civil Engineering*, pp. 187–196.

Gioffrè, M., M. Grigoriu, M. Kasperski, and E. Simiu (2000). Wind-induced peak bending moments in low-rise building frames. *Journal of Engineering Mechanics 126*(8), 879–881.

Gioffrè, M. and V. Gusella (2007). Peak response of a nonlinear beam. *Journal of Engineering Mechanics 133*(9), 963–969.

Grigoriu, M. (1984). Crossing of non-gaussian translation process. *Journal of Engineering Mechanics 110*(4), 610–620.

Grigoriu, M. (2006). A model for directional hurricane wind speeds. *NIST Government Contractor Report 06-905*.

Grigoriu, M. (2009). Algorithms for generating large sets of synthetic directional wind speed data for hurricane, thunderstorm, and synoptic winds. *NIST Technical Note 1626*.

Jensen, H. A. (2000). On the structural synthesis of uncertain systems subjected to environmenatal loads. *Structural and Multidisciplinary Optimization 20*, 37–48.

Kang, B. S., G. J. Park, and J. S. Arora (2006). A review of optimization of structures subjected to transient loads. *Structural and Multidisciplinary Optimization 31*, 81–95.

Kargahil, M. and J. C. Anderson (2006). Structural weight optimization of frames using tabu search. ii: Evaluation and seismic performance. *Journal of Structural Engineering, ASCE 132*(12), 1869–1879.

Kargahil, M., J. C. Anderson, and M. M. Dessouky (2006). Structural weight optimization of frames using tabu search. i: Optimization procedure. *Journal of Structural Engineering, ASCE 132*(12), 1858–1868.

Kim, C. K., H. S. Kim, J. S. Hwang, and S. M. Hong (2007). Development of drift design model for high-rise buildings subjected to lateral and vertical loads. *The Structural Design Of Tall And Special Buildings, John Wiley & Sons, Ltd*.

Simiu, E. and T. Miyata (2006). *Design of Buildings and Bridges for Wind*. Hoboken, New Jersey: John Wiley & Sons.

Simiu, E. and R. H. Scanlan (1996). *Wind effects on structures: Fundamentals and applications to design 3rd Ed.* New York: Wiley-Interscience.

Spence, S. M. J., M. Gioffrè, and V. Gusella (2008). Influence of higher modes on the dynamic response of irregular and regular tall buildings. In *Proceedings, 6th International Colloquium on Bluff Bodies Aerodynamics and Applications (BBAA VI)*. Milano, Italy, July 20–24.

Reliability and Optimization of Structural Systems – Straub (ed)
© 2010 Taylor & Francis Group, London, ISBN 978-0-415-88179-1

Reliability updating with measurements in spatially distributed systems using stochastic simulation

D. Straub
Engineering Risk Analysis Group, TU München, Germany

ABSTRACT: A recently developed method for reliability updating with equality information [Straub D., *Probabilistic Engineering Mechanics*, under review] is investigated for application to spatially distributed systems. In contrast to existing ones, the novel method enables the use of stochastic simulation (e.g. Monte Carlo or importance sampling) for reliability updating with any type of information. The main benefit of this approach over existing methods is that it is robust and calculations can thus be automated. This is of particular relevance for spatially distributed systems subject to uncertain information (e.g. though measurements, monitoring, observations of system performances), which require a large number of conditional reliability calculations. In the current paper it is shown that for such systems, stochastic simulation in combination with the novel method can provide accurate results for a large number of locations simultaneously.

1 INTRODUCTION

We are concerned with computing the probability of a rare event F (the failure event) conditional on uncertain information, e.g. as obtained from measurements, inspection or monitoring of structures. Such Bayesian updating has been applied in the context of structural reliability since the 1970s (e.g. Tang 1973). The conditional probability of failure given the information event Z is computed as

$$\Pr(F \mid Z) = \frac{\Pr(F \cap Z)}{\Pr(Z)} \tag{1}$$

In structural reliability, failure events F and information events Z are described by domains Ω in the outcome space of the basic random variables $\mathbf{X} = (X_1, X_2, \ldots, X_n)$. The failure domain Ω_F is defined in terms of continuous limit state functions $g(\mathbf{x})$. In the simplest case, it is

$$\Omega_F = \{g(\mathbf{x}) \leq 0\} \tag{2}$$

In the general case, Ω_F is defined in terms of a number of limit state functions (e.g., Der Kiureghian 2005), corresponding to systems of components that are defined by limit state functions. For the purpose of the present paper, the formulation in (2) is sufficiently general; extension to the system application is straightforward (Straub 2010).

Information obtained on the system, e.g. in the form of measurements, monitoring, inspection, observed system performance, is also described in terms of continuous limit state functions $h(\mathbf{x})$ and corresponding domains. Information is said to be of the inequality type if it can be written as

$$\Omega_Z = \{h(\mathbf{x}) \leq 0\} \tag{3}$$

and it is said to be of the equality type if it can be written as

$$\Omega_Z = \{h(\mathbf{x}) = 0\} \tag{4}$$

Structural Reliability Methods (SRM) solve Eq. (1) by computing integrals in the space of **X**:

$$\Pr(F \mid Z) = \frac{\int_{\mathbf{x}\in\{\Omega_F \cap \Omega_Z\}} f(\mathbf{x})\mathrm{d}\mathbf{x}}{\int_{\mathbf{x}\in\{\Omega_Z\}} f(\mathbf{x})\mathrm{d}\mathbf{x}} \tag{5}$$

If information is exclusively of the inequality type, Eq. (3), evaluation of the above integrals is straightforward using any of the available SRM. However, if the information event Z is of the equality type, the integrals result in zero, since these events have zero probability a-priori. Direct application of SRM is thus not possible in this case.

Solutions to overcome this problem have been suggested by Madsen (1987) and the group of Rackwitz (e.g. Schall et al. 1988). Madsen's solution is based on inserting a dummy random variable and then computing probability sensitivities with respect to this variable. The solutions of the Rackwitz group are based on computing surface integrals, using first- or second-order approximations of the surfaces $h_i(\mathbf{x}) = 0$. These solutions are implemented in the Strurel software (Gollwitzer et al. 2006). Both Madsen's and Rackwitz' methods are efficient and, in many cases, represent a sufficiently accurate approximation. However, in cases where FORM/SORM solutions are not sufficiently accurate or in which it is difficult to identify the joint design point, these methods should not or cannot be used. Furthermore, it is often difficult to appraise the error made by the first- or second-order approximation.

Recently, the author has proposed a novel method for solving Eq. (5) using SRM when information is of the equality type (Straub 2010). The method is based on transforming equality information into inequality information, which enables the direct use of Eq. (5) using any SRM. The aim of the present paper is to study the application of the methodology to reliability problems involving spatially distributed systems. For such systems, commonly a large number of reliability problems must be solved simultaneously, which requires that the applied algorithms are robust and efficient. Here, it is investigated if and how stochastic simulation approaches are suitable in this context, whereby particular focus is put on examining the trade-off between computational efficiency (measured in number of evaluations of the limit state functions) and robustness (i.e. the applicability of a method without the need for problem specific adjustments). First- and second-order reliability methods (FORM/SORM), which are highly efficient but not necessarily robust and accurate in all cases, are not investigated here. They are considered in Straub (2010). Additionally, Papaioannou and Straub (2010) utilize subset simulation in conjunction with the methodology for reliability updating in stochastic FEM.

2 BAYESIAN UPDATING WITH EQUALITY INFORMATION

This section presents a summary of the method described in Straub (2010), with slight modifications in view of the application of stochastic simulation methods.

2.1 *Describing information through likelihood functions*

We note that in statistics, information is not commonly described in the form of domains Ω_Z. Instead, the usual way to describe (uncertain) information on **X** is by means of the likelihood function, which is defined as

$$L(\mathbf{x}) \propto \Pr(Z \mid \mathbf{X} = \mathbf{x}) \tag{6}$$

As noted in Straub (2010), any domain Ω_Z can be translated into a likelihood function. However, it is often more convenient to directly identify the likelihood function. As an example, consider a measurement s_m of a system characteristic $s(\mathbf{X})$. The measurement has an additive error ε that is a zero mean random variable uncorrelated with **X**. The limit state function $h(\mathbf{X}, \varepsilon)$ describing this equality information as well as the corresponding likelihood function are given in the following, with $f_\varepsilon()$ being the PDF of ε.

$$h(\mathbf{X}, \varepsilon) = s(\mathbf{X}) - s_m + \varepsilon \tag{7}$$

$$L(\mathbf{x}) = f_\varepsilon(s_m - s(\mathbf{x})) \tag{8}$$

For the case of several observations $Z_1, \ldots, Z_m$ with corresponding likelihood functions $L_i(\mathbf{x})$, it is always possible to combine the likelihood functions into a single likelihood function $L(\mathbf{x})$. E.g., if measurements are uncorrelated for given $\mathbf{X} = \mathbf{x}$, it is simply $L(\mathbf{x}) = \Pi_{i=1}^{m} L_i(\mathbf{x})$. It is thus sufficient to consider only the case of a single likelihood function describing the combined observations $Z = Z_1 \cap \ldots \cap Z_m$ in the following.

2.2 *Transform equality information into equivalent inequality information*

Let P be a random variable with uniform distribution in the range [0,1] and let c be a constant that is selected so that $0 \leq cL(\mathbf{x}) \leq 1$ for all $\mathbf{x}$. In this case, the following identity holds for given values of $\mathbf{X} = \mathbf{x}$:

$$L(\mathbf{x}) = \frac{\Pr[P \leq cL(\mathbf{x})]}{c} \tag{9}$$

Let α denote the proportionality constant in the likelihood definition given in Eq. (6). By combining with Eq. (9), we obtain

$$\Pr(Z \mid \mathbf{X} = \mathbf{x}) = \alpha L(\mathbf{x}) = \frac{\alpha}{c} \Pr[P \leq cL(\mathbf{x})] \tag{10}$$

It follows that the probability of the information event Z is

$$\Pr(Z) = \int_{\mathbf{X}} \Pr(Z \mid \mathbf{X} = \mathbf{x}) f(\mathbf{x}) \mathrm{d}\mathbf{x} = \frac{\alpha}{c} \int_{\mathbf{X}} [P \leq cL(\mathbf{x})] f(\mathbf{x}) \mathrm{d}\mathbf{x} \tag{11}$$

Next, we define the event $Z_e = \{P \leq cL(\mathbf{x})\}$ through the limit state function

$$h_e(\mathbf{x}, p) = p - cL(\mathbf{x}) \tag{12}$$

and the corresponding domain $\Omega_{Ze} = \{h_e(\mathbf{x}, p) \leq 0\}$. This has the same form as the domains describing inequality information, Eq. (3). Equation (11) can now be rewritten to

$$\Pr(Z) = \frac{\alpha}{c} \int_{\mathbf{x}, p \in \Omega_{Ze}} f(p) f(\mathbf{x}) \mathrm{d}\mathbf{x} \mathrm{d}p = \frac{\alpha}{c} \int_{\mathbf{x}, p \in \Omega_{Ze}} f(\mathbf{x}) \mathrm{d}\mathbf{x} \mathrm{d}p \tag{13}$$

The second identity follows from $f(p) = 1$. Accordingly, Pr $(F \cap Z)$ is obtained as

$$\Pr(F \cap Z) = \int_{\mathbf{X}} \Pr(Z \mid \mathbf{X} = \mathbf{x}) \Pr(F \mid \mathbf{X} = \mathbf{x}) f(\mathbf{x}) \mathrm{d}\mathbf{x} = \frac{\alpha}{c} \int_{\mathbf{x}, p \in \{\Omega_F \cap \Omega_{Ze}\}} f(\mathbf{x}) \mathrm{d}\mathbf{x} \mathrm{d}p \tag{14}$$

The conditional probability of F given Z is therefore

$$\Pr(F \mid Z) = \frac{\int_{\mathbf{x}, p \in \{\Omega_F \cap \Omega_{Ze}\}} f(\mathbf{x}) \mathrm{d}\mathbf{x} \mathrm{d}p}{\int_{\mathbf{x}, p \in \Omega_{Ze}} f(\mathbf{x}) \mathrm{d}\mathbf{x} \mathrm{d}p} \tag{15}$$

Here, the proportionality constant α disappears. Both integrals in Eq. (15) can be computed using any SRM. The denominator corresponds to a component reliability problem, the nominator to a parallel system reliability problem.

As shown in Straub (2010), the limit state surfaces $h_e(\mathbf{x}, p) = 0$, which describe the information, are generally highly non-linear. For this reason, there it was recommended not to apply FORM/SORM for the evaluation, unless the shape of the limit state surface is well understood. This is not generally the case for problems involving more than a few random variables and few observations. This motivates the use of stochastic simulation methods for the evaluation of the integrals in Eq. (15).

3 STOCHASTIC SIMULATION FOR COMPUTING THE CONDITIONAL PROBABILITY

Based on the transformation of equality information into inequality information, basic stochastic simulation methods (Monte Carlo, importance sampling) are adapted to the computation of the conditional probability $\Pr(F|Z)$ in this section. Other methods are available and further improvements of the suggested methods are possible and are under investigation.

3.1 *Monte Carlo evaluation*

The use of Monte Carlo Simulation (MCS) for the evaluation of the integrals in (15) has two main advantages: It is easy to implement and it provides unbiased results, which do not depend on choices made by the analyst. For these reasons, it is suggested to use MCS whenever possible for the verification of the results obtained with other methods.

The disadvantage of MCS is the required large number of evaluations of the limit state functions $g(\mathbf{x})$ and $h_e(\mathbf{x}, p)$ when the events F and Z_e have small probabilities of occurrence. In structural reliability, $\Pr(F|Z)$ is commonly required to be small and MCS is therefore not applicable for problems where the evaluation of $g(\mathbf{x})$ is computationally expensive. Typically these are problems involving Finite Element Models (FEM), e.g. Papaioannou and Straub (2010). In addition, even for problems that have a parametrical function $g(\mathbf{x})$, MCS might not be feasible because of small values of $\Pr(Z_e)$. Nonetheless, there are a number of applications for which MCS represents a viable solution strategy. An example of MCS applied to reliability updating of a structural component subject to fatigue crack growth is presented in Straub (2010).

With MCS, the integrals in Eq. (15) are approximated by

$$\Pr(F \mid Z) \approx \frac{\sum_{i=1}^{n_s} I[h_e(\mathbf{x}_i, p_i) \leq 0] I[g(\mathbf{x}_i) \leq 0]}{\sum_{i=1}^{n_s} I[h_e(\mathbf{x}_i, p_i) \leq 0]} = \frac{n_{FZ}}{n_Z} \tag{16}$$

where n_s is the number of samples, $\mathbf{x}_i$ is the i th sample of $\mathbf{X}$, p_i is the i th sample of P and $I[\,]$ is the indicator function that takes value 1 if the condition holds and 0 otherwise. The denominator in Eq. (16), n_Z, is the number of sample that fall into the domain Ω_{Ze}; it can be interpreted as the effective number of samples that are available to compute the probability of F conditional on Z. Its expected value is $\mathrm{E}[N_Z] = n\Pr(Z_e)$. If $\Pr(Z_e) >> \Pr(\mathrm{F} \cap Z_e)$, we can neglect the variance of N_Z and the standard deviation of the MCS estimator is approximately

$$\sigma_{MCS} \approx \sqrt{\frac{\Pr(F \cap Z_e) - \Pr(F \cap Z_e)^2}{n\Pr(Z_e)}} \tag{17}$$

As the accuracy of the MCS estimator is a function of $\Pr(Z_e) = \Pr(h_e(\mathbf{x}, p) \leq 0)$, its applicability depends on the available information Z. As will illustrated later in the numerical example, $\Pr(Z_e)$ decreases with increasing number of observations, but is also a function of the dependence among the observations. Generally, the higher the information content of the observations with respect to the basic random variables $\mathbf{X}$, the lower $\Pr(Z_e)$ will become.

3.1.1 *Selecting the constant c*

$\Pr(Z_e)$ is directly proportional to the chosen value of the constant c in Eq. (9). To increase the efficiency of MCS, it is thus beneficial to select the largest possible value of c that still fulfills the condition $0 \leq cL(\mathbf{x}) \leq 1$.

If the location of the maximum value of $L(\mathbf{x})$ in the space of $\mathbf{X}$ is known (this is the Maximum Likelihood Estimator $\mathbf{x}_{MLE}$), the constant can be selected as $c = 1/L(\mathbf{x}_{MLE})$. When applying MCS, a simpler solution is to select c by computing the likelihood for all samples of $\mathbf{X}$ and setting it equal to the inverse of the largest observed value:

$$c = \frac{1}{\max[L(\mathbf{x}_i)]} \tag{18}$$

In doing so, the value of the likelihood function is truncated at $1/c$, as evident from Eq. (9). This has no influence on the result obtained with MCS, however, because the likelihood function $L(\mathbf{x})$ is not evaluated in the regions of $\mathbf{X}$ where $L(\mathbf{x})$ is truncated.

3.2 *Importance sampling*

MCS becomes inefficient for problems with more than a few random variables and a few observations. Importance sampling (IS) can provide a more efficient estimation of $\Pr(F|Z)$. General introductions to IS in reliability applications are given e.g. in (Engelund and Rackwitz 1993). The IS estimator for the conditional probability $\Pr(F|Z)$ in Eq. (15) is

$$\Pr(F \mid Z) \approx \frac{\sum_{i=1}^{n_s} I[h_e(\mathbf{x}_i, p_i) \le 0] I[g(\mathbf{x}_i) \le 0] \dfrac{f_{\mathbf{X}}(\mathbf{x}_i)}{\psi(\mathbf{x}_i, p_i)}}{\sum_{i=1}^{n_s} I[h_e(\mathbf{x}_i, p_i) \le 0] \dfrac{f_{\mathbf{X}}(\mathbf{x}_i)}{\psi(\mathbf{x}_i, p_i)}} \tag{19}$$

wherein the samples $\mathbf{x}_i$ and p_i are simulated from a distribution with sampling density $\psi(\mathbf{x},p)$. The efficiency of importance sampling is determined by the selected sampling density $\psi(\mathbf{x},p)$. It is not possible to generically define an optimal sampling density that is valid for all applications; however, it is possible to split $\psi(\mathbf{x},p)$ into

$$\psi(\mathbf{x}, p) = \psi_1(\mathbf{x})\psi_2(p \mid \mathbf{x}) \tag{20}$$

where $\psi_1(\mathbf{x})$ is the sampling PDF of $\mathbf{X}$ and $\psi_2(p|\mathbf{x})$ is the conditional sampling density of P given $\mathbf{X}=\mathbf{x}$. An optimal conditional sampling density $\psi_2(p|\mathbf{x})$ that is valid for any application of Eq. (19) is given as

$$\psi_2(p \mid \mathbf{x}) = \frac{1}{cL(\mathbf{x})}, \quad 0 \le p \le cL(\mathbf{x}) \tag{21}$$

If it holds that $L(\mathbf{x}) > 0$ for any $\mathbf{x}$, then $I[h_e(\mathbf{x}_i, p_i) \le 0] = 1$ for any value of p_i that is sampled from the above conditional density $\psi_2(p|\mathbf{x})$. In this case, Eq. (19) reduces to

$$\Pr(F \mid Z) \approx \frac{\sum_{i=1}^{n_s} I[g(\mathbf{x}_i) \le 0] \dfrac{f_{\mathbf{X}}(\mathbf{x}_i)L(\mathbf{x}_i)}{\psi_1(\mathbf{x}_i)}}{\sum_{i=1}^{n_s} \dfrac{f_{\mathbf{X}}(\mathbf{x}_i)L(\mathbf{x}_i)}{\psi_1(\mathbf{x}_i)}} \tag{22}$$

Note that the constant c disappears when using this sampling density.

The optimal choice of $\psi_1(\mathbf{x})$ is problem dependent. Ideally, $\psi_1(\mathbf{x})$ should be centered around the area of interest in the space of $\mathbf{X}$. If it is possible to identify the design point of the event Z_e and the joint design point of the event $\{Z_e \cap F\}$, the sampling densities $\psi_1(\mathbf{x})$ can be centered around these points. This approach was successfully applied in (Straub 2010), and for many problems such a strategy works well. For the identification of the design points, efficient algorithms exist and are implemented in commercial software (e.g., Strurel, Gollwitzer et al. 2006). However, for spatially distributed systems requiring the computation of the reliability probability at many locations, the approach might not be sufficiently robust in all cases, since the identification of the design points requires solving many optimization problems simultaneously.

As an alternative, a robust choice for $\psi_1(\mathbf{x})$ is the a-priori PDF of $\mathbf{X}$, $f_{\mathbf{X}}(\mathbf{x})$. If the prior probability of failure is relatively large (say $\beta < 3$), this sampling density can provide reasonably accurate results with limited number of samples. With $\psi_1(\mathbf{x}) = f_{\mathbf{X}}(\mathbf{x})$, the estimate of the conditional probability in Eq. (22) simplifies further to

$$\Pr(F \mid Z) \approx \frac{\sum_{i=1}^{n_s} I[g(\mathbf{x}_i) \le 0] L(\mathbf{x}_i)}{\sum_{i=1}^{n_s} L(\mathbf{x}_i)} \tag{23}$$

Table 1. Measurements considered for the reliability updating.

Case	n_o	Locations l_i	Observations y_i	$\Pr(Z_e)$
a	2	30, 70	1, −1	0.077
b	5	10, 30, 50, 70, 90	0.5, 0.5, 0.5, 0.5, 0.5	0.016
c	10	5, 15, 25, …, 85, 95	0.0, 0.5, 0.5, 0.0, −0.5, 0.5, 0.0, 0.5, 0.5, 0.0	0.0014

It should be evident that this estimate can be interpreted as a Monte Carlo estimate wherein each sample is weighted with its corresponding likelihood. The same result could have be obtained by noting that the posterior PDF of $\mathbf{X}$ is $f''_{\mathbf{X}}(\mathbf{x}|Z) = \alpha L(\mathbf{x}) f(\mathbf{x})$, with constant

$$a = \frac{1}{\int_{\mathbf{X}} L(\mathbf{x}) f(\mathbf{x}) d\mathbf{x}} \tag{24}$$

and that $\Pr(F|Z)$ can be calculated from $f''_{\mathbf{X}}(\mathbf{x}|Z)$ by

$$\Pr(F \mid Z) = \int_{\mathbf{X}} I[g(\mathbf{x}) \le 0] f''(\mathbf{x} \mid Z) d\mathbf{x} = \frac{\int_{\mathbf{X}} I[g(\mathbf{x}) \le 0] L(\mathbf{x}) f(\mathbf{x}) d\mathbf{x}}{\int_{\mathbf{X}} L(\mathbf{x}) f(\mathbf{x}) d\mathbf{x}} \tag{25}$$

The solution in Eq. (23) is the MCS estimator of the two integrals in Eq. (25).

4 EXAMPLE: OBSERVATIONS OF A GAUSSIAN PROCESS

This example is included to investigate and illustrate the performance of the proposed method and the different computation approaches. Let $\mathbf{X} = [X_1, X_2, \ldots, X_n]^{\mathrm{T}}$ be the values of a Gaussian process at $n = 100$ locations with coordinates $\mathbf{l} = [1, 2, \ldots, 100]^{\mathrm{T}}$. All X_i have zero mean and unit variance; the covariance structure is fully described by the autocovariance function

$$\mathrm{Cov}(X_i, X_j) = \exp\left(-\frac{|l_i - l_j|}{20}\right) \tag{26}$$

At each location l_i, the failure event i is defined as the variable X_i exceeding value 2, i.e.,

$$g_i(\mathbf{x}) = 2 - x_i \tag{27}$$

We consider n_o observations y_i of the process at different locations l_i, $i = 1, \ldots, n_o$. Each observation is associated with an uncorrelated Gaussian error, with zero mean and standard deviation 0.5. The likelihood function is thus

$$\mathrm{L}(\mathbf{x}) = \prod_{i=1}^{n_0} \frac{1}{0.5\sqrt{2\pi}} \exp\left[-\frac{1}{2}\left(\frac{y_i - x_{l(i)}}{0.5}\right)^2\right] \tag{28}$$

For the numerical analysis, we consider three scenarios, summarized in Table 1.

In the last column of Table 1, the probability of the equivalent information event $\Pr(Z_e)$ is provided. Here, $\Pr(Z_e)$ is calculated with MCS, whereby Z_e is defined with $c = 1/L(\mathbf{x}_{MLE})$.

4.1 *Monte Carlo Simulation*

The results obtained with MCS using 10^5 samples are summarized in Figure 1, together with the true reliability calculated analytically. In case (a), MCS provides useful results with reasonably close credible intervals (credible intervals are the Bayesian equivalent of confidence intervals),

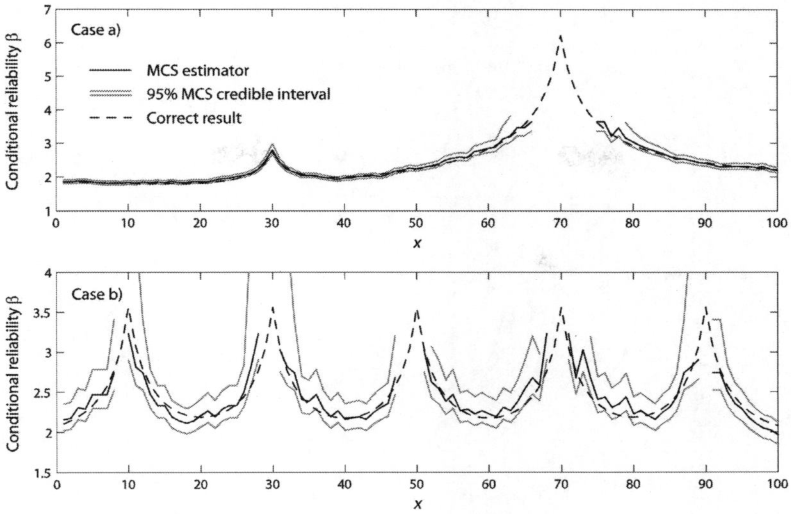

Figure 1. Results obtained with MCS with 10^5 samples.

with the exception of those areas that exhibit large posterior reliability. In case (b) MCS results are suitable for the purpose of verifying results obtained with importance sampling methods (or other methods, e.g. FORM/SORM), but are not sufficiently accurate for most other purposes. This is mainly due the fact that the probability of the equivalent information event $\Pr(Z_e)$ reduces with increased number of measurements, Table 1, and therefore the effective number of samples n_Z also diminishes according to Eq. (16). In case (c), due to the same reasons, no result can be obtained with MCS. Here, additionally, the increased number of measurements reduces the uncertainty and, therefore, increases the reliability, which leads to a reduction in the accuracy of MCS.

4.2 *Importance sampling*

We investigate importance sampling with the conditional sampling density $\psi_2(p|\mathbf{x})$ selected as the optimal one, Eq. (21), and with two different sampling densities $\psi_1(\mathbf{x})$, listed in the following.

1. The prior PDF of $\mathbf{X}$, $\psi_1(\mathbf{x}) = f_{\mathbf{X}}(\mathbf{x})$;
2. Multinormal distribution with mean equal to the design point corresponding to failure of element i and covariance matrix equal to that of $\mathbf{X}$, $C_{\mathbf{XX}}$.

With the first choice of $\psi_1(\mathbf{x})$, the sampling density is the same for the evaluation of the conditional reliability at all locations l_i. The second choice involves a different $\psi_1(\mathbf{x})$ for the evaluation of the conditional reliability at every location l_i. However, the design points applied here are readily obtained: It suffices to identify the design point u_i of the variable X_i (which is the same for every l_i and in the present example is simply equal to $u_i = 2$) and then determine the conditional mode of the other variables given $X_i = u_i$. Other investigated choices for $\psi_1(\mathbf{x})$ did not lead to significant improvement in accuracy over these two choices. As an example, the use of a sampling density $\psi_1(\mathbf{x})$ located around the MLE leads to similar accuracy than choice 1 above.

The results obtained with importance sampling are summarized in Figure 2 below. Not surprisingly, the choice of a different sampling density for the evaluation of the conditional reliability at every location l_i (choice 1) leads to higher accuracy in all cases. However, this choice does require a higher computational effort than the use of a common $\psi_1(\mathbf{x})$ for all locations. As evident from Figure 2, the accuracy decreases with increasing number of measurements. This effect is similar to that observed in the application of MCS.

Computation of the results presented here takes in the order of 50CPU seconds on a standard PC with a 1.6 GHz processor. The computation time is approximately linear to the number of simulations.

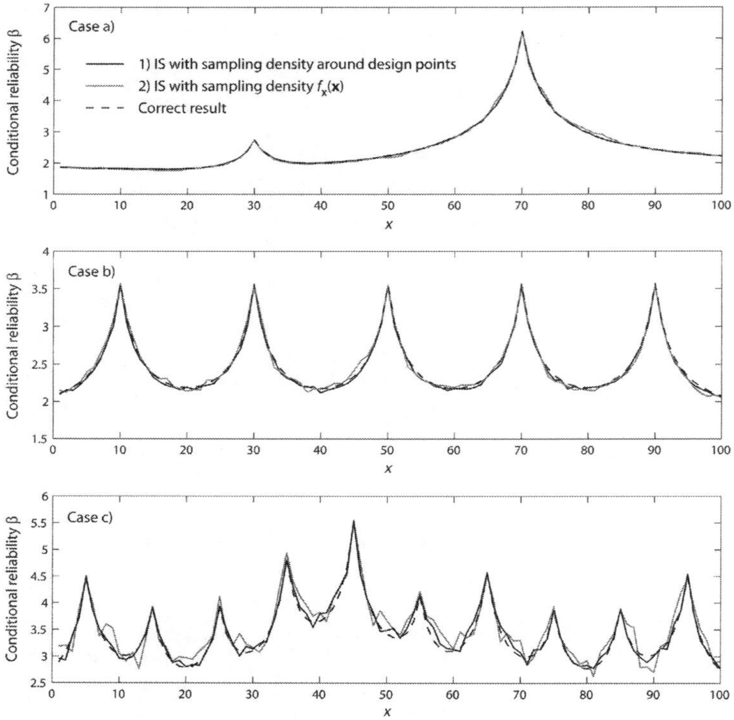

Figure 2. Results obtained with importance sampling with 10^5 samples.

5 CONCLUSIONS

With the novel approach proposed by the author in (Straub 2010), which is summarized in this paper, Bayesian updating in the context of structural reliability can be performed with stochastic simulation. In the present paper, the use of Monte Carlo simulation and Importance Sampling techniques is investigated. The approach has strong potential for reliability updating in spatially distributed systems, where robust algorithms are required, due to the large number of conditional reliability evaluations. Potential applications are manifold, including reliability updating of deteriorating large-scale engineering systems following inspections and monitoring results.

REFERENCES

Der Kiureghian, A. (2005), First- and second-order reliability methods. Chapter 14, in *Engineering design reliability handbook,* edited by E. Nikolaidis et al., CRC Press, Boca Raton, FL.

Engelund, S., and R. Rackwitz (1993), A benchmark study on importance sampling techniques in structural reliability, *Structural Safety,* **12**(4), 255–276.

Gollwitzer, S., B. Kirchgäßner, R. Fischer, and R. Rackwitz (2006), PERMAS-RA/STRUREL system of programs for probabilistic reliability analysis, *Structural Safety,* **28**(1-2), 108–129.

Madsen, H. O. (1987), Model Updating in Reliability Theory, in *Proc. ICASP 5,* Vancouver, Canada.

Papaioannou, I. and D. Straub (2010). Geotechnical reliability updating using stochastic FEM. *Proc. IFIP WG7.5 Working Conference on Reliability and Optimization of Structural Systems*, Munich.

Schall, G., S. Gollwitzer, and R. Rackwitz (1988), Integration of multinormal densities on surfaces, in *Proc. 2nd IFIP WG 7.5 Working Conference,* London.

Straub, D. (2010), Reliability updating with equality information. *Probabilistic Engineering Mechanics*, under review.

Tang, W. H. (1973), Probabilistic Updating of Flaw Information, *Journal of Testing and Evaluation,* **1**(6), 459–467.

Reliability and Optimization of Structural Systems – Straub (ed)
© 2010 Taylor & Francis Group, London, ISBN 978-0-415-88179-1

Structural optimization based on reliability index mapping functions

Wilson H. Tang & J. Zhang
Department of Civil and Environmental Engineering, The Hong Kong University of Science and Technology, Hong Kong SAR, China

ABSTRACT: Reliability Based Optimization (RBO) is often computationally intensive because the coupling of reliability evaluation and cost minimization. The mean first order reliability method (MFORM) is efficient for reliability evaluation, but it may not accurate. Nevertheless, the accurate reliability index is often highly correlated with the MFORM reliability index. In this study, we suggest that one can determine the relationship between the MFORM reliability index and the accurate reliability index through a regression analysis, which is called the reliability index mapping function. Based on the reliability index mapping function, an equivalent MFORM target reliability index can be determined such that MFORM can be used for reliability evaluation in RBO. Since the computational work involved in MFORM is minimal, RBO based on reliability index mapping functions is efficient. The reliability index mapping function determined via a regression analysis may only reflect the approximate relationship between the accurate reliability index and the MFORM reliability index. How to consider uncertainties in reliability index mapping functions in RBO is also described in this paper.

1 INTRODUCTION

Reliability based optimization (RBO) has received the attention of many researchers, for instance, Thoft-Christensen & Sørensen (1987), Madsen & Hansen (1991), Enevoldsen & Sørrensen (1994), Kuschel & Rackwitz (1997), Der Kiureghian & Polak (1998), Melchers (2001), Frangopol & Maute (2003), Kharmanda et al. (2003), Youn & Choi (2004), Straub & Faber (2005), Royset et al. (2006), Ching (2007), Liang et al (2007), Du (2008), Marti (2008), Bucher (2009), and Nguyen et al. (2010). A major challenge in solving RBO problems resides in the coupling between reliability evaluation and cost optimization, which could both be computationally intensive. The implementation of RBO can be eased if the computational work involved in the reliability constraints evaluation process is reduced. The mean first order reliability method (MFORM) is efficient for reliability evaluation, but it is not enough accurate for many problems. Nevertheless, as will be shown in this paper, the accurate reliability index is often highly correlated with the reliability index calculated with MFORM. Let β^M denote the MFORM reliability index and β^A denote the accurate reliability index. If one could find a functional form between β^A and β^M, it is then possible to evaluate the reliability constraint using MFORM in RBO, thus reducing the computational work in RBO. The relationship between β^A and β^M is called the *reliability index mapping function* in this study. The objective of this paper is to illustrate how to construct reliability index mapping functions and how to implement RBO through the use of reliability index mapping functions.

This paper consists of two parts. In the first part, we use a beam example to illustrate the basic idea about constructing reliability index mapping functions and implementing RBO based on reliability index mapping functions. It is possible that sometimes an accurate relationship between β^A and β^M cannot be found and there would be uncertainties in the reliability index mapping functions. In the second part of this paper, how to consider the uncertainties in reliability index mapping functions in RBO is described.

2 RBO BASED ON RELIABILITY INDEX MAPPING FUNCTIONS: A BEAM EXAMPLE

2.1 *Problem description*

A simple supported beam subjected to a uniformly distributed load w (Figure 1) is first studied. In this example, three failure modes are considered, i.e., bending failure, shear failure, and failure under combined effects of bending and shear. The limit state functions for the three failure modes are as follows

$$g_1(\mathbf{x}) = M_0 - wL^2/8 = 0 \tag{1a}$$

$$g_2(\mathbf{x}) = V_0 - wL/2 = 0 \tag{1b}$$

$$g_3(\mathbf{x}) = 1 - \left[wl^2/M_0 + wM_0/\left(2V_0^2\right)\right] = 0 \tag{1c}$$

where $\mathbf{x}$ = uncertain variables in the limit state functions; $M_0 = bh^2 f_y/6$; and $V_0 = 2\tau_y bh/3$. The uncertain variables include the load w, the tensile yield strength f_y, and the shear yield strength τ_y, i.e., $\mathbf{x} = \{w, f_y, \tau_y\}$. The three uncertain variables are assumed statistically independent and their statistics are summarized in Table 1.

Suppose the objective is to minimize the area of the cross section of the beam and the design variables are the width (b) and height (h) of the beam cross section, as shown in Figure 1. The design space is described by 0.2 m < b, h < 1 m and b/h > 1/3. The design requirement is that the reliability index of each failure mode is not smaller than 3. The RBO problem can then be denoted as

$$\begin{aligned} &\min A = bh \\ &\text{Subjected to} \quad \beta_i \geq 3, \quad i = 1,2,3 \\ &\qquad 0 \leq b, h \leq 1 \\ &\qquad b/h \geq 1/3 \end{aligned} \tag{2}$$

2.2 *Construction of reliability index mapping functions*

In general, the relationship between β^{A} and β^{M} not only depends on β^{A} but also the value of the design variables. To check if it is possible to find a relationship between β^{A} and β^{M} independent of the design variables, we randomly sample ten points from the design space, which result in ten possible designs. For each design, the reliability index of the first failure mode is calculated using both MFORM and the advanced first order reliability method (AFORM) (e.g., Hasofer & Lind 1974, Rackwitz & Fiessler 1978), resulting in ten pairs of β^{A} and β^{M}, as shown in Figure 1(a). Similarly, ten pairs of β^{A} and β^{M} for other two failure modes are obtained and shown in Figures 2(b)–(c), respectively.

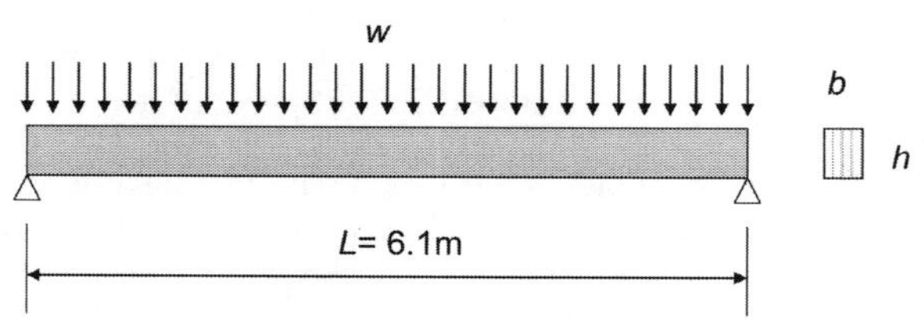

Figure 1. A simple supported beam subjected to a uniformly distributed load.

Table 1. Statistics of the uncertain variables in the beam example.

Distribution	w (kN/m) Lognormal	f_y (MPa) Lognormal	τ_y (MPa) Lognormal
Mean	40	300	150
COV	0.25	0.1	0.15

Figures 2(a)–2(c) show that there seems to be a one-to-one mapping between the two types of reliability indices. This implies that the relationships between the two types of reliability indices for the three failure modes are insensitive to the design variables. For each of the above three failure mode, the functional relationship between the two types of reliability indices, i.e., the reliability index mapping function, may be obtained accurately through regression analysis of the data points.

As the relationship between β^{A} and β^{M} is not strongly nonlinear, a second order polynomial function is used for regression analyses in this study to develop the reliability index mapping functions. The regression lines are also shown in Figures 2(a)–2(c). The mapping functions obtained for the three limit state functions are

$$\beta^M = 0.013\left(\beta^A\right)^2 + 1.2144\beta^A + 0.2557 \tag{3a}$$

$$\beta^M = -0.0496\left(\beta^A\right)^2 + 1.263\beta^A - 0.2347 \tag{3b}$$

$$\beta^M = 0.2311\left(\beta^A\right)^2 + 0.765\beta^A - 0.029 \tag{3c}$$

$$\beta^M = 0.2311\left(\beta^A\right)^2 + 0.765\beta^A - 0.029 \quad (3c)$$

2.3 *RBO through reliability index mapping functions*

Based on Equation 3(a), when $\beta^{\mathrm{A}} = 3$, $\beta^{\mathrm{M}} = 4.018$, implying that when MFORM is used for evaluating the reliability constraint, the target reliability index should be 4.018, which is called the *equivalent MFORM target reliability index* here. The equivalent MFORM reliability indices for the other two failure modes are 2.873 and 4.288, respectively. The RBO problem in Equation 1 can then be approximated as follows

$$\begin{aligned} &\min A = bh \\ &\text{Subjected to} \quad \beta_1^M \geq 4.018, \quad \beta_2^M \geq 2.873, \quad \beta_3^M \geq 4.288, \\ &\qquad 0 \leq b, h \leq 1 \\ &\qquad b/h \geq 1/3 \end{aligned} \tag{4}$$

Compared with Equation 1, the minimization problem in Equation 4 is easier to solve as the computational work involved in evaluating the reliability constraints based on MFORM is minimal. The Matlab routine *fmincon.m* is used here to find the minimization point in Equation 4. Based on Equation 4, the optimal values for b and h are 0.200 m and 0.568 m, respectively. Using AFORM, the reliability indices of the three failure modes are 10.831, 15.814, and 3.001, respectively, indicating that the optimal design is governed by the third failure mode. The reliability index of the most critical failure mode, i.e., the third failure mode, of the optimal design is not exactly the target reliability index. This implies that the MFORM target reliability index is not exactly equivalent to the original target reliability index. Nevertheless, the reliability index of the critical failure mode of the optimal design is very close to the target reliability index.

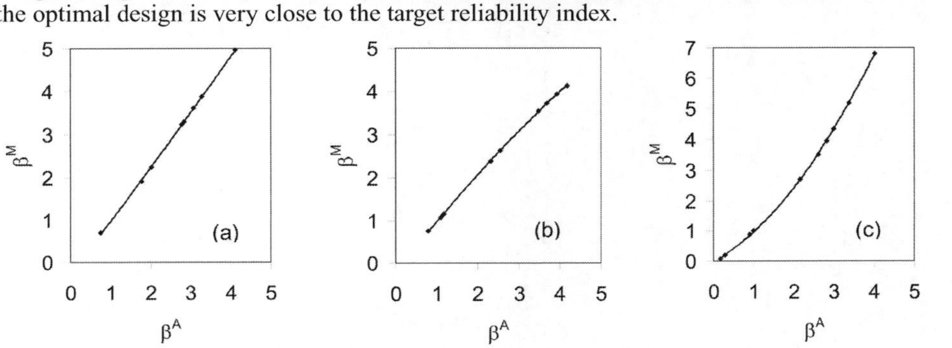

Figure 2. Relationships between β^{A} and β^{M} for the failure modes of the beam example.

3 RBO CONSIDERING UNCERTAINTIES IN RELIABILITY INDEX MAPPING FUNCTIONS

In the above beam example, the reliability index mapping functions can accurately reflect the relationship between the two types of reliability indices. There are cases where there are relatively larger uncertainties in the relationship between the two types of reliability indices. How to implement RBO considering the uncertainties in the reliability index mapping functions is described in this section.

3.1 *A numerical example*

A numerical example taken from Youn & Choi (2004) is examined. The RBO problem is as follows:

$$\begin{aligned} &\min \mu_1 + \mu_2 \\ &\text{Subjected to} \quad \beta_i \geq 3, \quad i = 1,2,3 \\ &\qquad\qquad\quad 0 \leq \mu_i \leq 10 \quad i = 1,2 \end{aligned} \tag{5}$$

The three limit state functions are:

$$g_1(\mathbf{x}) = x_1^2 x_2/20 - 1 = 0 \tag{6a}$$

$$g_2(\mathbf{x}) = (x_1 + x_2 - 5)^2/30 + (x_1 - x_2 - 12)^2/120 - 1 \tag{6b}$$

$$g_3(\mathbf{x}) = 80/(x_1^2 + 8x_2 + 5) - 1 = 0 \tag{6c}$$

where $\mathbf{x} = \{x_1, x_2\}$; and μ_1 and μ_2 = mean values of x_1 and x_2. The standard deviations of x_1 and x_2 are both 0.3. The design variables are μ_1 and μ_2. Both variables are normally distributed.

Figures 3(a)–(c) shows the relationships between β^A and β^M for the three failure modes, respectively. Compared with those in Figures 2(a)–(c), the relationships between β^A and β^M in Figures 3(a)–(c) are relatively uncertain in this example. Nevertheless, approximate reliability mapping functions can still be obtained through regression analysis. Based on equivalent MFORM reliability index obtained from the reliability index mapping functions, the optimal values for μ_1 and μ_2 are found to be 3.484 and 3.291, respectively. The AFORM reliability indices for the three failure modes are 3.143, 2.963, and 9.906, respectively. The reliability index of the most critical failure mode, i.e., the second failure model, is a bit smaller than 3.0. This implies around the optimal design point the actual equivalent MFORM reliability index is larger than the one determined directly based on the reliability index mapping function.

To make the reliability index of the third failure model at the optimal point closer to the target reliability index, the equivalent MFORM reliability index can be increased. Based on the data points in Figure 2(b), the standard deviation of the β^M around its best estimated value determined based on the reliability index mapping function is 0.017. Figure 2(b) shows that the uncertainty

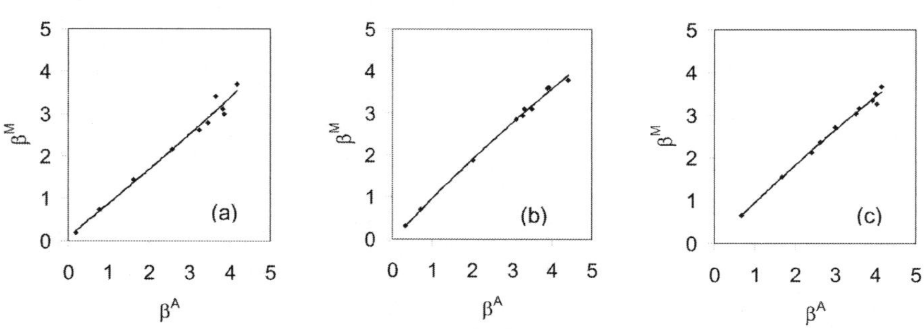

Figure 3. Relationships between β^A and β^M for three failure modes of the example adapted from Youn and Choi (2004).

in the reliability index mapping function increases with β^M. In the above calculation, the standard deviation of the β^M is assumed to be constant for simplicity. Let's try to increase equivalent MFORM reliability index by 0.017. Using this new equivalent MFORM reliability index, a new optimal design is obtained. The reliability index of the second failure mode at this new design is 2.974, which is closer to target reliability index, but still below the target reliability index. The equivalent MFORM reliability index is then further increased until the target reliability index of the second failure mode is larger than 3.0. Indeed, after adjusting equivalent reliability index for three times, the reliability indices of the three failure modes are all larger than 3.0, as also summarized in Table 2. Thus, when only approximate reliability mapping functions can be found, one can first locate the approximate location of the optimal design based on the MFORM reliability indices, and then adjust MFORM target reliability index until all reliability targets are met.

3.2 *Ten-bar example*

In the above numerical example, there is only one equivalent reliability index needed to be adjusted. When there are more uncertain variables and more limit state functions, multiple equivalent MFORM reliability indices may be adjusted simultaneously. This is illustrated with a ten-bar truss example adapted from Haftka & Gurdal (1992), as shown in Figure 4. The truss is subjected to two vertical loads, i.e., P_1, and P_2. Each bar may fail due to either excessive compression or tension. Assume the compressive and tensile strengths are the same. Let A_i and f_{yi} denote the area and yield strength of the ith bar, respectively. The design variables are A_1, A_2, ..., and A_{10}. The uncertain variables in this problem are $f_{y1}, f_{y2}, \ldots, f_{y10}$, P_1, and P_2. The yield strength of the ninth bar, i.e., f_{y9}, is assumed to be lognormally distributed with a mean of 516 MPa and a COV of 0.1. The yield strengths of the other bars are assumed to be lognormally distributed with a mean of 172 MPa and a COV of 0.1. P_1 and P_2 are assumed to be lognormally distributed with a mean of 300 kN and a COV of 0.3. All the uncertain variables are assumed to be statistically independent. The design requirements are that target reliability index for each bar against failure is not smaller

Table 2. Optimal values of design variables in the numerical example.

	μ_1	μ_2	β_1	β_2	β_3
Initial[1]	3.461	3.279	3.057	**2.950**	9.991
Adjusted[2]	3.454	3.295	3.056	**3.008**	9.983

[1] Obtained based on the equivalent reliability indices directly obtained from the reliability index mapping functions

[2] Obtained based on the adjusted equivalent reliability indices

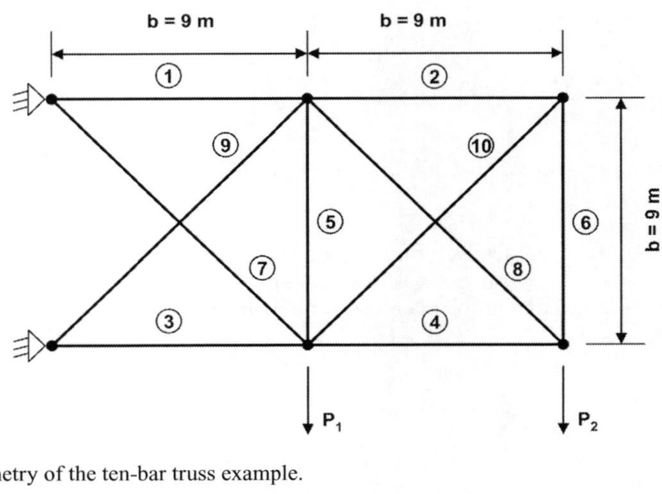

Figure 4. Geometry of the ten-bar truss example.

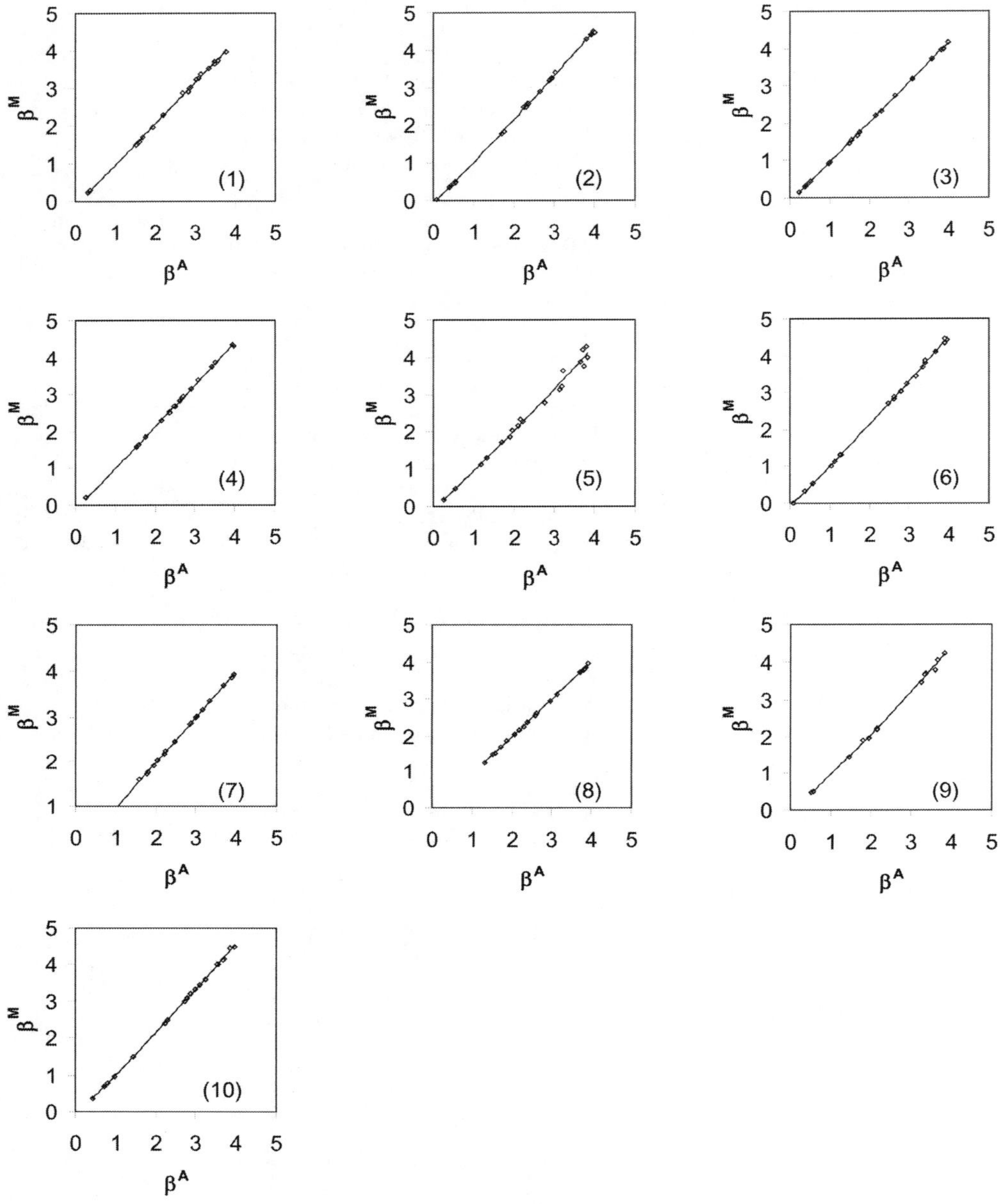

Figure 5. Relationships between β^{A} and β^{M} for the ten failure modes of the ten-bar truss example.

than 3.0. Supposing that the objective of the design is to minimize the volume of the materials to construct the truss and the design space is described by $0.0001\ \mathrm{m}^2 \leq A_i \leq 1\ \mathrm{m}^2 (i = 1, 2, \ldots, 10)$, the present optimization design problem can be formulated as

$$\begin{aligned} &\min \quad V = b\sum_{i=1}^{6} A_i + \sqrt{2}b\sum_{i=7}^{10} A_i \\ &\text{Subjected to} \qquad \beta_i \geq 3, \qquad i = 1,2,\ldots,10 \\ &\qquad\qquad\qquad 0.0001 \leq A_i \leq 1 \quad i = 1,2,\ldots,10 \end{aligned} \tag{7}$$

Table 3. Optimal areas of the ten bars.

	Area (cm^2)									
	A_1	A_2	A_3	A_4	A_5	A_6	A_7	A_8	A_9	A_{10}
Initial optimal[1]	61.89	1.00	59.51	32.18	7.66	1.00	41.94	41.04	35.40	1.42
Adjusted optimal[2]	63.04	1.00	59.39	32.53	7.88	1.00	42.63	42.00	35.96	1.41

[1] Obtained based on the equivalent reliability indices obtained directly from the reliability index mapping functions
[2] Obtained based on the adjusted equivalent reliability indices

Table 4. Reliability indices of the ten bars of the optimal design.

	β_1	β_2	β_3	β_4	β_5	β_6	β_7	β_8	β_9	β_{10}
Initial optimal	**2.935**	**2.938**	3.039	**2.967**	4.281	**2.936**	**2.893**	**2.900**	6.799	**2.942**
Adjusted optimal	**3.017**	**3.020**	3.039	**3.012**	4.317	**3.009**	**3.002**	**3.002**	6.867	**3.008**

where the ten limit state functions can be denoted as

$$g_i(\mathbf{x}) = A_i f_{yi} - |N_i| = 0 \quad i = 1, 2, \ldots, 10 \tag{8}$$

where $\mathbf{x} = \{f_{y1}, f_{y2}, \ldots, f_{y10}, P_1, P_2\}$ and N_i = axial force of the ith bar. Equations for calculating the axial forces of the ten bars can be found in Kumar et al. (2009).

The reliability mapping functions for the ten failure modes are shown in Figure 5. We can see that the two types of reliability indices are also highly correlated for the ten failure modes. Regression analysis are then used to obtain reliability index mapping functions for the ten failure modes. Based on the equivalent MFORM reliability index, an initial optimal design is found, as summarized in Table 3. The AFORM reliability indices of the ten bars of the optimal design are summarized in Table 4. Among the ten bars, there are seven bars with reliability indices smaller than 3.0. For each failure mode with reliability index smaller than the target reliability index, the standard deviation of β^M around its best estimated value determined based on the reliability index mapping function is determined. Then, the equivalent MFROM reliability indices of these failure modes are gradually increased with a step size equal to the standard deviations of β^M as estimated previously until the reliability indices of the ten failure modes are all larger than 3.0. For comparison, the adjusted optimal design is also summarized in Table 3, and the ten AFORM reliability indices of the adjusted optimal design are summarized in Table 4. It can be seen that all the target reliability indices are met. This demonstrates that the method of adjusting equivalent MFORM target reliability index introduced previously is also applicable when there are more than one equivalent MFORM needed to be adjusted.

4 SUMMARY AND CONCLUSIONS

The research work and findings from this study can be summarized as follows:

(1) Although the mean first order reliability method (MFORM) is not an accurate method for reliability evaluation, the accurate reliability index is generally highly correlated with the MFORM reliability index. The relationship between MFORM reliability index and that calculated using a more accurate method can be determined via a regression analysis, which is called the reliability index mapping function in this study.
(2) Through the reliability index mapping function, a target reliability index can be transferred to an equivalent MFORM target reliability index such that RBO can be implemented using MFORM

for reliability evaluation. As the computational work involved in MFORM is minimal, RBO based on reliability index mapping functions can be quite efficient.

(3) When the reliability index mapping function only reflects the approximate relationship between the MFORM reliability index and the accurate reliability index, the reliability of an optimal design found based on MFORM equivalent reliability indices may be below the target. To ensure all the reliability constraints are satisfied, one can gradually adjust the equivalent target MFORM reliability indices and repeat the optimization until all the reliability targets are met.

REFERENCES

Bucher, C. 2009. Probability-based optimal design of friction-based seismic isolation devices. *Structural Safety* 31(6): 500–507.

Ching, J. & Hsieh, Y.H. 2007. Approximate reliability-based optimization using a three-step approach based on subset simulation. *ASCE Journal of Engineering Mechanics* 133(4): 481–493.

Der Kiureghian, A. & Polak, E. 1998. Reliability-based optimal design: a decoupled approach. Reliability and optimization of structural systems. In: Andrzej S. Nowak (ed.), *Reliability and optimization of structural systems, Proc. 8th IFIP WG7.5 working conf.*, Cracow, Poland, 11–13 May 1998. Chelsea: Book Crafters.

Du, X.P. 2008. Saddlepoint approximation for sequential optimization and reliability analysis. *ASME Journal of Mechanical Design* 130(1): 011011-1-11.

Enevoldsen, I. & Sørrensen, J.D. 1994. Reliability-based optimization in structural engineering. *Structural Safety* 15(3):169–196.

Frangopol, D.M. & Maute, K. 2003. Life-cycle reliability-based optimization of civil and aerospace structures. *Computers & Structures* 81(7): 397–410.

Haftka, R.T. & Gurdal, Z. 1992. *Elements of structural optimization*. 3rd Edition. Dordrecht: Kluwer.

Hasofer, A.M. & Lind, N.C. 1974. Exact and invariant second moment code format. *ASCE Journal of Engineering Mechanics Division* 100(EM1): 111–121.

Kharmanda, G., Mohamed A. & Lemaire M. 2003. Efficient reliability-based design optimization using a hybrid space with application to finite element analysis. *Structural and Multidisciplinary Optimization* 24(3): 233–245.

Kumar, S., Pippy, R. J., Acar, E., Kim, N.H. & and Haftka, R.T. 2009. Approximate probabilistic optimization using exact-capacity-approximate-response-distribution (ECARD). *Structural and Multidisciplinary Optimization* 38:613–626.

Kuschel, N. & Rackwitz, R. 1997. Two basic problems in reliability-based structural optimization. *Mathematical Methods of Operations Research* 46(3): 309–333.

Liang, H., Haukaas, T. & Royset, J.O. 2007. Reliability-based optimal design software for earthquake engineering applications. *Canadian Journal of Civil Engineering* 34(7): 856–869.

Madsen, H.O. & Hansen, P.F. 1991. A comparison of some algorithms for reliability-based structural optimization and sensitivity analysis. In: Rüdiger Rackwitz & Palle Thoft-Christensen (eds.), *Reliability and optimization of structural systems, Proc. 4th IFIP WG7.5 Working Conf.*, Munich, 11–13 September 1991. Berlin: Springer-Verlag.

Marti, K. 2008. *Stochastic Optimization Methods*. Berlin: Springer-Verlag.

Melchers, R.E. 2001. Rational optimization of reliability and safety policies. *Reliability Engineering & System Safety* 73(3): 264–268.

Nguyen, T.H. Song, J. & Paulino, G.H. 2010. Single-loop system reliability-based design optimization using matrix-based system reliability method: theory and applications. *ASME Journal of Mechanical Design* 132(1): 011005-1-11.

Rackwitz, R. & Fiessler, B. 1978. Structural reliability under combined random load sequences. *Computers and Structures* 9(5): 484–494.

Royset, J.O., Kiureghian, A.D. & Polak E. 2006. Reliability-based optimal design of series structural systems. *ASCE Journal of Engineering Mechanics* 127 (6): 607–614.

Straub, D. & Faber, M.H. 2005. Risk based inspection planning for structural systems. *Structural Safety* 27(4): 335–355.

Thoft-Christensen, P. & Sørensen, J.D.1987. Recent advances in optimal design of structures from a reliability point of view. *Quality and Reliability Management* 4(4): 19–31.

Youn, B.D. & Choi, K.K. 2004. Selecting probabilistic approaches for reliability-based design optimization, *AIAA Journal* 42(1): 124–131.

Reliability and Optimization of Structural Systems – Straub (ed)
© 2010 Taylor & Francis Group, London, ISBN 978-0-415-88179-1

Fatigue life of wind turbine blades

P. Thoft-Christensen, R.R. Pedersen & S.R.K. Nielsen
Aalborg University, Aalborg, Denmark

ABSTRACT: The present paper analyses the possibility of reducing the expected damage accumulation during tower passage by modifying the wind turbine tower design from a traditional mono-tower to a tripod. Due to a narrow stagnation zone the stress reversals and hence the damage accumulation in the blades is substantial smaller in the tripod tower design compared to the mono-tower.

1 INTRODUCTION

Wind turbines are designed on basis of advanced modelling of the aerodynamic loading and of the strength of the involved materials. The most complicated parts of the wind turbines are the blades. The blades are made of composite materials because of the low weight and the high stiffness of these materials. However, the blades are highly fatigue critical. According to Shokrieh & Rafiee (2006), there are several factors which expose the wind turbine blades to fatigue phenomena:

- Long and flexible structures,
- Vibrations in its resonant mode,
- Randomness in the load spectra due to the nature of the wind,
- Continuous operation under different conditions,
- Low maintenance during lifetime.

The primary reason for fatigue damage accumulation in up-wind turbine blades is wind turbulence in the inflow. However, the stress reversal in the wind turbine blades during the blade passages through the tower shadow also contributes considerably to the fatigue damage accumulation. Therefore, it is of interest to investigate whether it is possible to reduce the fatigue damage accumulation by modifying the standard mono-tower design.

The tower shadow effect on the stresses in the wind turbine blades is due to the aero-dynamic disturbances of the wind because of the presence of the tower. During tower passage the blades of a horizontal axis upwind wind turbine hit a zone of stagnating. The resulting stress variation in the blades during tower passage depends on the size of the stagnation zone which again is determined from the radius of the tower and the clearance of the blade from the tower. Therefore, a reduced tower radius for the same clearance will reduce the stagnation zone and reduce the stress variations in the blade.

The effect of the tower shadow on the fatigue life of the wind turbine blades, and the influence of tower design, was studied recently by the authors in a preliminary study; see Thoft-Christensen et al. (2009). The present paper is to some extend based on a new study by the authors; see Pedersen et al. (2010).

Shadow effects have been studied in detail by several researchers; see e.g. Powles (1983), Wang (1998), Rosas et al. (2000), Wang & Coton (2001), Munduate & Colan (2000), Munduate et al. (2004), Dolan & Lehn (2005), Moriarty & Hansen (2005), and Crawford (2006).

2 WIND TURBINE FAILURES

Caithness Windfarms Information Forum (2009) has published a short report on documented wind turbine accidents. The report contains all documented cases of wind turbine related accidents up to 31 December 2009.

In figure 1 is shown how the accidents are distributed on years. The total number is 715, but this number is clearly a lower limit of the real number of accidents since a large number of non-documented accidents are not included in the report. Anyway, the figure shows a growing number of accidents, which is not strange since increasing number wind turbines are being built. In the report it is stated that the number of fatal accidents is 60 and that further 38 accidents regarding human injury are documented.

The dominant number of documented failure incidents is due to failure of the wind turbine blades, namely 167; see figure 2. Other types of structural failure are often more damaging than blade failure. However, blade failures are in general more risky to human health due to the fact that broken blades may be thrown far away from the wind turbines.

Although the number of wind turbine failures is not catastrophic high, the data above shows that it is still worthwhile and necessary to improve the safety of wind turbines. Especially, more research is needed regarding the safety of the blades since failure of the blades may have very severe consequences.

A typical example of wind turbine failure is shown in figure 3. The wind turbine is placed in Conisholme, Lincolnshire, UK. A 65 ft blade that flew off the turbine came loose after bolts attaching it to the hub failed. The turbine manufacturer Enercon has released an interim report identifying material fatigue as the cause of the accident. They are now carrying out further tests to establish what caused the bolts to come loose, focusing on the blade and hub components to which

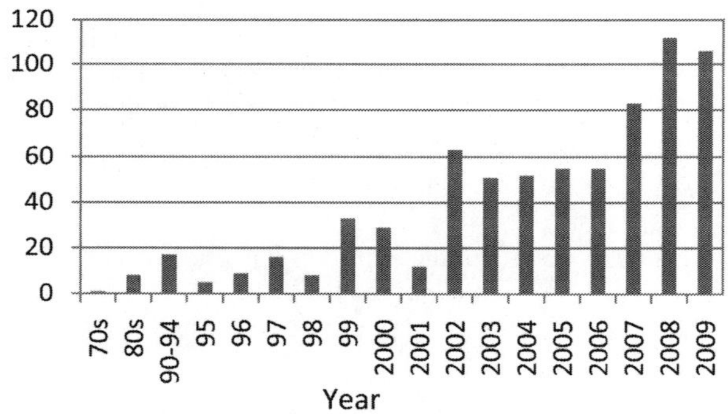

Figure 1. Number of documented wind turbine failure accidents.

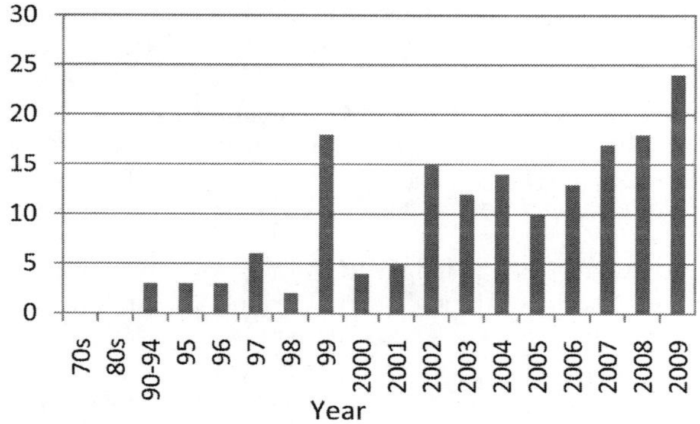

Figure 2. Number of wind turbine blade failures.

they were attached. If one of these failed, the bolts could have been put under unbearable stress; see Telegraph (2009).

3 DAMAGE ACCUMULATION

3.1 *Introduction*

In this section, the expected value and the standard derivation of the fatigue life of the wind turbine blades are estimated for a wind turbine with a tripod tower. The estimation is compared with a standard mono-tower; see Pedersen et al. (2010).

The stresses in the wind turbine blades are calculated from a dynamic mechanical model. The self-induced aeroelastic loading and the turbulence loading are modelled by means of a quasi-static model linearized around the operational point, assuming that the time-scale for readjustment of the non-stationary flows during changes of the angle of attack is small compared to the fundamental eigen-vibration period of the blade. Furthermore, the rotational effect of the turbulence hitting the blade is taken into consideration.

The analysis is based on Monte Carlo simulation where 10 minute realizations drawn from auto-spectral density of the rotational sampled turbulence process are generated. The equations of motion and the differential equations for the blade and state variables of the rational memory filter is solved numerically. The damage accumulation is calculated by rain-flow counting on the realizations of the bending stress.

A numerical example shows that the expected damage accumulation per unit of time in the wind turbine blades are reduced for the tripod when compared to a comparable mono-tower design. The mono-tower is shown in figure 4. To the left is shown the cross section of the circular tower and of the blade. In the middle is shown the dimensions of the tower. The blade dimensions are indicated to the right. The corresponding information on the tripod design is indicated in figure 5.

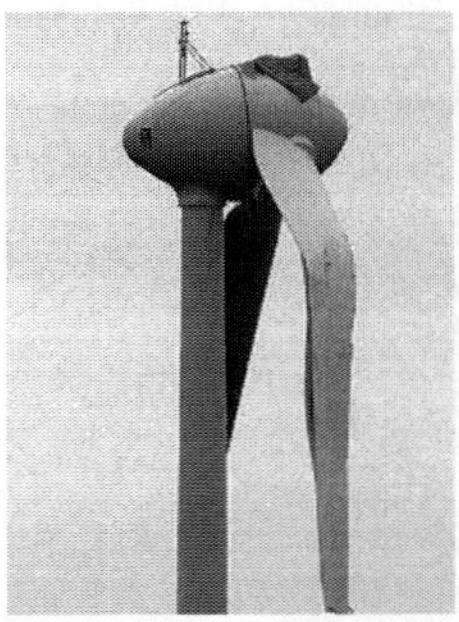

Figure 3. Wind turbine failure. From Telegraph (2009).

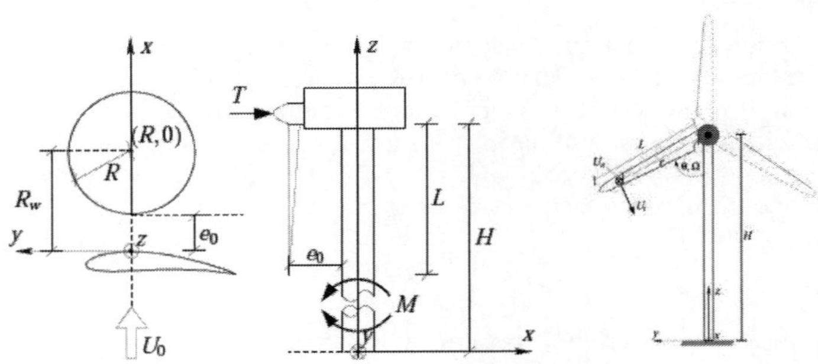

Figure 4. The mono-tower wind turbine. From Pedersen et al. (2010).

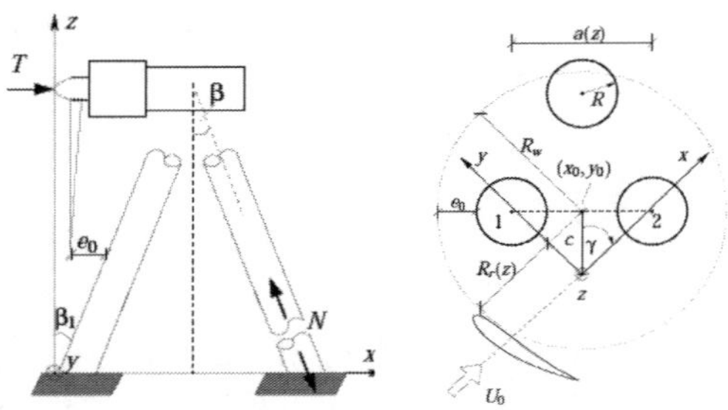

Figure 5. The tripod wind turbine. From Pedersen et al. (2010).

3.2 *Wind loading and turbulence*

The free mean wind speed U_0 at a height z is

$$U_0 = u_{10} k_t \ln(z/z_0) \tag{1}$$

where u_{10} is the 50-year return value of the 10-minute mean wind speed at 10 m height, and the terrain roughness parameters used here are $z_0 = 0.03$ and $k_t = 0.17$.

The turbulence model presented in Gilling et al. (2009) is adopted to generate the turbulence wind fields with a turbulence intensity $I_v = 0.16$ in Wind Turbine Class I; see IEC (2005). A frozen turbulence box is generated for each mean wind speed considered in the fatigue analysis, which is convected into the rotor with the mean wind at hub in agreement with Taylor's hypothesis of frozen turbulence. Next, the turbulence at a given cross section is obtained by interpolation among the grid points of the turbulence box.

The unsteady effects associated with a tower passage of the blades are important in the estimation of the fatigue life. The physic behind the phenomena is studied by Hand et al. (2001) using wind tunnel experiments to obtain the data necessary to validate the numerical models. During tower passage the mean bending stresses are reduced because of the diminished thrust on the blade. Due to the aerodynamic damping the dynamic amplification of the stress reversal is relatively small. This means that the shadow effect has minor influence on the extreme load case.

The load increment of the normal load per unit length due to turbulence and the velocity of the cross section is derived and added to the static load. I energy is dissipated from the system in case the load increment is in opposite direction to the velocity of the cross section. This is denoted aero elastic damping. Aero-elastic damping of the blades is an essential part of the structural behavior of the blades and has influence on the fatigue life. The wind loading and the turbulence is described in detail in Pedersen et al. (2010).

4 FATIGUE LIFE

4.1 *Constant life diagrams*

A detailed description of the fatigue load spectra and the fatigue behavior of the blade material are needed to evaluate the fatigue damage of a wind turbine blade. The fatigue behavior of the composite material is often described by a Goodman diagram. In a Goodman diagram the cycles to failure are plotted as a function of mean stress and range along lines of constant R_f-values, see Sutherland (1999) and Sutherland & Mandell (2005). R_f is defined as the relation between the minimum stress and the maximum stress in a fatigue cycle

$$R_f = \sigma_{\min} / \sigma_{\max} \tag{2}$$

Fatigue properties of wind turbine blades made of fiberglass composite materials are discussed in details by Sutherland & Mandell (2005), and an optimal constant life diagram (mean Goodman diagram) is proposed. The constant life diagram is further developed by Toft & Sørensen (2008).

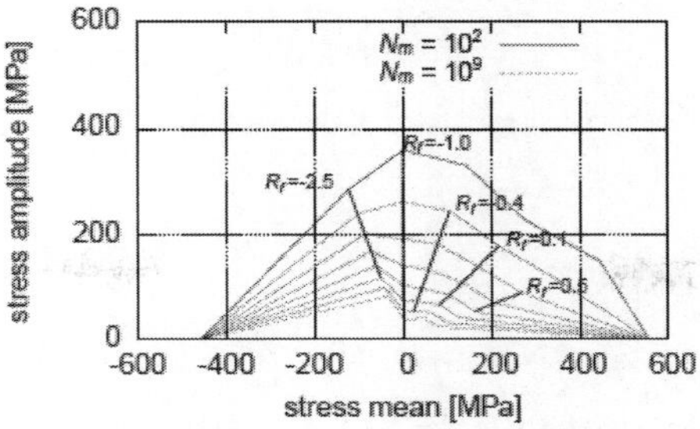

Figure 6. Constant life diagram. $N_m = 10^2$ and $N_m = 10^9$ indicate the lower and upper bound for the number of cycles. From Toft & Sørensen (2008).

Table 1. Main wind turbine data for the tripod and the mono-tower.

Eigen-frequency	ω_1, ω_2	6.32, 30.54 rad/s
Hub height	H	90 m
Mono tower radius	R	3.0 m
Tripod radii	R	1.0 m
Tubular thickness (mono)	t	0.09 m
Tubular thickness (tripod)	t	0.03 m
Total amount of steel (both)		25.5 m^3
Blade length	L	62 m
Hub diameter of blade	D	3.8 m
Clearance	e_0	4.0 m
Leg distance (tripod)	a_0	8.0 m
Rotational speed	Ω	1.26 rad/s
Life time		30 year
Number of samples	N_c	20
Sampling time	T	600 s
Young's modulus of blade	E	0.05 10^{12} N/m^2
Damping ratios	ζ_1, ζ_2	0.01; 0.01
Cut-in; cut-out wind velocity	u^{in}, u^{out}	5;25 m/s
Yaw angle inverval	γ	$[-\pi/2, \pi/2]$ rad

The fatigue constant life diagram used here is taken from Toft & Sørensen (2008) and is shown in figure 6. The procedure in the IEC norm (2005) is followed to compute the fatigue life.

4.2 *Numerical results*

The data, results, tables and figures in this section are all taken from the analysis by Pedersen et al. (2010). The main data concerning the mono-tower and the tripod are shown in table 1. The thickness and radii in the two designs are adjusted so that the amount of steel used in the two designs is identical.

The dynamic response of the blade during tower passage is primarily quasi-static and the dynamic effect is confined to the two lowest blade eigen-modes. The blade characteristics are taken from a 5 MW NREL reference wind turbine; see Jonkman et al. (2008). The two lowest eigen-modes are shown in figure 7, and the corresponding eigen-frequencies are listed in Table 1.

In figure 8 is shown the variation of the normal stress in the hub of the blade, when the blade passes a mono-tower and tripod tower in case of a yaw angle $\gamma = 0$ deg and the free wind velocity $u_{10} = 15$ m/s.

The Miners sum linked to $\gamma = 0$ rad and $u_{10} = 15$ m/s is presented in Figure 9, where each sample corresponds to a random turbulence field. This result indicates that the fatigue damage is lower when a tripod tower is used.

Table 2 shows the results from a complete analysis, where all yaw angles and mean wind velocities are included for both the tripod and mono tower. The final unconditional mean damage value μ_D and

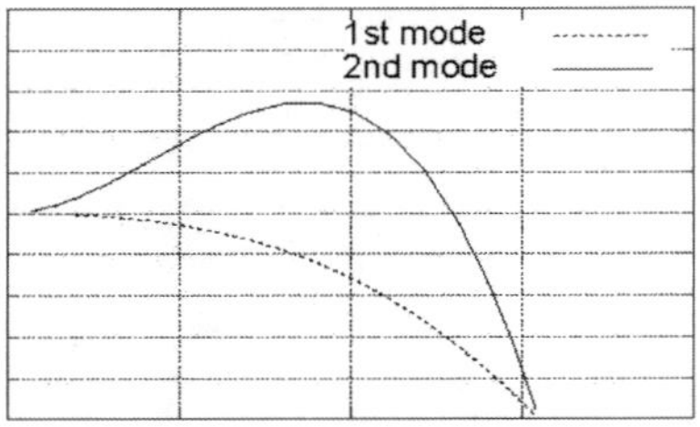

Figure 7. Eigen-modes for the wind turbine blades.

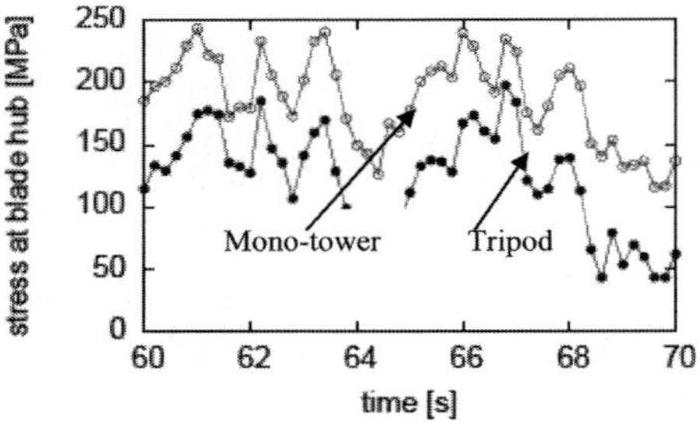

Figure 8. Normal stress in the blade hub when the blade passes a mono-tower and a tripod tower.

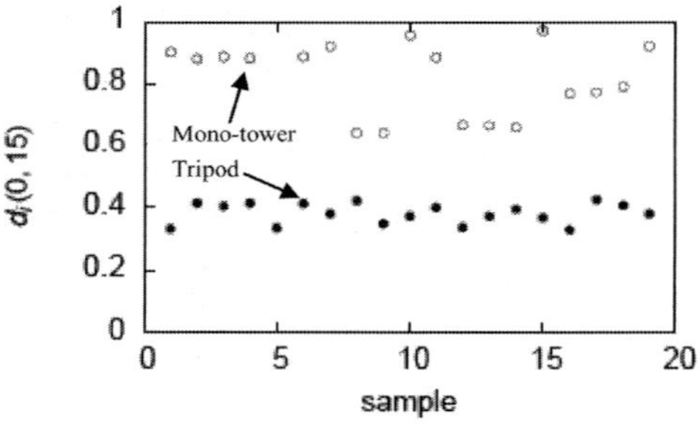

Figure 9. Samples of Miners sum.

Table 2. Mean and standard deviation of the damage.

	Mono tower	Tripod tower
μ_D	0.532	0.324
σ_D	0.078	0.042

standard derivation σ_D is presented in Table 2. The damage state is lower for the tripod configuration. Therefore, with the adopted assumptions and simplifications made in this example the lifetime is increased for a tripod design compared with a mono-tower design.

5 CONCLUSIONS

Evaluation of fatigue life of wind turbine blades due to shadow effects is discussed in this paper. A tripod design of wind turbine is compared to the standard mono-tower design. The shadow zones are decreased for the tripod design since the tripod tubular legs have a small radius compared to the radius of the mono tower. The wind turbine blade experiences more disturbances in a tripod design because of the three tripod legs. However, in the considered example the total fatigue damage is decreased.

The results indicate that the fatigue damage is significantly lower when the wind turbine blade is attached to a tripod configuration. The practical aspects concerning the construction and implementation of the tripod solution is not considered.

REFERENCES

Caithness Windfarms Information Forum (2009) Summary of wind turbine accident data to 31 December 2009. *www.caithnesswindfarms.co.uk.*

Crawford CA. (2006) *Advanced engineering models for wind turbines with application to the design of a coning rotor concept.* Dissertation, Trinity College Department of Engineering, University of Cambridge.

Dolan DSL & Lehn PW. (2006) Simulation model of wind turbine 3p torque oscillations due to wind shear and tower shadow. IEEE Transactions on Energy Conversion, 21: 717–724.

Gilling L, Nielsen SRK & Sørensen NN. (2009) Generation of synthetic turbulence in arbitrary domains. *Proceedings of the 10th International Conference on Structural Safety and Reliability*, Kansai University Japan.

Hand MM, Simms DA, Fingersh LJ, Jager DW, Cotrell JR, Shreck SJ & Larwood SM. (2001) *Unsteady Aerodynamics Experiment Phase VI: Wind Tunnel Test Configurations and Available Data Campaigns.* National Renewable Laboratory United States; NREL/TP-500-29955 2001.

IEC. 2005. *Wind turbine - part 1 Design requirement.* IEC 614001:2005(E) 3rd edition.

Jonkman J, Butterfield SS, Musial W & Scott G. (2008) *Definition of a 5-mw reference wind turbine for offshore system development.* NREL/TP-500-38060, Golden, National Renewable Energy Laboratory (to be published).

Moriarty PJ & Hansen AC. (2005) *AeroDyn Theory Manual.* National Renewable Energy Laboratory United States; NREL/TP 2005.

Munduate X & Coton FN. (2000) Identification of dynamic stall regions on horizontal axis wind turbines. *19th ASME Wind Energy Symposium*, Reno; Paper: 2000-0039.

Munduate X, Coton FN & Galbraith RAM. (2004) An investigation of the aerodynamic response of a wind turbine blade to tower shadow. *Journal of Solar Energy Engineering*; 126: 1034–1040.

Pedersen, RR, Nielsen SRK & Thoft-Christensen P. (2010) Stochastic analysis of the influence of tower shadow on fatigue life of wind turbine blade. *Submitted to Structural Safety.*

Powles SRJ. (1983) The effects of tower shadow on the dynamics of a horizontal-axis wind turbine. *Wind Engineering*; 7: 26–42.

Rosas PAC, Sørensen P & Binder H. (2000) Fast wind modelling for wind turbines. Proceedings (on CD-ROM), *Wind power for the 21 Century: EUWEC Special Topic Conference and Exhibition,* September 2000: 184–187.

Shokrieh MM & Rafiee R. (2006) Simulation of fatigue failure in a full composite wind turbine blade. *Composite Structures*; 74: 332–342.

Sutherland HJ. (1999) *On the fatigue analysis of wind turbines.* SAND99-0089, Sandia National Laboratories, Albuquerque, NM.

Sutherland HJ & Mandell JF. (2005) Optimized constant-life diagram for the analysis of fiberglass composites used in wind turbine blades. *Journal of Solar Energy Engineering-Transactions of the ASME*; 127: 563–569.

Telegraph. (2009) UFO wind turbine "broke due to mechanical failure not collision with flying object". http://www.telegraph.co.uk/news/newstopics/howaboutthat/4578998/UFO-wind-turbine-broke-due-to-mechanical-failure-not-collision-with-flying-object.html.

Thoft-Christensen P, Pedersen RR & Nielsen SRK. (2009) Numerical estimation of fatigue life of wind turbines due to shadow effect. *Proceedings of the 10th International Conference on Structural Safety and Reliability,* Kansai University Japan.

Toft HS & Sørensen JD. (2008) Stochastic models for strength of wind turbine blades using tests. *Proceedings of EWEC, Europe's Premier Wind Energy Event*, Brussels, Belgium.

Wang T. (1998) An examination of two tower-shadow modelling strategies for downwind wind turbines. In: *AIAA, Aerospace Sciences Meeting and Exhibit, 36th, and 1998 ASME Wind Energy Symposium, Reno, NV, Jan. 12–15, 1998, Collection of Technical Papers* (A98-16844 03-44), Reston, VA/New York, American Institute of Aeronautics and Astronautics, Inc./A.

Wang T & Coton FN. (2001) A high resolution tower shadow model for downwind turbines. *Journal of Wind Engineering and Industrial Aerodynamics*; 89: 873–892.

Reliability and Optimization of Structural Systems – Straub (ed)
© 2010 Taylor & Francis Group, London, ISBN 978-0-415-88179-1

Load level and target column over-design factors of frames under a specific reliability level

Yan-Gang Zhao
Department of Architecture, Kanagawa University, Yokohama, Japan

Mohammad Sharfuddin
Nagoya Institute of Technology, Nagoya, Japan

ABSTRACT: The objectives of the present paper are to probabilistically evaluate the load level of column over-designed frame structures under a specific reliability level and to present the target values of the Column Over-design Factor (COF) by which the undesirable story mechanisms can be avoided in probabilistic means. The load levels for target COF evaluation are analyzed, and an empirical formula describing the relationship between the load and COF is proposed. The COF requirements for probabilistically avoiding story mechanisms of frame structures are investigated based on the comparison of occurrence probabilities between entire beam-hinging failure mode and the most likely story mechanisms.

1 INTRODUCTION

In earthquake resistance design of frame structures, to ensure enough capacity of structures to absorb earthquake energy before a failure, the entire beam-hinging pattern mode in which the yielding of all the beams in flexure happens prior to possible yielding of the columns is generally considered to be a preferable. In order to ensure frame structures collapse according to the beam-hinging mode, structures are generally designed with a column over-design factor larger than unity to make the columns relatively stronger than beams. However, Due to the uncertainties in member strengths and external loads, it becomes uneasy to absolutely ensure the structures collapse according to the preferable failure modes, the better way may be to ensure the structures collapse according to the preferable mode probabilistically, namely, to ensure an occurrence probability of a preferable failure mode larger than the probabilities of undesirable modes.

Many studies have been performed on various aspects of COF requirements of structures. In a study by Kuwamura et al. (1989), the variation of the yield strength was taken into account in COF evaluation. Kawano et al. (1998) investigated COF by a finite element method and some results of the basic properties of the COF were obtained. Nakashima et al. (1999) simplified a frame structure into a fish bone shape for performing dynamic analysis with earthquake motions as input, and indicated that the necessary COF value ensuring no plastic hinges forming in columns, increased approximately linearly with the increase of peak velocity of the earthquake wave. Dooley et al. (2001) stated that the minimum COF requirements should be about 2.0 to have a significant probability of satisfying the building performance objectives during a design-basis seismic event of the reinforcement concrete frame. This value is much greater than 1.2, which is recommended by American Concrete Institute for concrete building design (ACI 1999). Almost all the aforementioned investigations were conducted on some specific structures using specific earthquake inputs, and it is required to develop a general approach that can be easily implemented to evaluate the target COFs for structures with arbitrary numbers of stories and bays.

Ono and Zhao (2000) proposed a probabilistic evaluation method to evaluate the target COF that controls the failure probabilities of undesirable failure modes within a specific tolerance. However, since the criterion used in the investigation is to ensure the frame structures collapse strictly according to the beam failure pattern, and any hinging in columns except the first-story column bases are unallowable, a much larger value of the COF is needed in the case where both the load and the member strength are uncertain than in the case of deterministic values of those parameters.

The objectives of the present paper are to probabilistically evaluate the load level of column over-designed frame structures under a specific reliability level and to present the target values of the Column Over-design Factor (COF) by which the undesirable story mechanisms can be avoided in probabilistic means. The load levels for target COF evaluation are analyzed, and an empirical formula describing the relationship between the load and COF is proposed. The COF requirements for probabilistically avoiding story mechanisms of frame structures are investigated based on the comparison of occurrence probabilities between entire beam-hinging failure mode and the most likely story mechanisms.

2 DEFINITIONS AND ASSUMPTIONS

2.1 *Definition of COF*

The COF used in the investigation is defined for each beam-column node as the ratio of the sum of the moment capacity of columns to the sum of moment capacity of beams as:

$$C_{of}(k) = \sum_i \mu_{mci} / \sum_i \mu_{mbi} \tag{1}$$

where k is the node number, and μ_{mci}, μ_{mbi} are the mean values of the ultimate moment strength of the column and beam connected to the kth node, respectively.

2.2 *Computational assumptions*

The following basic assumptions are used in this study:

(1) The strengths of beams and columns are designed to make the structure have identical COF at all beam-column nodes.
(2) The external loads consist of only the lateral earthquake loads triangularly distributed over the structures. This assumption is often used in simple aseismic design.
(3) Plastic moment capacities of sections are statistically independent of one another, and independent of the earthquake loads.
(4) All the random variables are assumed to follow a lognormal distribution. Because the coefficient of variation (COV) of earthquake load is generally taken to be 0.6–0.7 (Kanda 1993; AIJ 1993), and considering the other uncertainties included in load model, the coefficient of variation for the lateral forces is assumed to be 0.8.

3 DETERMINATION OF LOAD LEVEL

In entire beam-hinging mode, plastification occurs at all beam edges and the bases of the first-story columns as shown in Fig. 1. For an n-story, m-span frame structure, the performance function of entire beam-hinging mode $G(\mathbf{X})$, of which $\mathbf{X}$ is the vector of random variables of structural system, is generally given by:

$$G(\mathbf{X}) = 2\sum_{i=1}^{m} M_{bni} + 2\sum_{j=1}^{n-1}\sum_{i=1}^{m} M_{bij} + \sum_{l=1}^{2} M_{csl} + \sum_{l=1}^{m-1} M_{cl} - \sum_{j=1}^{n} jhP_j \tag{2}$$

where M_{bni} is the moment strength of the beam of the top story, M_{bij} is the moment strength of the beam of the ith span and jth story, M_{cl} is the moment strength of the interior column, and M_{csl} is the moment strength of the exterior column, P_j is the load acting on the jth story of structure, n and m are the number of stories and spans, respectively, and h is story height. The member strength and load are considered as random variables, while n, m and h are assumed to be deterministic variables.

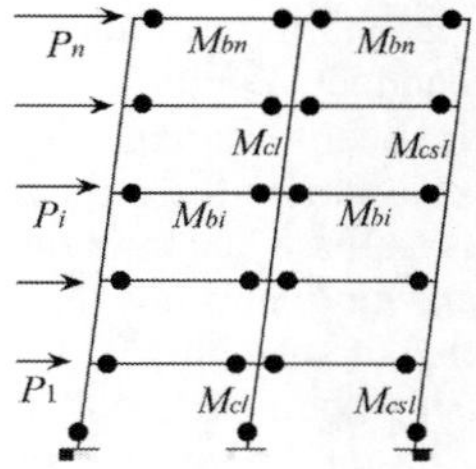

Figure 1. Entire beam-hinging pattern

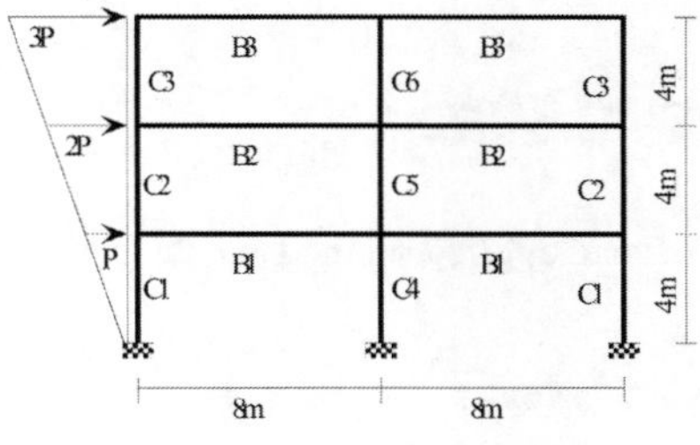

Figure 2. 2-span-3-story frame

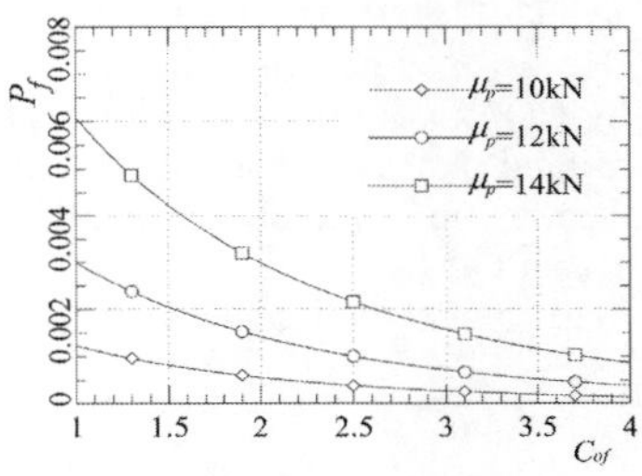

Figure 3. Failure probabilities failure of the entire beam-hinging mode

The mean strengths of structural members are determined through Eq. (3) to make the structure designed with the same COF in all beam-column nodes. Equation (4) gives the coefficient of variation of each random variable.

$$\mu_{bni} = \mu_b; \mu_{bij} = 2\mu_b;$$
$$\mu_{csl} = C_{of}\mu_b; \mu_{cl} = 2C_{of}\mu_b; \mu_{pj} = j\mu_p \tag{3}$$

$$V_{bni} = V_{bij} = V_{cl} = V_{csl} = V_1; V_{pj} = V_2 \tag{4}$$

In Eq. (3), μ_b is the mean yield strength of top beam, μ_p is the mean value of load applied on the first floor of structure. V_1 and V_2 in Eq. (4) are coefficients of variation of member strength and load, respectively. The mean value and COV defined for structural members correspond to those parameters with same subscript in Eq. (2).

The relationships between the load and other parameters of a two-span three-story structure as shown in Fig. 2 are investigated. The mean strength of top beam is assumed to be 106.2 kN·m; the mean strengths for other members can be obtained by applying Eq. (3). For example, given $C_{of} = 1.1$, the mean strength of 212.4 kN·m can be obtained for the beams B2 and B1, 116.8 kN·m for the exterior columns C1, C2 and C3, and 233.6 kN·m for the columns C4, C5 and C6. For other cases of COF, the member strengths can be similarly designed.

Using the first order reliability method (FORM) (Ang & Tang 1984), the failure probability of the entire beam-hinging mode is obtained and shown in Fig. 3, from which one can see that the load level has a great effect on the failure probability of the entire beam-hinging mode. In the present study, the target COF investigation is conducted under a specific reliability level, which means that for a given reliability index of the entire beam-hinging failure mode the load levels are adjusted to ensure the first order reliability index β_F remain the same as the target reliability index β_T for the frame structures designed with various COFs. The target reliability index β_T reflects the reliability of structure with respect to entire beam-hinging mode, and it should be determined according to the significance and safety requirements of structures.

In order to understand the relationships between μ_p–C_{of}, μ_p–μ_b and μ_p–m, consider the reliability index of the entire beam-hinging mode with performance function described in Eq. (2). When V_1 is so far smaller than V_2 that it can be ignored, the second moment reliability index can be approximately given by (Zhao et al. 2002):

$$\beta_{SM} \approx [2m\mu_b(2n-1+C_{of})-\mu_p h\sum_{j=1}^{n} j^2]/(V_2\mu_p h\sqrt{\sum_{j=1}^{n} j^4}) \tag{5}$$

From Eq. (5) the mean value of load applied on the first story of structure is obtained as:

$$\mu_p \approx 2m\mu_b(2n-1+C_{of})/h(\beta_{SM}V_2\sqrt{\sum_{j=1}^{n} j^4}+\sum_{j=1}^{n} j^2) \tag{6}$$

Under a specific reliability level of FORM, because β_{SM} is unknown, Eq. (6) cannot be directly used to determine μ_p. However, since the higher order moments of the performance function Eq. (2) is not affected by COF under the condition that V_1 is far smaller than V_2 (Zhao et al. 2002), one can easily understand that μ_p is basically a linear function of COF. Meanwhile, β_{SM} is also not affected by the mean value of member strength or the number of bays, so μ_p also changes linearly with respect to μ_b and m.

Therefore, the mean value of load can be given by the following formula as a linear function of m, μ_b and C_{of}.

$$\mu_p = m\mu_b(a+bC_{of}) \tag{7}$$

where a and b are constants independent of μ_b, m and C_{of} but related to β_F and n. Comparing Eq. (6) with Eq. (7), one can see that:

$$b = a/(2n-1) \tag{8}$$

then the mean value of load can be given as:

$$\mu_p = am\mu_b(1+\frac{1}{2n-1})C_{of} \tag{9}$$

From larger number of computation, a is empirically expressed as.

$$a = \frac{3.3\beta^2\text{-}29.8\beta\text{+}73.5}{n\text{+}0.45} - \frac{n^{1.5}\text{-}(n/4)^3}{\beta^4\text{-}\beta} \tag{10}$$

Comparison between the values obtained from Eq. 10 and accurate values of a is shown in Fig. 5, one can see that Eq. 10 approzimate a wery well.

Assuming $V_1 = 0.1$, the mean values of load corresponding to different COFs are shown in Fig. 5(a) in terms of solid line, from which a linear relationship between μ_p and C_{of} can be recognized. In the same way, the μ_p–μ_b curves and μ_p–m curves are obtained in Fig. 5(b) and Fig. 5(c), from which the linear relationships between μ_p and μ_b, μ_p and m are also observed.

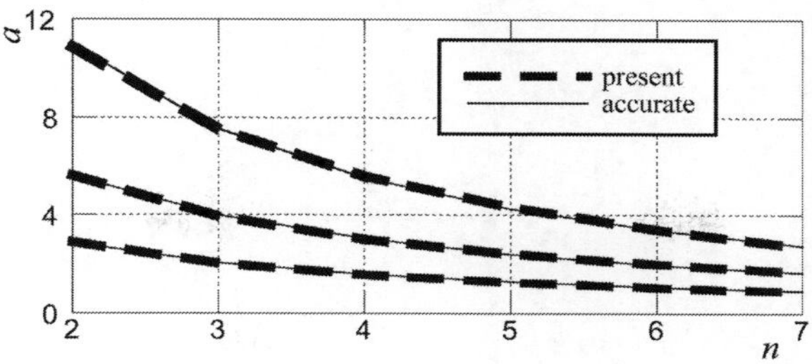

Figure 4. Accraacy of Eq. 10

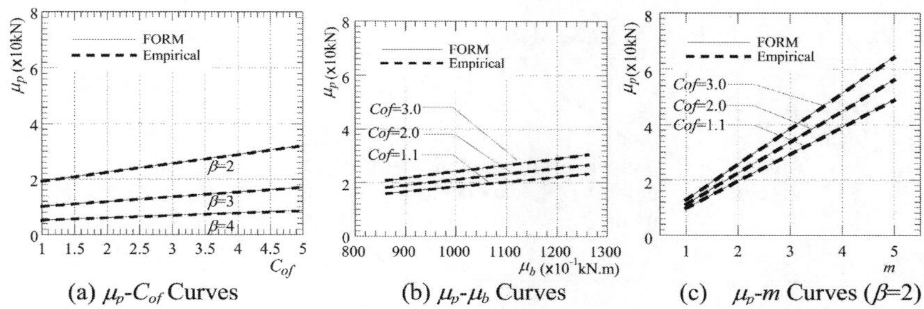

Figure 5. Relationship curves

4 EVALUATION METHOD FOR TARGET COF

4.1 *Evaluation method*

Since the structure is deterministically designed to collapse according to the entire beam failure mode, the most likely failure mode is generally the preferable beam-hinging mode, and all the likely story failure modes are undesirable. To probabilistically avoid the story mechanisms, the relative occurrence ratio of the most likely story mechanism to the entire beam-hinging mode γ should be limited to a specific allowable level γ_0 lower than or equal to 1 as follows:

$$\gamma = P_{f2} / P_{f1} \leq \gamma_0 \leq 1 \tag{11}$$

where P_{f1} is the occurrence probability of the desirable beam-hinging mode, and P_{f2} is failure probability of the most likely story mechanism. It should be noted that what is defined for P_{f2} is not the joint probability of all possible story mechanisms but the probability of individual story mechanism. The joint probability, which means the probability of appearance of any story mechanism, may be some different from that of individual failure mechanism. The evaluation with consideration of other story mechanisms needs another seperate research. P_{f1} is computed from the target reliability index β_T. Obviously, the larger the value of the COF, the smaller the value of the relative occurrence rate of the story mechanism. By conducting failure mode analysis and reliability analysis on frames designed with various COFs, a γ–C_{of} curve can be obtained and the value of target COF satisfying Eq. (11) can be determined.

4.2 *The most likely story mechanisms*

The number of the potential story mechanisms is large for multi-story frames, and it increases rapidly with the increase of the number of stories. In the target COF evaluation, however, only the most likely story mechanisms are required. An earlier study on the story failure modes of frame structures (Zhao et al. 2007) showed that the upper story mechanism with the most failure stories and all the lower story mechanisms would occur the most likely.

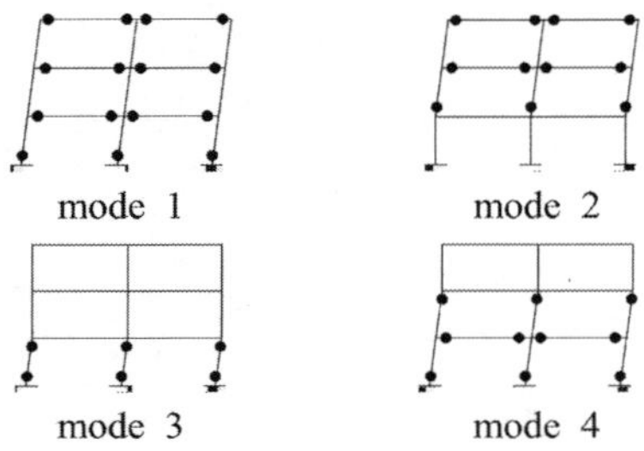

Figure 6. Failure modes of 2-span-3-story frame

Table 1. The Occurrence Order of Collapse Modes

COF Range	1.0~1.09	1.09~1.155	1.155~1.24	1.24~1.245	1.245~1.547	1.547~5.0
Occurrence Order	mode 3	mode 4	mode 4	mode 4	mode 1	mode 1
	mode 4	mode 3	mode 1	mode 1	mode 4	mode 2
	mode 1	mode 1	mode 3	mode 2	mode 2	mode 4
	mode 2	mode 2	mode 2	mode 3	mode 3	mode 3
Mode for P_{f2}	mode 3		mode 4			mode 2

4.3 *Evaluation example*

The target COF evaluation of the two-bay, three-story structure shown in Fig. 2 is presented to demonstrate the method. As the results of failure mode analysis, the failure modes that are necessary for COF evaluation are shown in Fig. 6, where mode 1 is the preferable entire beam-hinging failure mode; mode 2 is the most likely upper story mechanism; and mode 3 and mode 4 are the most likely lower story mechanisms. In order to conduct the COF evaluation at a same reliability level, the external load level should be adjusted to ensure that the reliability index corresponding to the entire beam-hinging pattern mode of the structure remains to be the same even though the value of COF is changed. From Eq. 10 one obtains that $a = 7.54 \times 10^{-4}$ for $\beta_F = 2$. Then the load level can be expressed as:

$$\mu_p = 7.54 \times 10^{-4} m \mu_b (1 + \frac{1}{5} C_{of}) \tag{12}$$

Using the obtained load level, the probabilistic analyses for the story mechanisms are conducted. Figure 7(a) shows the $P_f - C_{of}$ curves. Table 1 gives the occurrence order of failure modes shown in Fig. 6. In Table 1, since mode 1 is the preferable mode, the occurrence probability corresponding to mode 1 is P_{f1}. P_{f2} will be the maximum occurrence probability corresponding to the likely story mechanisms, i.e., mode 2, 3 and 4. When the COF ranges from 1.0 to 1.09, the occurrence order is mode 3-4-1-2, and mode 3 is the most likely story mechanism, and therefore P_{f2} is the occurrence probability corresponding to mode 3. Similarly, when COF ranges from 1.09 to 1.547, P_{f2} corresponds to mode 4 and if COF is larger than 1.547, P_{f2} corresponds to mode 2. By computing the ratio of P_{f2} to P_{f1} at each COF value, the γ–C_{of} curve is obtained as shown in Fig. 7(b). The target COF is the COF value where γ equals the allowable level γ_0. Given an allowable level γ_0 the corresponding target COF values can be determined from Fig. 7(b). For example, target values of 1.323 and 1.422 can be obtained for $\gamma_0 = 0.9$ and $\gamma_0 = 0.8$.

5 BASIC COF AND OPTIMUM COF

The analysis above indicated that the allowable level γ_0 denotes the safety requirement and the target COF is sensitive to γ_0. The smaller the value of γ_0, the larger the target COF, and the lower

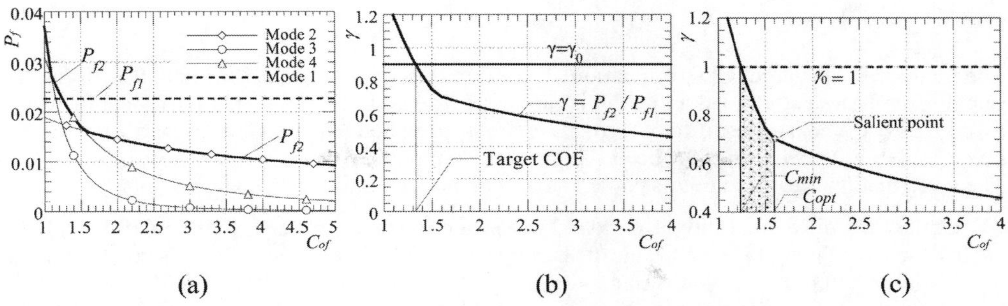

Figure 7. (a): P_f–C_{of} Curves; (b): γ–C_{of} Curve; (c): Basic and optimum COF

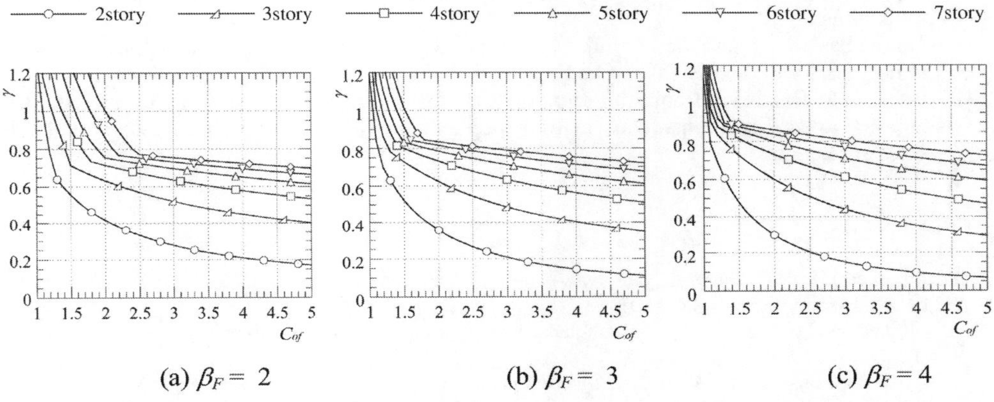

Figure 8. The γ–C_{of} curves for multi-story frames

Table 2. Basic COF and Optimum COF

		2story	3story	4story	5story	6story	7story
$\beta = 2$	Basic.	1.098	1.234	1.377	1.568	1.792	2.011
	Optimum.	1.279	1.543	1.784	1.992	2.174	2.464
$\beta = 3$	Basic.	1.055	1.108	1.187	1.284	1.399	1.522
	Optimum.	1.146	1.253	1.377	1.489	1.596	1.758
$\beta = 4$	Basic.	1.039	1.072	1.092	1.142	1.198	1.270
	Optimum.	1.046	1.122	1.177	1.235	1.301	1.388

the occurrence probability of undesirable story failure modes will be. However, if COF increases, it means a higher cost of structural construction. The target COF in application must satisfy the requirements of both safety and economy. At present, what level of evaluation index γ_0 should be used is still in confusion. In order to provide COF values applicable for design, the basic COF and the optimum COF are proposed in this section.

Figure 8 shows the evaluation curves of two-seven story frames under a reliability level of $\beta_F = 2$, 3 and 4, respectively. The basic COF and optimum COF values obtained from the evaluation curves are listed in Table 2, from which one can see that the basic COF and optimum COF increase with the increase of the number of stories and decrease with the increase of target reliability level of frame.

6 CONCLUSIONS

The load levels for target COF evaluation are analyzed, and an empirical formula describing the relationship between the load and COF is proposed. The COF requirements for probabilistically avoiding story mechanisms of frame structures are investigated based on the comparison of occurrence probabilities between entire beam-hinging failure mode and the most likely story mechanisms. The investigation presented in this paper can be summarized into the following points.

(1) Under a specific reliability level, load level is approximately a linear function of the column over-design factor, the mean value of strength and the number of bays. The presented empirical load function provides results in good agreement with that obtained by FORM.
(2) The basic COF and optimum COF are proposed. The basic COF is the lowest limit that ensures the occurrence probability of the story mechanism is at least not larger than that of the preferable beam-hinging pattern, and the optimum COF can be used to avoid story mechanisms most efficiently. The COF between basic COF and optimum COF is recommended to use in design applications.
(3) Both the basic COF and the optimum COF increase with the increase of the number of stories and decrease with the increase of target reliability level of frame.
(4) An example frame with column strength varying for different stories was evaluated. It is suggested to take into consideration all the potential story mechanisms before the most likely modes are known.

REFERENCES

ACI Committee 318 1999. *Building code requirements for structural concrete (ACI 318-99) and commentary (318R-99)*, Farmington Hills, MI, 391.

Anderson, J. C., and Gupta, T. P. 1972. Earthquake resistant design of unbraced frames. *J. Struct. Div., ASCE*, 98(11), 2523–2539.

Ang, A. H-S. and Tang, W. 1984. *Probability concepts in engineering planning and design*. Vol.II-Decision, Risk and Reliability. J. Wiley & Sons, New York.

Architecture Institute of Japan. 1993. *Recommendations for loads on buildings*.

Clough, R. W., and Penzien, J. 1982. *Dynamics of structures*, McGraw-Hill, Singapore, 597–602

Dooley, K. L., and Bracci, J. M. 2001. Seismic evaluation of column-to-beam strength ratios in reinforced concrete frames. *Structural Journal*, ACI, Vol. 98, No. 6, 843–851.

Kanda J. 1993. *Introduction to limit state design*. KentikuGijyutu, Tokyo (in Japanese).

Kawano, A., Matsui, C. and Shimizu, T. 1998. Basic properties of column over-design factors for steel reinforced concrete frames. *J. Struct. Constr. Engrg.*, No. 505, 153–159.

Kuwamura, H., Sakai, M., and Kato, B. 1989. Mechanism-based seismic design considering random yield strength. *J. Struct. Constr. Engrg.*, No. 401, 151–162.

Lee, H.-S. 1996. Revised rule for concept of strong-column weak-girder design. *J. Struct. Engrg., ASCE*, 122(4), 359–364.

Nakashima, M. and Sawaizumi, S. 1999. Effect of Column-to Beam Strength Ratio on Earthquake Responses of Steel Moment Frames (Part 1: Column-to-Beam Strength Ratio Required for Ensuring Beam-Hinging Responses). *J. Steel Struct.* 6(23), 117–132.

Ono, T., Zhao, Y.G. and Ito, T. 2000. Probabilistic Evaluation of Column Overdesign Factors for Frames. *J. Struct. Engrg., ASCE*, 126(5), 605–611.

Park, R., and Paulay, T. 1975. *Reinforced concrete structures*, Wiley, New York, 554–568.

Zhao Y.G., Ono T., Ishii K. et al. 2002. An investigation on the column over-design factors for steel framed structures. *Journal of Structural Construction Engineering*. No. 558, 61–67.

Zhao Y.G., Pu W.C., Ono T. 2007. Likely story mechanisms of steel frame structures. *Journal of Structural Construction Engineering*. No. 613, 35–42.

Reliability and Optimization of Structural Systems – Straub (ed)
© 2010 Taylor & Francis Group, London, ISBN 978-0-415-88179-1

How to improve the dissemination of structural reliability methods in engineering practice?

Summary of the second panel discussion at the 15th IFIP WG7.5 working conference

Summarized by Luis Esteva and Bruno Sudret

1 REFERENCE FRAMEWORK

The session started with a presentation of both the concepts in engineering practice where it would be advantageous to make explicit use of structural reliability methods, and the actors involved, namely those professionals who should be involved in different stages or activities in the process:

- **Lines of application**: engineering design, construction quality control, vulnerability and risk assessment;
- **Actors involved**: code writers, design engineers (structures, foundations, etc.), construction managers and supervisors, educators.

2 ACHIEVEMENTS, LIMITATIONS, CHALLENGES

At the technical level, it seems that the lack of information often appears as a hindrance to the development of probabilistic modeling and associated risk analysis. The contrast between "basic" rules for initial design and complex models for structural assessment (*e.g.* advanced finite element models) impede the consistent modelling of uncertainty at the various steps of a project.

At the educational level, there is a need for creating "consciousness of implicit uncertainty" among engineers-to-be as well as established engineers that are used to deterministic design rules. In particular, the distinction between aleatoric and epistemic uncertainty and the way they may be addressed should be understood by the lay engineer.

There was consensus about the fact that significant technical advances in models and methods have been achieved during the last years, including their use by groups responsible for the preparation of normative documents. However significant worries were expressed about some difficulties or inconveniences restraining their direct application in the design practice of specific structural systems. Some of those hindrances may be associated with the professional demand for "as-simple-as-possible" design criteria and methods. Others arise from the insufficient knowledge and expertise of most practical engineers to handle probabilistic concepts and models. In order to cope with these limitations, actions along the following lines were recommended:

a. Focusing efforts on the education of younger generations, providing them with a solid theoretical background on probability and statistics. In order to enhance interest on these concepts and their understanding for practical applications, teaching of basic concepts should be integrated into the engineering curriculum and taught by engineers with illustrative problems of their practical applications.
b. Continuous education courses for professionals.
c. Gradual introduction of explicit reliability-based design criteria into codified design rules. Adequate controls should be imposed in those rules, aiming at avoiding their inadequate application, which may occur either by ignorance, misunderstanding or intentional violation oriented to reducing initial construction costs, at the expense of increasing vulnerability.
d. Developing adequate practical design aids.

Reliability and Optimization of Structural Systems – Straub (ed)
© 2010 Taylor & Francis Group, London, ISBN 978-0-415-88179-1

Author index